MEN IN ARMS

MEN IN ARMS

A HISTORY OF

WARFARE AND ITS

INTERRELATIONSHIPS

WITH WESTERN SOCIETY

FIFTH EDITION

RICHARD A. PRESTON
DUKE UNIVERSITY

ALEX ROLAND
DUKE UNIVERSITY

SYDNEY F. WISE
CARLETON UNIVERSITY

Harcourt Brace Jovanovich College Publishers
Fort Worth   Philadelphia   San Diego
New York   Orlando   Austin   San Antonio
Toronto   Montreal   London   Sydney   Tokyo

Publisher	Ted Buchholz
Acquisitions Editor	David Tatom
Developmental Editor	Martin Lewis
Senior Project Editor	Mark Hobbs
Copyeditor	Wanda Giles
Production Manager	Ken Dunaway
Art & Design Supervisor	Vicki McAlindon Horton
Cover & Text Designer	CIRCA 86, Inc.

Library of Congress Cataloging-in-Publication Data

Preston, Richard Arthur.
    Men in arms : a history of warfare and its interrelationships with
Western society / Richard A. Preston, Alex Roland, Sydney F. Wise
—5th ed.
    Includes bibliographical references and index.
    1. Military history. I. Roland, Alex, 1944–   . II. Wise, S. F.
(Sydney F.), 1924–   . III. Title.
    D25.P7   1991
    909—dc20   90-28315
    ISBN: 0-03-033428-4   CIP

Requests for permission to make copies of any part of the work should be mailed
to the Permissions Department, Harcourt Brace Jovanovich, Inc., 8th Floor, 6277
Sea Harbor Drive, Orlando, Florida 32887

Address for editorial correspondence: Harcourt Brace Jovanovich, Inc., 301 Commerce Street, Suite 3700, Fort Worth, TX 76102

Address for orders: Harcourt Brace Jovanovich, Inc., 6277 Sea Harbor Drive, Orlando, Florida 32887. 1-800-782-4479, or 1-800-433-0001 (in Florida)

PRINTED IN THE UNITED STATES OF AMERICA

 2 3 090 9 8 7 6 5 4 3 2

Harcourt Brace Jovanovich, Inc.
The Dryden Press
Saunders College Publishing

CONTENTS IN BRIEF

PREFACE TO THE
FIFTH EDITION

The thesis of this book, that military history can be properly understood only when it is set in its relationship with political, economic, social, and technical change, was adopted as the theme for two new military history courses in the Royal Military College of Canada when it re-opened after World War II. They were an advanced seminar for history specialists and a lecture course for third-year engineers. These courses were in line with similar innovations in other fields of history, where students were trained to set their special interests in a broader background of social development.

Events since that time, and also the publication of new scholarship, especially that of William McNeill, Martin van Creveld, John Keegan, and Paul Kennedy, have fully justified this approach. They also pointed to a need to bring this book up to date.

Since the first edition in 1955, three later editions added chronological chapters to cover more recent events. This fifth edition synthesizes and updates those additions in three new topical chapters that cover the whole period from the Second World War to 1990. At the same time, this new edition revises earlier chapters to bring them in line with modern scholarship.

We continue to be indebted to our previous co-authors, Sydney F. Wise and the late Herman O. Werner, and also to those thirty persons whose constructive criticism was acknowledged in prefaces to the earlier editions.

For this fifth edition we wish to mention our indebtedness to Dr. Everett Wheeler, a scholar in residence in the Classics Department at Duke University, who made many valuable suggestions for the amendment of the chapters on classical and Byzantine warfare; to Colonel Joe B. Sharpe, Professor of Military Science in the Duke Army Reserve Officer Training Corps, and to the members of the Duke Military History Colloquium in 1989 and 1990, especially Lt. Timothy Budd, U. S. Air Force, Lt. Col. David E. Johnson, U. S. Army, Major Rob Owen, U. S. Air Force, and Major Gordon V. Rudd, U. S. Army.

We must also record our indebtedness to Johannah Sherrer, Head of Duke's Perkins Library's Reference Department, and to her informed and dedicated staff, especially Kenneth W. Berger, Joseph Rees, Jane Vogel, and Bessie Carrington, who gave us much help with difficult bibliographic problems. Stuart Basefsky and his able colleagues in the Public Documents Department of Perkins Library were always willing and ready to guide us through their vast holdings. Dr. John L. Sharpe and J. Samuel Hammond of the Rare Book Room helped us find illustrations for the book.

At the Royal Military College, Kingston, the Chief Librarian, Samuel Alexander, furnished us with illustrations from rare works in the Massey Library Collection. Benoit Cameron and Suzanne Burt helped with the location of other material.

We must add our indebtedness for the typing assistance by the History Department staff at Duke University, Grace Guyer, Vivian Jackson, Thelma Kithcart, and Dorothy Sapp.

We note that David Tatom and Martin Lewis of Holt, Rinehart and Winston not only foresaw the need to bring the book up to date but also courteously helped us to overcome many difficulties.

Richard A. Preston
W. K. Boyd Professor Emeritus, Duke University, and Honorary Professor of History, Royal Military College of Canada

Alex Roland
Professor of History, Duke University

November 1990

CONTENTS

LIST OF ILLUSTRATIONS

LIST OF MAPS

MEN IN ARMS

INTRODUCTION

The history of human society has been thoroughly permeated by war, but studies of military history have often been narrowly specialized, or written as if war existed in a vacuum. One reason for this is that by far the greater number of those who studied war in the past had a professional interest in it. Battlefield experience, training maneuvers, war games, and tactical exercises without troops (TEWTs)—important ways of learning about war—were seen as not enough to provide the understanding that the practice of war required. Only by reading history could the mind of the future commander be trained. So the military student turned to military history.

Within the course of the present century, soldiers and military historians discovered and enunciated certain basic precepts that appeared to affect the outcome of battles and that were apparently pervasive and permanent despite changing circumstances. Applicable especially at the level of grand tactics and strategy, these came to be regarded as useful guides at other levels of operations. Originally worked out in relation to land operations, the "principles of war," being general propositions, could be adapted to war on the sea and in the air. Not surprisingly, they came to interest and dominate the attention of some of those who studied military history for professional reasons.

Instruction in military history, especially in army programs, still often centers on an exposition of the principles of war. In view of the prominence they are thus given, all those who read military history should be aware of their nature, content, value, and use. The principles of war presently authorized by the United States Chiefs of Staff are as follows.

PRINCIPLES OF WAR
FM 100-1, *The Army* (29 August 1986)

1. OBJECTIVE: Direct every military operation toward a clearly defined, decisive, and attainable objective.
2. OFFENSIVE: Seize, retain, and exploit the initiative.
3. MASS: Concentrate combat power at the decisive place and time.
4. ECONOMY OF FORCE: minimum essential combat power to secondary efforts.
5. MANEUVER: Place the enemy in a position of disadvantage through the flexible application of combat power.
6. UNITY OF COMMAND: every objective, insure unity of effort under one responsible commander.
7. SECURITY: Never permit the enemy to acquire an unexpected advantage.
8. SURPRISE: Strike the enemy at a time or place or in a manner for which he is unprepared.

9. SIMPLICITY: Prepare clear, uncomplicated plans and clear, concise orders to insure thorough understanding.

These principles of war draw the attention of military readers to those aspects of campaign histories that can help them to absorb their lessons. But they should be used solely for that purpose. They are the distillation of current wisdom, not universal laws. They are not rules of thumb for the conduct of operations. Rather they are warnings that if certain factors are neglected, failure may follow.

New principles and new definitions have from time to time been added. Lists prepared in different countries and by different staffs vary in content and in wording. Thus, in 1942 Joseph Stalin promulgated what he called five "permanently operating factors," which were similar in purpose but different in detail from the principles taught in the West. The nature of some of the American principles differs from one to the other, some being aids to planning and others being conditions which, if not met, may bring defeat. Some principles qualify others to such an extent as to require the application of individual judgment, that is, common sense. Each principle does not necessarily apply to each and every military situation. As they are primarily pedagogical devices to assist in learning, the principles of war must not become substitutes for thought. To be able to repeat them by rote, even with a supply of appropriate examples, has very limited value.

This book, not being campaign history, will not refer in the text to the principles of war. It has a wider focus. Military history, especially campaign history, has too often been written with inadequate attention to such closely related factors as supply, communications, administration, and organization, or even weapon development. It is in fact only in recent times, with the rise of air power and the renewed importance of amphibious operations, that operational history has become regularly treated as a whole, instead of as separate military and naval occurrences. Furthermore, the political, social, economic, psychological, and technological circumstances that often decided the outcome of wars were also regularly ignored. Military historians today can no longer afford to make that mistake. Military history must be set in a broader picture. Finally, there is good reason to believe that, with the development of weapons of potentially all-destructive capacity, the purpose of war and the function of the soldier has been so much altered that some of the traditional lessons of military history are likely to be no longer fully adequate.

Military history is no longer the interest only, or primarily, of soldiers. Although there is much talk of its abolition, war has, in fact, become ever more closely a part of the life of the community, rather than the plaything of kings. Just as war has been more and more affected by other aspects of society's evolution, so it in turn has affected the course of events far and wide. Therefore, general history cannot be understood without a knowledge of military history, nor military history without a knowledge of general history. These facts have not always been appreciated. In the period between the two world wars of this century, distaste for things military, growing out of the bitter experiences of 1914–18, led to the relegation of military history to a very minor place in the curriculum. There are indeed some academics who still appear to believe that by not thinking about war, they may be able to induce it to go away.

This is unfortunate. In modern war, civilians make decisions that are vital to a war effort. Indeed, in modern times the higher direction of war in most countries is in the hands of civilian leaders, not of soldiers. Furthermore, preparation for defense in times of peace needs an understanding of war's purposes and functions. The degree to which civilian life, even in peace, is affected by war, and the amount of civilian energy and resources that it absorbs, make it essential that all, whether politicians, businesspeople, engineers, manufacturers, or teachers—indeed, all voters—know military history and understand warfare. What they need to grasp especially is the relation between warfare and the way society has developed in the past and presumably will continue to develop in the future. For this purpose the military history of their own country is not enough. They must study war as a universal institution.

There has been some disagreement whether it is better, for both military and civilian students, to concentrate on a few wars, campaigns, battles, or leaders, or, on the other hand, to make a wide survey. Narrow intensive studies can come closer to recreating reality, including what has been called "the fog of war"; but such specialized studies provide relatively few illustrations, examples, and experiences. On the other hand, surveys, which give a varied experience by the production of many examples, tend to oversimplify the story. The ideal system is to use both approaches. What is most important is that the student should read over a number of years not limited to the duration of an academic course or to the preparation for a professional examination. The essential first step is to gain the interest of the student, whether military or civilian, so that he or she will continue reading after the first period of acquaintance with military history ends. It can be argued that that first critical acquaintance can best be made through an overall survey.

This book has a dual aim: to provide a background to warfare by presenting the ancillary factors that have affected its evolution and to facilitate an understanding of the way in which military events and decisions have affected the evolution of society. It is restricted to the concerns of Western society and of those societies from which its military practices, methods, and traditions were drawn. Byzantium is included because it was an alternative diversion derived from the same classical forerunners of warfare as was the medieval world, and because of the extent to which it simultaneously confronted the feudal society of Europe's Middle Ages and, at the same time, that society's long-term enemy, Islam.

In sum, this book traces the development of warfare and relates it to the development of Western society, showing how political, economic, social, and technological developments governed the decisions achieved in wars, and vice versa. Placed in this frame, the story of war looms large in the history of Western civilization.

1

WARFARE AND SOCIETY

PRIMITIVE WAR

The warfare of modern times may be defined as organized societal violence. It includes every conflict between rival groups, by force of arms
or other means, which has claims to be recognized as a legal conflict.
Under this definition there may be a state of war without actual violence or
clash of arms. It excludes riot and acts of individual violence but includes
insurrection and armed rebellion, especially that which is of sufficient extent to be regarded as civil war. Wars between nation-states, political entities that monopolize armed force within their borders, are its most usually
recognized form today.

The conflicts of primitive peoples may be classed as warfare if they were
the result of organized group activity with some continuity of purpose and
action; but if they consisted only of incidental and casual acts of violence,
they cannot be considered as warfare. Primitive societies include many stages
of development, from the simple social organization of the peaceable Eskimo
(now known as Inuit, "The People") to the more complex and differentiated
societies of the warlike Zulu or of the Indians of the Pacific Northwest. The
basic feature of primitive societies is that, having no written form of language,
they rely upon oral tradition for the transmission of knowledge. Primitive life
is ruled by custom; social relationships within the group and the techniques
for providing the necessities of life have been sanctioned by the unvaried experience of generations. Primitive warfare is conditioned by the same rule of
custom that governs the other activities of the group. Since society is static,
the techniques and weapons of war are static also and in this way differ from

5

those of civilized societies in which methods of making war are subject to
constant change.

Primitive warfare is different from that waged by civilized peoples in
almost every major respect. Specialization of function in primitive societies
is almost nonexistent; each member of the group participates in all its activities. Thus there is no professional, or "regular," military class whose
only function is that of waging war. Just as there is no ruling class in most
primitive groups, so in war there is no command structure, no means of
enforcing orders or discipline. The war chiefs of the North American Indians were those warriors who had distinguished themselves by their individual exploits; their prowess attracted others to follow them on the warpath;
but they did not act as commanders in our sense of the term. The lack of
any principle of authority, other than custom, means that the tactical organization of civilized armies is never achieved. The simplest form of attack
is to ambush an enemy, preferably by employing missile weapons from a
distance. Many primitive groups never get beyond this technique, and very
few of them are prepared to engage in a stand-up pitched battle or to attack
fortified places. Usually primitive war parties break up after the initial engagement, and individuals fight for themselves.

Rational planning for war, like that undertaken by military staffs or political bodies in civilized nations, does not exist in the primitive approach to
conflict: the war plan may be no more than the auspicious dream of a war
chief, as among the Crow people of the Plains. Because its food resources are
too scanty, because its warriors are also the hunters and food suppliers to the
whole community, and because the members of its war parties are free to go
home if the spirit so moves them, no primitive society is capable of extended
campaigning. Although casualties in primitive war are usually light, because
a primitive society can ill afford to lose one of its vital food producers, the
loss of a few men usually causes the abandonment of fighting. The Seneca,
joining with the British against the Americans in the battle of Oriskany
(1777), quite understandably regarded the death of thirty-three braves as a
tribal disaster.

Finally, the motives for primitive war only vaguely resemble those of organized, civilized states. The economic motive is not usually dominant. Most
primitive groups have little to tempt an attacker. Wars for the acquisition of
territory or the domination of another people are equally unusual. War is
generally undertaken by a group of individuals, or by the whole tribe, as a
matter of prestige; that is, war is fought for the glory which attaches to the
outstanding warrior, for revenge, for wives to perpetuate the group, or for
food.

The nature of primitive warfare is closely related to the state of the social, political, and economic organization of primitive society. Change in the
methods of making war may occur if there are changes in the organization of
the group, either as the result of internal innovations or as the result of influences from other cultures. Such change might come about through the
mastering of a technique, such as metallurgy, which makes specialization

possible. It may occur through the evolution of political institutions, e.g., a kingship. When the royal principle was established among the Zulu, it resulted in the emergence of an individual who had the power to command. The Zulu, organized into regiments *(impis)*, and led by powerful kings, especially Chaka, broke through the barrier of custom which impedes change and evolved a system of warfare intermediate between that of more primitive groups and that practiced by civilized societies. Similarly, most primitive peoples, upon encountering the European, have imitated Western techniques of war, with varying degrees of success. The Plains Indians obtained horses from the Spaniards and became cavalry warriors; firearms added to their effectiveness; but although the Sioux succeeded in vanquishing Custer's troops at the battle of the Little Big Horn (1876), Indian command and tactics remained rudimentary, Indian progress in armament was not matched by an advance in social organization.

WARFARE AND CIVILIZATION

In civilized societies warfare is a condition which is distinguishable from many other forms of violence by the fact that it is an accepted form of behavior on the part of certain groups within the community and is also legal. The endemic disorder of the Middle Ages, although frequently regarded by authorities in church and state as illegal because it threatened the integrity of the Christian polity in Europe, was in fact warfare. The political unity of medieval Christendom was largely theoretical; conflict between groups was a recognized method of settling differences; and military operations were planned and carried through in an organized fashion.

In modern times the type of conflict that predominates is warfare between nation-states. Such conflict must be classed as warfare even though it is often alleged to be unjust or illegal.

Philosophers have wrestled, although not very successfully, with the problem of distinguishing between just and unjust wars on the basis of the morality of causes. Today, under the charter of the United Nations, it is claimed by some that warfare of the old type between nation-states is illegal and that the only legal military action is an international police action to prevent or punish aggression. But powers that have opposed the United Nations forces have claimed that they were fighting for legitimate ends and that their cause was both just and legal. It is thus still impossible to find any significant distinguishing characteristics between wars which are legal or just and those which are illegal or unjust. Therefore, any organized armed conflict between national or ideological groups must be regarded as warfare.

The requirement that to be classed as warfare a conflict must be in some degree an organized activity might suggest that war is a condition which develops with the development of civilization. Indeed, some philosophers, like Jean-Jacques Rousseau, and some anthropologists, like W. J. Perry and the school of "diffusionists," have implied that war and civilization had a common origin. Some anthropologists have pointed to certain stone-age tribes, like the Eskimos, to suggest that primitive man was "peaceful"; and it is possible to infer that war has been intensified as man has become more "civilized." From these sources, it is possible to draw an inference that war is "unnatural." If it is "unnatural," then there is powerful support for the belief that war can be eliminated from society; for if man is not by nature a combative animal, then it is conceivable that a cure can be found for a disease that plagues civilizations. On the other hand, the majority of anthropologists disagree with the belief that the most primitive man did not know war because they regard his food-gathering and wife-seeking raids as a primitive form of warfare.

Clearly the march of civilization, while it has intensified war at certain levels—for instance, in the resounding clash of nations—has in fact abolished its manifestations at other levels, such as within the area of the nation-state itself. Few will deny that the nation-state has become one of the most important man-made instruments for the advancement of civilization. War is, however, not the product of the nation-state and extreme nationalism, as is sometimes said. It occurred before the nation-state existed. Hence, from one point of view, modern war between nations may be looked upon as a reversion to barbarism which has not yet been controlled by the growing civilizing forces that have restrained men within the state. Quincy Wright, in *A Study of War,* published in 1942, argues that since war in the modern technical and legal sense developed along with civilization, it may be assumed that it can be contained.

Whether one accepts the thesis that war is the child of civilization or the contrary one that war stems from human nature, it is clear that development in warfare has been closely related to the process of historical change. Man's social, political, economic, and cultural progress has been affected both for good and for ill by the incidence and impact of armed conflict. The verdict of war has been, time and again, the deciding factor in the process of historical change. The Persian Wars have been said to have saved Europe from Asiatic tyranny. The Roman Empire was established by warfare, and warfare contributed to its destruction. William the Conqueror, as his name indicates, exercised his influence upon the history of England because of a successful war. And so it has gone on through the centuries. War has thus always been the arbiter when other methods of reaching a decision failed. But the judgment it has given is based on might rather than on right; it is never a moral judgment. At times right has prevailed, but whether this was accidental or due to inherent moral strength is a question upon which agreement is impossible.

But war has also been much more than a crude trial by combat deciding the course of history. It is intimately involved with the whole historical process. The nature of war itself has been fashioned by social factors and by technical development; war and organizations for war or for defense have affected social and technical progress or retrogression.

That war has been affected by social and technical change needs little elucidation. Weapons are the products of contemporary technology. Armies reflect the society from which they spring. When social or industrial revolutions take place, when power is passed from one economic class to another, when new techniques of administration or of distribution are discovered, warfare is automatically affected. The ever-accelerating rate of scientific discovery in our own time has made this clear as never before, so much so that one British writer on military affairs, General J. F. C. Fuller, has gone so far as to say that weapons account for 99 percent of victory. While few would agree with so extreme a statement, it is a fact that "superior arms favor victory," as I. B. Holley, Jr., wrote in *Ideas and Weapons* in 1953. Therefore, because of the present acceleration in invention, success in war depends more than ever before on the facility with which soldiers adapt their organization, tactics, and doctrine to the use of improved weapons. It may equally depend upon the extent to which military leaders bring their forces into line with the society from which they come and which they are designed to defend. Armies that are anachronisms will be swept away like the *condottieri* of Renaissance Italy. In the past, partly through a conservatism inherent in their craft, soldiers have often been peculiarly slow to adopt weapons and methods that were ready to hand. Resistance to improvements in military efficiency has sometimes resulted from militarism, which includes among other things a distortion of military values by an overemphasis upon the superficialities of military traditions and the insulation of the military craft from society at large. Only in very recent times has scientific research been accepted by the soldier as an important part of defense; and only slowly is it becoming understood that an army cannot stand aloof from the rest of the community.

The other thesis, that war shapes society, raises more contentious issues. Stanislav Andreski, in his *Military Organization and Society,* argues that all societies reflect their armies because societies that do not develop adequate defensive systems inevitably disappear. Some thinkers have actually concluded that war has been a constructive force in social and technological progress. The German economist Werner Sombart argued that war fostered the modern economic system and therefore modern society: that the medieval knight was the earliest example of the specialization of labor; that the growth of professional armies developed the spirit of discipline and the organizing spirit essential to modern capitalism; that the cost of war led to the expansion and development of credit; and that the demands of modern armies for standardized products on a huge scale compelled the introduction of the techniques of mass production in the basic metal-working and textile industries. The American social philosopher Lewis Mumford contended that the machine was propagated by war; that the invention of gunpowder stimulated the production of the basic element of modern civilization, iron; that since the gun was itself a primitive single-chambered combustion engine, it inspired also the invention of power-engines; that war

produced the military engineer who was the prototype of the industrial director and something very different from the simple craftsman of the Middle Ages; and that it was in the professional army that there was elaborated
the ideal form of organization for a purely mechanical system of industrial
production. Furthermore, a host of writers have dwelt upon the military
virtues as they are revealed by individuals, especially in time of crisis; and
some, like Friedrich Nietzsche, have argued that war must therefore be an
ennobling experience for society as a whole.

A conflicting view was advanced by A. J. Toynbee in his multi-volume
Study of History, which, along with due recognition of the importance of the
martial virtues, showed that war was the "proximate cause" of the breakdown
of every civilization in the past. Toynbee's description of the death-symptoms
in the "time of troubles" of previous civilizations was uncomfortably similar
to the nationalist wars and proletarian revolutions of the early twentieth century. Although he wrote about many ancient civilizations, he was thinking
primarily of the classical periods of Greece and Rome and parallels with contemporary problems.

A great many other writers have held that war is a great destroyer,
both of materials and of moral standards. Some have attempted to distinguish between militarism and other military characteristics and have come
to the conclusion that there is a strong tendency, perhaps an inevitable one,
for the efforts expended in necessary defensive organization and operations
to lead to excesses of the military spirit which, in the end, tend to destroy
the society that it was the original intention to defend. Many books and
articles, inspired by the revelations made by scientists about the atom and
hydrogen bombs, and by the impact of total warfare, have resurrected the
theme of Armageddon and of the destruction of civilization, perhaps even of
man himself.

Published in 1950, J. U. Nef's *War and Human Progress* is a most detailed investigation of the impact of war upon society, written to refute
Werner Sombart. In it, many of the claims of the "constructive" school about
the contributions made by war to society were shown to be either false or
exaggerated. Nef contended that it is the "limitations" on war, rather than
war itself, that led to social and technical advance.

War may be limited in several different ways. It may be limited in duration, in space or location, in intensity or mode of fighting, in its impact
upon the contending peoples, or in objective. All of these forms of limitation
need not be present at the same time. The origin of such limitation is diverse. Shortness of the duration of wars is likely to be brought about by a
preponderance of strength on one side rather than by lack of zeal of the
contestants; limitation in location may be caused by the localized nature of
the issues in dispute; restraints in the mode of fighting, lack of impact upon
the civilian populations, and limitation of objective can be caused by physical limitations as well as by a climate of opinion which places restrictions
on the nature of warfare. Limitations of this latter kind existed in theory in

the Middle Ages and were a powerful restraint in the eighteenth-century
Age of Enlightenment.

TOTAL WAR

It is obvious that the concept of limitation is opposite to the concept of
totality in war. Furthermore, just as limitation can never be absolute, so
also totality in war is a relative concept rather than an absolute one. Total
warfare in the most complete sense would mean fighting with all resources
and all kinds of weapons without any restrictions imposed by humanity or
by expediency, killing all prisoners and civilians without respect for age or
sex, disregarding completely the rights of neutrals, and using psychological
techniques to wipe out individual personality and to obliterate all standards. No warfare has yet reached this stage. Clausewitz said that total war
of this kind is an abstraction that exists only in theory; in practice, all
war is limited by the political ends it seeks. The conqueror does not wish to
find at his feet a pestilence-ridden lazarhouse. Absolute totality in warfare
could only mean chaos and a return to barbarism. Hence, when we speak of
limitations upon warfare, we mean that the growth of civilizing influences,
or the lack of ability to overcome physical barriers, has exercised restraint
upon warfare and so has lessened its impact upon society, and when we
speak of total warfare, we really mean approximation to totality.

Man's increasing power of construction has been closely parallelled by
the growth of his power of destruction. Similarly, his greater capacity for
destruction has been matched by, and to some extent has produced, a desire
to place limitations on warfare in the name of humanity. Through the centuries, the forces moving toward totality, on the one hand, and toward limitation, on the other, have been roughly in balance with a tendency first to
one side and then to the other. Just as the Middle Ages and the eighteenth
century were periods of limitation, so the ages of the Reformation and of
the French Revolution saw increasing approximation to totality. The nineteenth century was a period when it seemed as if man had a choice between
two alternative paths. In 1910 Sir Norman Angell, in *The Great Illusion,*
showed that war profited neither victor nor vanquished. A logical conclusion
was that rational man would eventually desist from resorting to it to settle
differences. However, the twentieth century showed a swing toward total
warfare. It was also marked by more conscious efforts than ever before to
control the scourge of war.

The invention of the atomic bomb seemed to threaten civilization and
perhaps even the very existence of man. Although it showed that war had
stimulated the rate of scientific and technical advance seldom achieved in
peacetime, war had now become a destructive force of boundless potential-
ity. Failure to restrict atomic power to peaceful uses accentuated the danger which was suspended, like the sword of Damocles, above man's head.

Movements to abolish war, which had already been stimulated by the horrors of modern strife, were greatly strengthened. Pacifism, limitation of armaments, collective security, and even the creation of a super-state by general consent have been advocated as means by which war might be ended. The annihilating power of the new hydrogen bomb is so vast that some responsible statesmen, scientists, and military men have begun to say that the frightfulness of the weapon will, in itself, be a deterrent against the onset of another major conflict.

It is just possible that the human instinct for self-preservation will eventually save mankind from extermination; however, it is also possible that it will not. Observers have been surprised to find that the development of a Soviet nuclear capability, which was at one time feared as a possible precipitant of a third world war, has so far had a quite different effect. The nuclear equilibrium, or stalemate, of the two great powers, achieved about 1959, established a state of mutual deterrence—a balance of terror. It also unfolded a nuclear umbrella under which lesser powers continued to wage war while the superpowers, although sometimes giving clandestine support, held back. Even when they became more involved, the two superpowers accepted restraints on their operations and shunned the temptation to resort to nuclear weapons. Moreover, both revealed a reluctance to share atomic secrets even with their closest friends, and both preached against the dangers of proliferation. The net result is that the world saw forty years of arms control negotiations between the superpowers, along with measures for what is called "strategic defense."

Because the superpowers are beyond control by general international agreement, and because statecraft has been incapable of abolishing or containing war, except fortuitously by the fragile nuclear balance, some scholars have turned to a more intensive study of war and its causes in order to learn more about its prevention. The earlier debate about whether war was a result of human nature or of developing civilization now seems too simplistic. Attempts to maintain peace by arbitration, disarmament, international organization, and collective security have proved equally frustrating. In 1958–59, the year when the possibilities of the balance of terror were first realized, new American research institutes sprang up at Duke, Harvard, Northwestern, Princeton, Michigan, Stanford, and elsewhere to explore what came to be called the science of peace. One of the most promising lines of approach in these programs was the study of conflict resolution and crisis management, in the belief that, if, under the nuclear umbrella, antagonisms between superpowers could be prevented from passing the point of no return, war could be contained by action planned in advance. However, the negotiations between the superpowers about nuclear arms limitation, which began in 1969, led only to a hope-inspiring treaty in 1988 to abolish intermediate-range nuclear missiles (INF treaty) in Europe, leaving the superpowers still divided on the more difficult problems of extending the ban to the short- and long-range arsenals.

The balance of terror fostered a more scientific approach to the study of the origins of warfare by social scientists and other scholars from a variety of disciplines. No one expects easy solutions. Even if an overall theory of the causes of war could eventually be produced from work done in several widely scattered fields of endeavor, there would still remain the serious problem of persuading statesmen, and also peoples, to apply the theory to practical situations. Nevertheless, such studies constitute promising steps toward peace.

William McNeill, in *The Pursuit of Power* (1982), argued that throughout the ages the course of history has been determined by technological development that affected the nature of warfare and of society. Technology led to the commercialization of the exercise of violence and to the professionalization of armies. The exercise of command of the technical developments and armed forces by monarchs or elites drastically changed the societies that they controlled. McNeill sees no evidence of an end to this process in the foreseeable future.

More recently Paul Kennedy, in *The Rise and Fall of the Great Powers* (1988), which like Toynbee's *Study of History,* was derived from a contemplation of the contemporary scene, argued that Western Europe's success in extending its sway across the face of the whole globe led to a changing internal balance of power among the national states on the continent and to the prevention of the creation of a single imperial power like those that dominated all earlier civilizations. Kennedy argued that prolonged wars lead to economic collapse and therefore to social and political change. As a result of the two world conflicts of this century, two superpowers have emerged, the United States and the USSR, states which originally were peripheral to Western civilization. He says that with the emergence of Japan, China, and the European Economic Community, there are now five such potential centers of power. His interpretation suggests that the leadership gained by midcentury by the United States having quickly waned, it may be followed by intense conflict between all five.

All these various scenarios warn against the uncontrolled and unlimited conflicts that could bring disaster to the human race; but at the same time they show the need for the vigilant maintenance of technically competent and up-to-date armed forces and defense strategies.

The study of the history of warfare, its relation to society, and the impact of social and technical development on it, which is the purpose of this book, is necessary to help us to speculate fruitfully upon the search for a means to control or limit conflict and so to eliminate a present danger to mankind and civilization. It is, at the same time, also essential for an adequate understanding of the way in which armed forces and conflict developed in the past and also the way in which they are likely to adjust in the future to changing circumstances and changing needs.

2

CLASSICAL WARFARE: THE AGE OF THE PHALANX

EARLY GREEK WARFARE

"Civilization begins, because the beginning of civilization is a military advantage," wrote Walter Bagehot in his *Physics and Politics;* "progress is promoted by the competitive examination of constant war." Most modern historians believe that the explanation of the origin of civilization is much more complex than this purely military interpretation. Civilization is associated with the growth of urban life, of a complicated social structure, and of extensive control over the physical environment. To achieve civilization, the oral transmission of knowledge must be superseded by a written language so that successive generations may build upon the accumulated knowledge of their predecessors. These conditions arise only out of a combination of circumstances: favorable topography, soil, and climate; the presence of mineral resources; knowledge obtained from other peoples; and the perfecting of new techniques, institutions, or ideas in response to certain stimuli. In this process, warfare must play a part.

All these influences seem to have contributed to the emergence of Greek civilization, the primary source of the Western military tradition.

Early in the second millennium B.C., Indo-Europeans settled in what is now mainland Greece, called Hellas by the Greeks. By about 1600 B.C. a distinct civilization had emerged from the settlement. The Mycenaens combined earlier Hellenic culture with strains of Minoan culture, independently centered in the capital of Knossos on the island of Crete. The warlike nature of the Mycenaens is recorded in the Homeric epics, in their grave furniture, and in their great citadels of Mycenae, Tiryns, and Pylos. They fought in war chariots and were armed and armored with bronze.

Following the collapse of Minoan civilization in the middle of the fifteenth century B.C., the Mycenaens occupied their capital and established themselves as the most powerful force in the Aegean. Their dominance lasted through the siege of Troy, which historian Michael Wood calls "the last fling of the Mycenaen world." Toward the end of the thirteenth century B.C., Mycenaen cities began to suffer a succession of disasters similar to those that overtook Minoan civilization. Greek literary tradition ascribes these reverses to Dorian invaders from the north, supposedly armed with iron weapons, but there is no archeological evidence for this view. Whatever the cause, the breakdown of Mycenaen civilization precipitated a comparative dark age in Greek history.

By about 750 B.C., the successors of the Mycenae had become literate, adopted the Phoenician alphabet, and initiated a political transformation that would alter the course of history. Gradually they evolved the city-state, the characteristic Greek form of political organization, and with it a complex social structure in which, in many states, simple tribal kingships gave way first to wealthy oligarchies and then, by the sixth and fifth centuries B.C., to democracies. Political uniformity, however, was hardly a Greek characteristic; Sparta, for example, remained a monarchy to the end, though the limited number of full citizens retained some powers over its dual kings. During the same period the Greeks produced a "civilized" fighting body, the phalanx, the germ of all future European military development. The story of Western civilization, and of organized warfare in the West, must thus begin with ancient Greece.

While it is possible to speak of the civilization or culture of Greece and, despite its many dialects, of a Greek language, politically Greece was marked by extreme diversity. Even in times of great peril, consciousness of a common identity as "Hellenes" never transcended strong local loyalties and a jealous regard for local political independence. The political unit was the city-state. This molecular political structure of Greece was inherited from the original tribal fragmentation of the Hellenic peoples at the time of their migration and was preserved by the geography of the Greek peninsula. Settlement concentrated on the infrequent coastal plains and in the river valleys, which were geographically isolated from neighboring areas by high mountain barriers. The early Greek city-state was the limit of its inhabitants' horizon. Particularism bred of isolation was, in the course of time, strengthened by distinctive historical and religious traditions which made for a powerful group identity. This sense of community and locale was the most important barrier against the establishment of a single state embracing the whole Greek mainland.

Greek unification was therefore not accomplished by a gradual historical process, since that process was working against it. It was eventually imposed by force of arms, but until the fourth century B.C. no single state was powerful enough to achieve it.

Moreover, Greek armies and military techniques were peculiarly unsuited to the act of conquest. Despite their great political differentiation, all the Greek states, with the exception of Thessaly and Aetolia, fought in the same way and with the same formation. The phalanx was a solid rectangle of heavily armored infantrymen carrying shield and spear and drawn up eight deep.

Except in Sparta, the heart of the phalanx was a citizen militia, not a professional standing army. Always before the fourth century B.C. and usually thereafter, the soldiers themselves paid for their own arms and armor, made available in quantity by the relatively advanced state of Greek metal-working crafts. Though slavery was well established in Greek society, the army was drawn from the citizenry. Drill of the phalanx was not frequent by modern standards. Its strength was moral as much as physical, based on common cause, volunteerism, and self-preservation. Only discipline, endurance, and commitment could prevent a breach in the phalangeal line.

A battle between phalangite armies was a test of weight and stamina, with the two formations, locked closely together, contending until one side or the other broke ranks and fled. There was no place in combat of this nature for generalship; once battle was joined, the individual was submerged in a mass of sweating bodies and, since reserves were unknown before the late fifth century B.C., there was no chance of outside assistance. Outflanking as a conscious maneuver was never used, except by the Spartans, beginning in the fourth century B.C. In battles between the citizen-soldiers of other states, outflanking occurred by accident or through the tendency of each man in the phalanx to sidle to the right in an effort to avoid exposing his unshielded right side. The highly trained Spartans took advantage of this tendency by wheeling, outflanking their enemy's left, and then rolling up his line from that flank. Victory in phalangeal warfare was limited in extent, for the phalanx, by itself, was incapable of achieving a crushing success. It was too unwieldy and immobile to be adapted to pursuit. Casualties in early Greek warfare were light, and most hoplites (heavily armored infantrymen) lived to fight another day.

Greek warfare remained stationary for several centuries. The phalanx was the standard military technique from the late eighth century B.C. until after the Peloponnesian War (431–404 B.C.). The explanation for this extraordinary conservatism lies only in part with the unwillingness to alter what was known and proven. Greece was not exposed to attack from an outside power employing different techniques until the Persian invasion in the early fifth century. Neither example nor incentive was present to compel innovation. Moreover, the social prestige of the infantry soldier has perhaps never been so high as in the ancient Greek world. The heavily armored hoplite, with crested helmet and massive shield, was for the Greeks the embodiment of military attainment, sanctioned by the triumphs of the past and the hero-image of an Achilles. It was an honor to serve in the phalanx; the most desirable position was in the front rank; and military service was the privilege as well as the duty of every free citizen. Only the aristocrats and the middle class fought as hoplites, however, because only they could afford the cost of the armor. The bulk of the free population served as lightly armed troops or rowed in the fleet. Moreover, there was no competing tradition of a cavalry elite, because most of Greece, owing to poor pasturage, was unsuitable for horse breeding, and the mountainous terrain was poorly suited to calvary tactics. Only Thessaly, in the north, and Boeotia developed significant cavalry forces. Neither state was able to use cavalry effectively away from its homeland plains.

The predominance of infantry can be accounted for on grounds of social prestige, military tradition, and political structure, and also the lack of strong competing traditions. But the paradox remains that in a country four-fifths of which was mountainous, the Greeks were employing a formation which, like cavalry, could be effective only on level ground. Only on the infrequent plains of Greece could the phalanx retain that solidarity which was its essential attribute. Enhancing the paradox is the fact that although most of the towns in Greece were fortified in one way or another (the Athenian Acropolis is an outstanding example), there were few sieges and no developed siege procedure.

Greek military development was highly specialized, both in the methods employed and in the terrain chosen for battles. This was so because warfare in Greece was profoundly affected by the fundamental economic problem of Greek history. The cultivable area of Greece was very small; through terracing and other artificial methods of coping with an unfavorable environment, Greece probably supported a somewhat larger population in the fifth century than in modern times. But food shortage was chronic, with the attendant evil of overpopulation. In part, these difficulties were overcome by the great colonizing activity of the seventh and sixth centuries B.C., which eased the population pressure by emigration.

The thin line between subsistence and deprivation meant, however, that agriculture became a primary focus of military conflict between city states. Recent scholarship has cast doubt upon the traditional view that invading armies attacked crops in order to force defenders to leave their walled cities and come out and fight; the crops were not all that easy to destroy. Still, it remains true that fighting was generally restricted to the autumn harvest period, when the crop was most vulnerable, and battles almost always took place in open fields.

Greek warfare was controlled by an unwritten warrior code, a part of the so-called laws of the Greeks. This early international law limited war and aggression within the community of Greek city-states. Destruction of a city-state was viewed as a barbarian act and prohibited. This unwritten warrior code inhibited the development of strategy, since the rules of the game, at least between Greeks, were already fixed.

At the beginning of the fifth century B.C., Greek warfare was running in channels already ancient, its frequency conforming to the rhythm of the seasons, its techniques adapted to the conditions of economic life, and its continuance guaranteed by military conservatism, social custom, and democratic politics. In democratic Athens, for instance, commanders could be called to account by the Assembly when they lost a battle. They therefore tended to be orthodox in their manner of fighting, as the least blameworthy course. For this reason, it is hardly surprising that no state had been able to dominate Greece politically. Leaving aside the obstacles of limited resources, rugged terrain, and complicated diplomacy, the conventionalized, severely restricted military methods pursued by the Greeks set sharp bounds to the ambitions of any city-state. In the fifth century, however, Greece was required to face the greatest power of the Mediterranean world. This challenge was survived, but it brought changes that were to alter the nature of Greek civilization.

THE PERSIAN WARS

The army of imperial Persia reflected the social and political organization of the empire; it was composed of men from many nationalities and language groups. Its core, however, was the Persians themselves, and those closely affiliated to them in race and religion. The best infantry, whether the Immortals of the Persian royal guard or the Medes and Elamites, consisted of regular professional troops. Although they carried daggers or short spears for close fighting, their basic weapon was the bow, and their standard tactic was to launch their arrows at an enemy from a distance, behind the protection of wicker shields planted in the ground. Other than their shields, they wore little or no protective equipment, a fact that astonished the Greeks. The cavalry of the Persian homeland probably came from the landowning aristocracy and their adherents; the Medes, Elamites, Bactrians, and Sakai also produced excellent horsemen. The basic weapon of cavalry, as well as infantry, was the bow; although some body armor was worn and the Sakai had axes, the function of the cavalry was not shock but the fixing of an enemy by a combination of speed and missile fire, so that he could then be destroyed by the infantry. These tactics had won Persia an Asian empire, but they proved unsuitable in Greek terrain against heavy infantry fighting on ground of its own choosing.

The Persian offensive against Greece was an attempt to prevent the Greek states from further helping their sister communities of the islands and eastern coastline of the Aegean Sea. The decisive engagement of the first Persian expedition, the battle of Marathon (490 B.C.), is known to us almost entirely from Athenian sources. According to them, the Athenian phalanx closed swiftly with the Persian infantry, and, although the Sakai broke the deliberately weakened Greek center, the great superiority of the heavy infantry in close combat was decisive on the wings and brought victory. There appears to have been little Persian cavalry at Marathon. In the subsequent battles of the second Persian expedition, Greek infantry successfully withstood the Persian combination of cavalry and infantry missile fire that had proved so lethal in Asia. In the stand of Leonidas and his Spartans at the pass of Thermopylae (480 B.C.), where the terrain was unsuitable for cavalry, the Persian infantry understandably refused to close with the hoplites, and the Spartans were eventually surrounded and overwhelmed by numbers and a hail of arrows from both front and rear. At Plataea (479 B.C.), in a preliminary engagement, the allied Greek phalanx was able to protect its flanks from cavalry assault by resting them against a mountain spur and a wall and indomitably kept ranks under steady missile fire. In the main battle, some days later, the Persian general, Mardonius, caught the Greeks dispersed during a retreat, and his cavalry halted the Spartan contingent. Spartan discipline held up under prolonged missile fire; then, when the Persian infantry was fully committed, the Spartan commander, Pausanias, launched his hoplites in an attack that smashed through the wickershield wall and put the Persians to flight. The Greek victories against the Persians were defensive in character, won by fighting on ground of their own choosing. It is noteworthy that the Greeks never took the offensive or offered battle in areas like the plains of Boeotia, where their shortcomings would have been revealed.

The truly decisive battle of the war, however, was fought not on land but at sea. The Athenian statesman Themistocles was aware that the Persian menace would persist as long as the Persian fleet was able to act as flank guard and support to a land army invading Greece. He succeeded in persuading his fellow citizens to devote the profits from a newly discovered vein in the state-owned silver mines at Laureion to the building of a fleet of 200 triremes, galleys of exquisite design and unmatched power. In the decade after Marathon, Athens by this supreme effort emerged as a naval power. The outstanding naval victory of Salamis (480 B.C.), won chiefly through a combination of ramming and boarding, resulted in Xerxes' withdrawal, with part of his army, from Greek soil. Mardonius, left with the bulk of the army and acutely sensitive to the threat to his communications now posed by the dominant Greek fleet, was driven to seek a swift victory. The result was the decisive Greek triumph at Plataea.

It has become commonplace to say that the thwarting of Persian imperialism allowed Greek civilization to reach its fullest flower and thus permitted a great cultural legacy to be bequeathed to Western society. As far as the Greeks could ascertain, however, the threat from the East had not ended, and the continued existence of this supposed threat had the widest political and military consequences. The Persian intrusion into the homeland had exposed a serious Greek weakness: Owing either to long-standing jealousies and traditions, or to the unwillingness of some states to sacrifice their own territories to Panhellenic requirements, the various states were unable to cooperate effectively against an outside power.

The necessity for coordination of effort in time of war drove certain of the Greek states into an alliance, the Delian League, so called because the treasury of the League was established on the sacred island of Delos. Members of the League were required either to contribute ships to the fleet or to render an equivalent in cash to the League treasury. Because Athens was the premier naval power, it was perhaps natural that she should head the League and be made responsible for its administration, although, had Sparta been more enterprising, the leadership would clearly have gone to her.

Under her great leader Pericles, Athens proceeded to transform the League from a defensive alliance of freely associated states into an empire based upon the coercive force of the Athenian navy. States wishing to secede from the League were brought to heel by military action and their contributions were increased; democratic governments on the Athenian model (and dependent upon Athenian support for their existence) were set up in all member-states; and in 454 B.C. the treasury was moved from Delos to Athens upon the feeblest of pretexts. With that, the Delian League essentially was converted into the Athenian Empire. League members expected that their contributions would end at the termination of hostilities with Persia. Instead, on the legitimate ground that her navy was performing a Panhellenic function by patrolling the Aegean, Athens insisted that their contributions should continue. Thus she irrevocably committed herself to a policy of naval imperialism. At the same time, Sparta was following much the same course in the Peloponnesian League she headed, except that she tended to support oligarchies, not democracies.

The middle years of the fifth century saw the splendor of Athens at its height: the building of the Parthenon and of other magnificent structures, the sculpture of Phidias, the poetry and drama of Aeschylus and Sophocles, the teaching of Socrates, the flowering of a great intellectual center. In this period Athens had a liberal democratic form of government, in which, in the words of Pericles, "power rests with the majority and not with a few; in private disputes all are equal before the law; and we ourselves personally either decide policy or at least form a sound judgment on it." At the same time Athens was enjoying an age of great commercial prosperity. A changeover from subsistence agriculture to the cultivation for export of the vine and the olive and the creation of large-scale pottery and metal-work manufactures had made Greece the premier trading nation of the Mediterranean, and Athens the wealthiest and most important commercial center in Greece.

To all these achievements the Athenian navy made an important, if not decisive, contribution. Its dominance of the eastern Mediterranean and Aegean safeguarded the channels of Athenian trade and guaranteed Athenian control of the vital wheat-producing areas around the Black Sea. Commercial wealth made possible the flourishing of Athenian culture and a leisure class to support it. It is scarcely necessary to point out the significance of the construction of the Long Walls linking Athens with its port of the Piraeus (456 B.C.). The fortunes of Athens were tied so intimately with sea power that she was attempting to make herself into an island.

The operation of the Athenian navy required thousands of men. Because the poorer citizens rowed in the fleet, the navy was always the stronghold of Athenian democracy; the army was a more likely source of potential oligarchs. Naval-induced prosperity brought certain problems. Manufacturers resorted to slave labor on a large scale, with the result that individual artisans were unable to compete and many free citizens were unable to find employment. The navy thus became a means of easing economic distress.

Although Pericles might have justified Athenian imperialism on the grounds that "our city as a whole is an education to Greece and our individual citizens excel all men," it might equally be shown that Athens was forced to convert the Delian League into an empire based on money contributions because of social and economic dislocation stemming originally from her very success as a naval-commercial power. Whether or not this is so, Athenian statesmen realized that, for Athens, there could be no return to the orbit of a parochial city-state, set predictably within the constellation of its fellows. Her seaborne commerce demanded an expensive fleet, which only commerce-derived wealth and the Delian contributions could maintain; inexorably, therefore, she was driven along the road to empire.

THE PELOPONNESIAN WAR

The threat posed by Athens to the autonomy of other Greek states caused them to cluster about the only state in Greece which had the military power to check her. Sparta was unquestionably the most formidable land power in Greece; but she owed this preeminence to a peculiar set of political and social arrangements comparable to nothing in history except perhaps the "national socialism" of modern totalitarian states. In the eighth century B.C., when other Greek states solved the problem of a steadily increasing population and a relatively diminishing food supply by colonization, Sparta turned instead to the conquest of neighboring Messenia. The food shortage was relieved by dividing arable Messenian land into equal allotments for each Spartan family. Since the land was cultivated by the Messenians, Spartan males were left free to engage their energies in the one great object of their state, the maintenance of an iron rule over the Messenians and other subject populations, collectively known as the Helots. Every Spartan, female as well as male, was trained, in the appropriate age group, to physical perfection through a competitive program of athletics; perpetuation of the best stock was achieved by placing eugenic control in the hands of the state, which rejected unfit infants, encouraged the marriage of fine physical specimens, and expelled the cowardly and the idle. From the age of seven (when he was taken from his mother) to the age of sixty, no Spartan male was free from military discipline and service. From this system emerged the best heavy infantry of the age, greatly superior even to other Hellenic hoplites. To this end was sacrificed the many-sided diversity of Greek culture: a narrow preoccupation with the forging of a remarkable military instrument meant the stifling of all other activities as mere distractions. For this reason, after the Messenian Wars there is no tradition of Spartan development in the plastic arts, poetry, drama, or philosophy. The political structure congealed into a dictatorship of the minority as early as the seventh century B.C. It is the great irony of Spartan history that, having labored at enormous cost to construct an irresistible army, the Spartan people were unable to use their creation in large-scale ventures outside their own territory. They were living atop a volcano. As Aristotle put it, "the Helots may be described as perpetually lying in wait to take advantage of their masters' misfortunes." Thus, even in the Persian Wars, Spartan participation was limited and unwilling.

The rise of these two great powers, Athens and Sparta, on the same peninsula, made a clash all but inevitable. The conflict arose within fifty years of the defeat of the Persian threat.

A prime cause of the Peloponnesian War (431–404 B.C.) was the anxiety of the commercial state of Corinth over Athenian trade expansion. Corinth succeeded in dragging a reluctant Sparta into war when Athens chose to apply an economic boycott to the city of Megara, a clear demonstration to all Greek states that Athens was prepared to use her naval dominance to control the distribution of food imports.

The Peloponnesian War was a death struggle between the two most powerful states of the Greek world, but neither Spartan heavy infantry nor Athenian naval preeminence was, by itself, sufficient to bring victory. The result of an early stalemate was the beginning of innovations which were to change the nature of Greek warfare. Learning from the wars with Persia, Athens added cavalry and light-armed troops (*peltasts*) to its order of battle, and Sparta reluctantly followed suit. These changes occurred because it was slowly recognized that the phalanx alone could not achieve war-winning results and because light troops frequently held their own against heavy infantry. The war dragged on because the Periclean strategy of the Athenians was to avoid land battle. Eventually Sparta found it necessary to fight Athens on its own element, the sea.

Sparta's victory was made possible by an Athenian naval and military disaster at Syracuse in 413 B.C. In Sparta's acceptance of Persian money to build a fleet lies the explanation of her final victory; in return, Persia was permitted to repossess the Ionian states lost by the Empire in 479 B.C. The war left in its wake a trail of devastation, especially in Athenian territory. In addition, Greek society proved too frail to withstand the stresses of fratricidal war. Everywhere, but particularly at Athens, there occurred a struggle for power between contending classes, resulting in violent political upheavals, tyrannies, and the rapid alternating of power from class to class.

ALEXANDER THE GREAT

The period between the end of the Peloponnesian War and the establishment of Macedonian hegemony in Greece was one of political chaos, social disintegration, and economic distress and, at the same time, of the most rapid military development. Victory in the war proved lethal for Sparta: it forced her to become, as the only state capable of giving order to Greece, an imperial power; but Spartan governors and garrisons, men trained exclusively in military matters, were unable to achieve the political flexibility that their new position demanded, and the harshness and rapacity of Spartan rule soon led

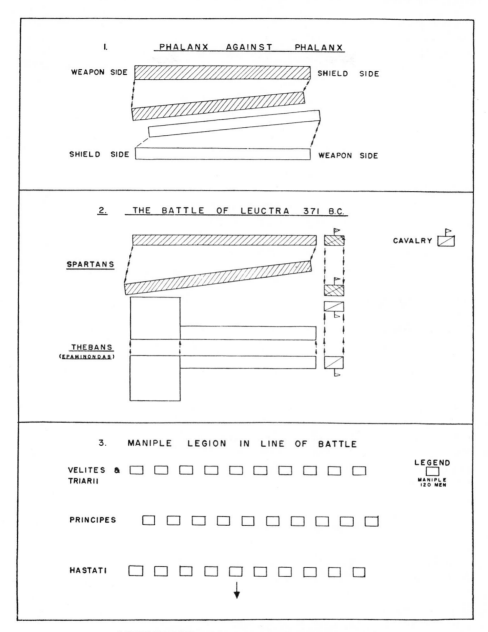

MILITARY FORMATIONS OF THE ANCIENT WORLD

to resistance. In addition, the underlying contradiction of Spartan might re-
asserted itself: the maintenance abroad of garrisons weakened Sparta's grip
on the Helots and rendered its imperial policy half-hearted. Even more important, the necessity of providing garrisons drastically weakened the core of
Spartan strength, the army. Thus, when the brief period of Spartan domination was ended by Thebes at the battle of Leuctra (371 B.C.), the proportion
of Spartans to Helots in the battle order had dropped from the normal four
in ten to one in ten.

Leuctra illustrates an evolution in Greek warfare which had begun during the Peloponnesian War. Thebes had achieved considerable military success by deepening the phalanx and advancing in heavy column, heightening
the shock effect, though on a narrow frontage. The Spartan commander, seeking to take advantage of his longer line, attempted to outflank the Thebans by
extending his right wing behind a screen of cavalry. The Theban commander,
Epaminondas, discerning the Spartan intentions, sent his Boeotian cavalry to
drive the enemy horse into the outflanking force, and immediately behind his
cavalry the Theban phalanx, advancing obliquely, smashed into the disordered
Spartan infantry. The issue might still have been in doubt, since some elements of the Spartan flanking force remained on the right, but they were
broken up by the Sacred Band, a picked body of heavy infantry which
Epaminondas had held in reserve. The coordination of different arms had
made possible effective offensive tactics and had also placed a new importance on the role of generalship.

But Thebes was no more successful than Athens or Sparta in holding
Greece together as one political unit. In part, Theban power was curbed by
the same shortage of manpower that had restricted Sparta and by the same
failure to create a form of government transcending the concept of the city-
state. The devastation and unemployment resulting from the Peloponnesian
War and subsequent conflicts heightened political instability in every Greek
state, including Thebes, and rendered military success fleeting and mean-
ingless. It was left for a state the Greeks considered barbaric, or at least not
quite "Greek," to impose upon the peninsula the centralized rule of a conquering power.

In constructing an empire, Macedonia did not have to break through the
city-state habit of mind, for it had clung to the institution of monarchy, a
form of government long abandoned in Greece itself. Therefore, extensions of
Macedonian rule were merely extensions of the personal power of the Macedonian king. The basis of Macedonian strength was an army which was perhaps the most powerful military force constructed before the coming of
gunpowder. Although its creator, Philip II, had borrowed inspiration during
an enforced stay in Thebes, the army was founded ultimately upon the social
resources of Macedonia itself. The country was predominantly rural; from its
free peasantry came the phalangite, differing from his Greek counterpart in
that he wielded a somewhat longer spear; the Macedonian phalanx was sixteen deep armed with a sarissa of fifteen feet or more, compared to the traditional Greek phalanx of eight to twelve ranks armed with a pike of ten
to twelve feet. From Macedonia's land-holding aristocracy came the prime

striking force of the army, the cavalry of the King's "Companions," some two thousand strong. To these basic elements Philip added mercenary cavalry components drawn from the neighboring state of Thessaly, and also infantry units of different kinds, including slingers and javelin throwers, whom he had probably seen in action in Greece.

The result was the first balanced army of historic times, an army equipped to fight over any kind of terrain and against any enemy. It could operate defensively and offensively with equal efficiency. Philip must also be regarded as the founder (at least for the Greek world) of scientific siege warfare; his army contained a full siege apparatus, including towers, rams, and torsion catapults, the latter invented shortly before Philip's reign began. The army was kept in the field by selective conscription for set periods; in this way the Macedonians obtained the first standing army for year-round campaigning. With the long campaigns of Philip's son, Alexander, the Macedonian army transformed itself from a national conscript army into a professional army with a history and *esprit* of its own.

Militarily, the dissension-torn Greek states could not resist the new army of Macedonia. In the battle of Chaeronaea (338 B.C.) the irresistible Macedonian horse supplied the margin of victory and made Philip master of Greece. It was Philip's ambition to lead a united Greece against the traditional Persian enemy, and to this end he had himself appointed Captain-General of the Hellenes. His assassination left the project to his son Alexander, who assumed the title and the ambition after quelling Greek insurrections. There were factors besides royal ambition, however, which encouraged the idea of an expedition against Persia. Such an attack against the "barbarians" had been preached for many years by political publicists like Isocrates, whose propaganda, truly imperialist in nature, was based on the superiority of the Greeks as a people and hence their undoubted right to govern others. Although Panhellenism as a unifying force had never been strong, Isocrates' arguments fell on fertile soil, for, in the fourth century, the Greek export economy was being slowly strangled by the rise of competitors like Carthage and by the growth of home manufacturing in Italy, the Black Sea areas, and the Persian Empire itself. As a result, population once more began to press upon food supply; and as before, many Greeks began to think in terms of overseas expansion as the solution of their difficulties. This situation explains one of the most marked phenomena of the military history of the fourth century: the presence on foreign soil of large numbers of Greek mercenary troops. As Isocrates put it, "many, through lack of daily sustenance, are compelled to serve as mercenaries, and to die fighting for the enemy against their own friends." In 401 B.C. 10,000 Greeks took service with Cyrus the Younger in his abortive attempt to seize the Persian crown from his brother Artaxerxes II; their exploits, described by Xenophon in his *Anabasis,* are the best known of many such adventures during the century.

At the moment in Greek history when social and economic exigencies were forcing Greeks to look abroad, a channel for the release of pent-up energies was provided by the perfected Macedonian war machine, with a military genius at its head and a rich and unwieldy empire as its target. It does not seem that at the outset Alexander contemplated the overthrow of the Persian Empire and the construction of another in its place, but an unbroken string of victories led him on beyond the limits of the world previously known to the Greeks. Syria, Egypt, Babylonia, the Persian heartland itself, the fringes of the great Russian steppes, the western ranges of the Himalayas saw his army pass, until, in northwest India, weary and homesick after eight years of campaigning (334–326 B.C.), the Macedonian veterans refused to go farther and Alexander was forced to turn back. In this unprecedented march (which the Romans failed to equal) each component of the army was tested in turn, as different terrain was entered and as different enemies were encountered. The light-armed peltasts proved their worth in arduous mountain fighting, while the siege train met the difficult challenge of the island fortress of Tyre successfully in a long and hazardous siege. The conquest of Tyre (332 B.C.) was the most spectacular of Alexander's successes in siege warfare; it was emulated many times in the capture of other strongholds.

The greatest element in the Macedonian technique of conquest, however, and that which constituted the most significant advance on traditional Greek warfare, was the combination of the rocklike phalanx with light and heavy cavalry. The union of an always dependable infantry base with the mobile shock supplied by cavalry was too much for the valorous but heterogeneous Persian masses, whose chief advantage, numbers, was vastly outweighed by steadiness, missile fire, cavalry shock, and generalship.

More than any previous commander, Alexander possessed the attributes which, in modern times, are associated with generalship; and it is this imponderable of generalship, the uniformly skillful wielding of an almost perfect military instrument, which is the most difficult factor to measure in accounting for the conquest of Persia. The qualities of Alexander as a general emerged most clearly in the most critical battle of the whole campaign at Gaugamela on the river Tigris (331 B.C.). Alexander, adhering to the old custom, always entered the battle personally at the head of his cavalry, but never until he had concluded that the decisive moment was at hand. Thus at Gaugamela he withheld his Companion cavalry until the Persians had been totally engaged and then launched an assault which crumpled the Persian flank, drove it in upon the center, and turned an apparent Persian victory into a rout. Alexander showed the qualities of greatness in other ways. He invariably followed up his victories on the field with a relentless pursuit; and he fought in all weather and in every season. As a strategist, he was the master of his age. When the Persian campaign began, the Persian fleet dominated the eastern Mediterranean and menaced Alexander's communications. Rather than risk a defeat at sea, Alexander systematically worked his way down the Mediterranean coastline, capturing and garrisoning every seaport until Persian naval superiority had been nullified.

The great empire constructed by force of arms survived Alexander's death by only a few years. Its enormous extent and the consequent communications problem made centralized administration almost impossible; the empire tended to break down into well-defined regions or national blocs.

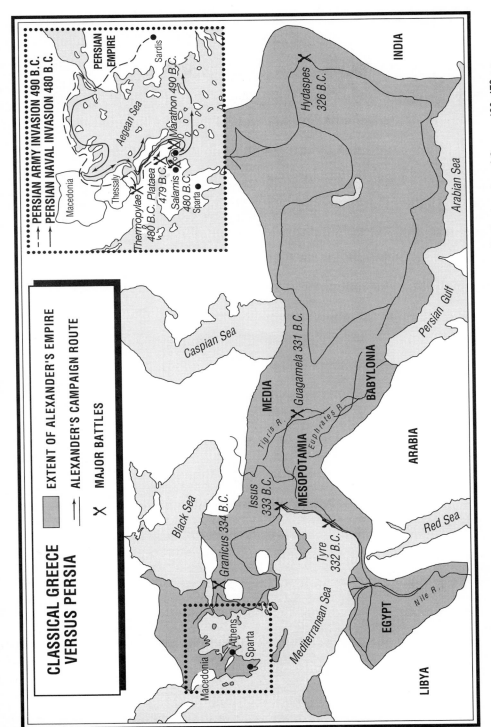

THE MACEDONIAN EMPIRE, 330 B.C., showing Alexander's campaigns, 334–323 B.C. Inset: Persian Invasion of Greece, 480–479 B.C.

The Greeks, especially the Macedonians, were too conscious of their own virtues to mix freely with the Persians and with other groups, and so there could be no supranational basis for the empire. It required more than an invincible military machine to win and hold this great territory; the failure of the Graeco-Macedonian conquerors to institutionalize their military triumphs into a positive political system meant that once the arresting personality of Alexander was removed, the empire rapidly disintegrated.

But the career of Alexander was much more than a transitory tour de force. His conquests were made possible by an army which itself was the product of centuries of Greek development. This army broke down barriers that could never be restored, vastly widened the area of international commerce open to the Greeks, and, in place of the old particularism, substituted the concept of the Mediterranean world. The Macedonian army served as the transmitter of Greek civilization. The "islands" of Greek culture left in its wake Hellenized the Mediterranean world and provided the foundation, not only for the Roman Empire, but for the coming of Christianity.

3

CLASSICAL WARFARE:
THE AGE OF THE LEGION

THE CONQUEST OF ITALY

"We find that the Romans owed the conquest of the world to no other cause than continual military training, exact observance of discipline in their camps, and unwearied cultivation of the other arts of war. Without these, what chance would the Roman armies, with their inconsiderable numbers, have had against the multitudes of the Gauls . . . or the Germans? The Spaniards surpassed us not only in numbers but in physical strength. We were always inferior to the Africans in wealth and unequal to them in deception and stratagem. The Greeks, indisputably, were far superior to us in skill in arts and all kinds of knowledge." These words, set down by the military writer Vegetius when night was falling on the Roman Empire of the West, indicate how much that empire, the most durable and extensive of ancient times, owed to the meticulous Roman approach to war. And, notwithstanding the deference of Vegetius to superior attributes of other peoples, the Romans alone possessed the political and administrative capacity to consolidate the gains of their generals and thereby bridge the gap from city-state to empire.

When Greece, having repulsed the Persians, was just entering upon its greatest days, Rome was a barely civilized city-state fighting for existence on a peninsula dominated by Etruscan warlords and the Greek cities of southern Italy and Sicily. By 265 B.C. Rome had become the master of Italy and was preparing to embark upon the conquest of the Mediterranean world. Favored by geography, Rome was sure of a place of some importance in Italy. The city straddled both the Tiber River and the main north-south trade route; and the rich plain of Latium supported a dense population which provided an ample reservoir for Roman armies. Early Rome was a farming community; its people were conservative, unostentatious, and practical in outlook; the society was marked by a singular toughness and cohesion despite its rigid division into patricians and plebeians. All the Roman virtues tended to involve obedience to some established authority: to the family, to the state and the laws, and to the gods. On this solid base the Roman republic was founded, its chief organ of government being the conservative, patrician Senate. The two chief magistrates, the consuls, were elected annually; in addition to their civil duties, they also shared the command of the army. This peculiar dual command, designed to lessen the possibility of military tyranny, was often a military drawback. The annual changeover made continuity of policy difficult and the election of inexperienced commanders possible; the day-to-day rotation of command resulted in deadlock when the consuls disagreed, a deadlock that could be broken only by the election of a dictator for an emergency period.

As the Roman army mirrored the society of which it was a part, it was a levy of the citizen-body, or rather of those citizens who could meet the property qualification which conveyed the right (and obligation) to serve. Each citizen-soldier was responsible for providing his own equipment; grading within the army was determined by wealth; and the mass of the common soldiery was composed of small farmers. The early Roman army was not much different from the armies of its neighbors in that it was an infantry force using a phalanx close to the Greek model. Its great strength lay in the simple patriotism and rigid discipline characteristic of Rome and in the attention the Romans seem always to have paid to matters of training and drill. From very early times, and certainly by 400 B.C., soldiers were paid by the state.

The first stride in Rome's climb to the headship of Italy was the development of the legion, the standard infantry formation of all later Roman armies. Roman tradition assigns this innovation to the Second Samnite War (326–304 B.C.), and there is no cause to question it. The Samnites were people of the mountainous region bordering on Latium, and in meeting them on their own ground, the Romans found that the phalanx could neither maintain ranks nor achieve the mobility required to come to grips with the enemy. It was therefore broken down into self-contained units called maniples ("handfuls") of 120 men. The legion was an open formation of thirty maniples, plus five in reserve, arranged in a checkerboard pattern. The mature legion had a high degree of flexibility and range of tactical employment; the first line of maniples, for example, could retire through the openings in the second line without throwing it into confusion, and the second line could similarly advance. For the single shock of the heavy phalanx the Romans substituted a series of smaller shocks by units infinitely more maneuverable and versatile, a tactical system that made much greater demands upon the individual soldier and that was attained only by the strictest training and discipline.

Rome was primarily an infantry power, and this was never more so than in the period of expansion on the Italian peninsula. One legion (i.e., one field army) numbered roughly 6,000 men, half of them heavy yeoman infantry and the remainder lightly armed skirmishers (*velites*) and a small body of cavalry to cover the flanks. In battle order, the first two lines of maniples were composed of the heavy infantry (*hastati* and *principes*), equipped with shields, helmets, body armor, short, broad-blade thrusting swords (*gladii*), and throwing spears (*pila*). Although the commander had many methods of attack at his disposal, fundamental tactics were simple: the front line let fly its *pila* and charged at the run, to fight at close quarters with the sword. If this initial wave was repulsed, the second line took over, using the same procedure. The third line, a mixture of heavy and light infantry, was held in reserve. Later Roman generals were to exploit the potentialities of this reserve to the fullest extent.

The perfecting of the legion as an instrument of war was a gradual process which took place over many years; but even in its semideveloped state it was so vastly superior to the crude tactics of Rome's neighbors that by the middle of the third century B.C. she controlled almost the whole of peninsular Italy. Military success, moreover, had been turned to political advantage by an imperial policy at once enlightened and practical. Each Italian state was treated not as a subject of Rome, but as an ally; there was no interference with local autonomy. Rome merely reserved the control of foreign policy to the Roman Senate and, in time of war, required contingents of troops from each dependent state. This liberal policy, coupled with the eventual extension of Roman citizenship and Roman culture throughout Italy, secured for Rome the attachment of other Italian states, an attachment cemented by the building of admirable roads of military as well as commercial importance. The roads were tied in with Roman colonies to form a military network throughout the peninsula. Unlike Greek colonies, the Roman colonies remained units of citizens under Roman direction and were to form an important part of the resistance to Hannibal.

THE PUNIC WARS

It would have seemed that in 265 B.C., with the reaching of her "natural" frontiers, Rome had also reached the limit of her legitimate aspirations; but at this point she was pulled into the wider theater of Mediterranean affairs. Rome, it is fair to say, had conquered Italy almost in self-defense and against the inclinations of the Senate. (At times, to evade its own regulation against the waging of aggressive war, the Senate trumped up fictitious aggressions by others.) Rome now collided with the mercantile state of Carthage, in North Africa, whose navy controlled the western Mediterranean and therefore menaced Roman interests in southern Italy. At this point the political effect of the long wars of expansion came into play: the continual demands upon Roman manpower had forced the patrician Senate to make concessions to the plebians. In 265 B.C. the democratic assembly (*plebiscites*) was able to override the misgivings of the Senate and force a war against Carthage.

The Romans found that, to deal effectively with Carthage, to strike at the African heart of her empire with land forces, they must first beat her on her own element, the sea. Although the Romans at no time felt at home on the sea or ever gave much consideration to sea power, they met the Carthaginian challenge by constructing a fleet, using as a model a Carthaginian trireme that had been driven ashore. At first, because of lack of experience in naval tactics and the mishandling of ships in bad weather, Roman losses were enormous, as many as 500 ships and 200,000 men. Losses, however, were made up by drawing upon the plentiful resources of Italy in men and materials; and naval shortcomings were surmounted by turning battles at sea into boarding encounters through the use of grappling hooks and swinging gangplanks. Thus Rome overcame her difficulties and henceforward was a naval power.

The First Punic War (265–241 B.C.) had brought Rome her first offshore possession, Sicily. Although Rome did not at this time maintain a permanent fleet, she had, in effect, deprived Carthage of the naval leadership of the western Mediterranean. Under the leadership of Hamilcar Barca, and later of his son Hannibal, in the interval between the First and Second Punic wars, Carthage attempted to restore her position by absorbing Spain. When Carthage and Rome clashed in this area in 218 B.C., the possession of sea power should have given the advantage of the offensive to Rome. Hannibal, however, forced his enemies to conform to his own strategical conceptions. He saw that the Roman navy was no more than the outwork of a state whose true vigor was on land and that his only chance to crush Rome lay in entering Italy and endeavoring to detach the Italian confederates. He crossed the Pyrenees, the Rhone River, and the Alps, thereby outflanking the Roman navy. Then, by a series of smashing victories, he brought Rome almost to her knees.

Hannibal's victories were won by a conglomerate army made up from the many subject peoples of Carthage. Through the force of his personality, and as a result of long training, he had welded this army into a sensitive, disciplined weapon which would respond to his guidance even in the heat of battle. Cavalry formed a high proportion of its total strength and was its most potent arm. When directed by the consummate tactical sense of Hannibal, the mobility and shock-power of the Carthaginians were too much for the mechanically efficient but poorly led Roman legions. At the River Trebia, hidden cavalry forces fell upon the rear of the Roman legionaries, who had hitherto been more than holding their own; at Lake Trasimene the Roman army in marching order blundered into an ambush and was slaughtered before it could deploy; at the crowning disaster of Cannae (216 B.C.) the largest army Rome had ever put in the field was destroyed through the outstanding generalship of Hannibal and the stupidity of the Roman commander. The consul at Cannae, C. Tarentius Varro, a Roman businessman whose day it was to be general, discarded the open order of the legion and crowded his 80,000 men in a phalangeal formation on a plain ideally suited for cavalry action. When engaged by the Carthaginian infantry, the huge Roman block lost its coherence; when Hannibal's cavalry swept in to envelop it on both sides, the Roman army became a tightly packed huddle of struggling men, powerless to use their arms.

In three calamitous battles Rome had lost well over 100,000 men and an immeasurable amount of prestige. That it did not collapse is a tribute not only to its own toughness and resilience, but also to its imperial and military power. Many slaves were freed and, along with underage youths, were enlisted to restore the shattered legions; 173 senators were appointed to take the place of those who had fallen at Cannae; and a policy of disengagement was substituted for the aggressive policy which had brought such unfortunate results. Under the leadership of Fabius Maximus, called *Cunctator* (the Delayer), Roman units practiced what came to be called Fabian tactics, hanging about the Carthaginian camps and practicing the familiar techniques of guerrilla warfare, nibbling at Hannibal's limited manpower but refusing at all times to come to grips. Much depended, however, upon the conduct of the Italian states and their response to the proclamation of Hannibal: "I have come not to fight against you but to attack Rome in your behalf . . . I have come to restore freedom to the Italians and to assist you to recover the cities and lands that you have one and all lost to Rome." Rome's past imperial policy was vindicated when Hannibal failed to receive any significant help from the Italians. Without their assistance he could menace Rome but not besiege it, and he had to content himself with methodically devastating the Italian countryside. Meanwhile, because of Roman sea power, reinforcements from Carthage had to march one-third of the way around the Mediterranean to reach him. For thirteen years this war of mutual frustration continued, although Hannibal's chances of success had really vanished by 207 B.C., when a Carthaginian army of reinforcement was totally defeated at the River Metaurus in northern Italy.

In Scipio Africanus, Rome finally produced a general of stature comparable to Hannibal. He did not achieve military command in the old way, through the ascension of the prescribed political ladder. Appointed by the people to command in Spain, he had trained his army in a series of hard campaigns to a high standard of excellence. Under Scipio the maniple became more than ever before the true combat unit. Each maniple had objectives and methods of its own; through the perfection of its parts the legion became one of the few infantry forces of the pre-gunpowder period to hold its own with cavalry. Scipio borrowed Hannibal's tactics of envelopment and applied them to his infantry by masking his third-line reserves behind skirmishers and shooting them out from either wing upon the flanks of his enemy. He also repaired somewhat the Roman deficiency in cavalry by increasing its numbers and by recruiting horsemen among peoples more accustomed to that form of war than were the Romans. In 205 B.C. he was given the direction of an army, not for a year, but until Carthage was defeated. Taking full advantage of control of the sea, he carried the war to North Africa, forced Hannibal to return to protect his own homeland, and defeated him in the decisive battle of Zama (202 B.C.), where his balanced and flexible army outmatched Hannibal's medley of mercenaries and conscripts.

This surprisingly swift end to a long war had been brought about by a fundamental recasting of the old Roman citizen army. Roman military organization had been shifted from its base to begin a new line of development which was to culminate in the professional army of Julius Caesar. Scipio's army did not change its composition yearly as one annual levy replaced another. Although the qualifications for service remained the same, in practice Roman citizens were henceforward enrolled for long service that, by the second century B.C., seems to have averaged six years. The only limitation on length of service was the sixteen-year maximum civic obligation. Rome owed her victory over Carthage to the qualities of discipline upon which the legion was based, to her facility in appropriating the techniques of other societies, including the use of sea power, and to the imperial policy that had brought her the nearly unanimous support of the Italian confederacy.

EXPANSION OF THE REPUBLIC

Carthage was the most formidable opponent the Romans were ever called upon to face. When, in the second century B.C., Rome turned her face to the eastern Mediterranean, there was no comparable power to block the way. The Macedonian successors of Alexander, who did most of their fighting in Greece, had successively discarded the complexities of his magnificent army, until nothing was left but the original phalangeal nucleus, now even more unmanageable than its predecessor, and with its flanks guarded only by a few inferior horsemen. This array was adequate to the conditions of Greek warfare but failed repeatedly in combat with the legion. The superior Roman cavalry drove in its flanks, or the speedy maniples took it in the rear. At the decisive battle of Pydna (168 B.C.), where the wings of the phalanx happened to be well guarded, the Greek infantry drove irresistibly forward but, whether through poor marching or uneven ground, a small gap opened in the front rank, the maniples poured in, and the phalanx was split as a brittle stone is split by ice. With Macedonia disposed of, it was not many years before every state bordering the eastern Mediterranean had acknowledged the Roman republic as suzerain.

Rome's rapid expansion from city-state to empire in all but name set in motion broad social and economic processes which altered completely the nature of the Roman state and society and profoundly affected Roman military organization. Although many of the territorial gains of the late republican period were justified in the rhetoric of the traditional defensive policy, more than a century of responsibility for foreign and military affairs was working to transform the Senate into an arbitrary body drawing its members from one exclusive class, the war-wealthy *nobiles,* and was changing its goals from a quest for defensible frontiers into an undisguised adventure for riches and territorial aggrandizement. This rapacious policy was symbolized by the ruthless sacking of Carthage in 146 B.C. The Senate had no imperial program similar to the sensible federal policy followed on the Italian peninsula. As new territories were added, governors (proconsuls) were sent out from Rome, nominally under the centralized authority of the Senate but actually having a largely free hand. Because the proconsul was commander of the army in his province, he presented a potential danger to the security of the state.

Domestic problems went much deeper. Several factors stemming from the long wars had conspired to weaken gravely the old base of the Roman state, the class of small farmers. Hannibal's widespread devastation of the rural areas had caused Italian agriculture to stagnate. The Italian population dropped by 17 percent during the Punic Wars. Much of this decrease was due to the casualties suffered by the farmers who had formed the mass of the Roman armies. For many veterans, the task of reestablishing their holdings was too great, especially for those near Rome, who had now to compete with the quantities of cheap grain pouring in from new provinces like Sicily and from Egypt. The *nobiles,* prevented by law and social taboo from engaging in commerce, found an outlet for their war profits by converting the small peasant holdings into vast ranches, or *latifundia,* and changing from grain cultivation to vines and olives, crops much more suitable for the lands south of the Apennines. Such products involved a long-term capital outlay that the small peasant could not undertake, nor could he compete as labor with the slaves available to rural entrepreneurs as a by-product of Roman conquests.

These social and economic movements drove many Roman farmers into the cities to swell the rootless urban proletariat and be manipulated by politicians prepared to exploit their bitterness. In social terms, a great wedge had sundered the connection between the governing class and the citizen body; in military terms, the property qualification for military service could no longer be depended upon to fill the legions. The old military framework was further weakened by the tendency of many members of the upper classes, who had hitherto officered the legions, to buy their way out of service in order to enjoy a life of ostentatious leisure which would never have been tolerated in earlier Rome. For the solid virtues of the antique Roman state were being corroded as well; in the train of Roman armies came not only slaves and tribute but also exotic cults from the east, much more exciting than the formal state religion, which broke down the sense of duty to the state. It could be argued that strongly individualistic codes of conduct such as Stoicism and Epicureanism had a similar effect, yet Stoicism reinforced the sense of duty to the state, and Epicureanism was espoused by many leading figures of the late Republic, including Julius Caesar. Whatever the proclivity of his leaders, the typical legionary remained a countryman, believed in the old Roman religion, and responded to traditional civic and patriotic appeals.

During the second century B.C., the demands made upon Roman manpower were immense. The wars with Macedonia; the expansion into Asia Minor, Egypt, and Gaul; the Third Punic War; and, above all, the terrible wars of pacification in Spain, which continued virtually throughout the century, meant continual levies upon the small landholders, who formed the backbone of the army. The new provinces required permanent garrisons, beginning with the stationing of two legions in Spain from 197 B.C. By the end of the century, the garrison army numbered eight legions, or about 42,000 men. Because garrison service tended to be even longer than the service performed by those called up "for the duration" to meet a war crisis, and an absence of even two or three years could ruin the small farmer, military service became increasingly unpopular among the very class that formed the bulk of the armies. The attempts by Tiberius and Gaius Gracchus, during their respective tribunates (133 B.C., 123 B.C.), to meet the distress of the small holders and discharged veterans through land reforms met with failure but were a precedent for the land-allotment policies of the political generals in the last century of the Roman Republic.

The time was ripe for the emergence of men who knew how to use social unrest to promote their own careers. Those who could summon the help of armed force were naturally in the best position, and the first century B.C. witnessed the marriage of politics and the army in the persons of four outstanding generals. The first of these, Gaius Marius, was repeatedly elected to the consulship (contrary to law) on the basis of vigorous demagogic appeals. Because in 107 B.C. he threw the legions open to volunteers from the whole citizen body, Marius has usually been considered the creator of the wholly professional and proletarian Roman army. But this measure was simply a reflection of the desperate manpower crisis that had caused similar measures in the past, and of the social and economic distress that had brought Marius the political support of those upon whom the draft bore hard. A century before, Scipio Africanus had been required to raise his expeditionary force exclusively from volunteers; throughout the second century, volunteers had repeatedly been called for from among those eligible for the levy; and in 152 B.C. the draft had been by lot rather than by selection, because the magistrates tended to pick those who already had some military experience. The inflation caused by trade expansion and vast imports of specie as the result of Roman conquest had had the effect of greatly lowering the property qualification in real terms; during the century, that qualification was further lowered by legislation from 11,000 *asses* to 4,000. Marius, then, was scraping the bottom of the barrel. Moreover, although, because of the Marian reform, the proportion of volunteers in the army greatly increased, particularly from propertyless citizens attracted by the promise of a land grant upon discharge, the compulsory levy based upon property continued to operate. While the professional character and quality of the army rose after Marius, the citizen-draftee was still to be found alongside the long-service volunteer.

Marius also was once regarded as the inventor of the cohort, but it is now thought that this tactical innovation evolved gradually during the second century to meet the onrush in battle of such peoples as the Spanish and Gallic tribes, who did not use the phalanx. Whatever the explanation, the result was that the maniple was replaced as the tactical unit by the cohort of 360 heavy infantrymen.

This innovation did not greatly reduce the efficiency of the legion; if anything, discipline was stricter than before and foot and arms drill was kept at a high level by the use of professional gladiators as instructors. To compensate for the loss in mobility sustained by the creation of the cohort-legion, Marius placed on a regular footing the bands of mercenaries who had long been associated with Roman armies. He abolished the velites and legionary cavalry and replaced them with Balearic slingers, Numidian and Gaulish horsemen, and other auxiliaries gathered from the provinces. Henceforth, these troops were permanent and indispensable parts of the Roman field force. The professional approach to war may be seen in the careful attention paid to such procedures as a vigilant order of march, the equipping and provisioning of the individual soldier, and the exemplary use of camp and field entrenchments.

Because tangible rewards of loot or the grant of a parcel of land upon discharge could come only from the general, the immediate result of the reform of Marius was to divorce the army from its allegiance to the state and make it almost the personal property of its general. The army's attachment to the Roman state was further weakened when, as an outcome of the Social War (91–88 B.C.), the inhabitants of the Italian provinces were admitted to full Roman citizenship. The integration into the legions of the Italian *socii,* men who had little cause to love the Senate or the old constitution, meant that no longer did an army under Roman command flinch from marching upon the city itself. The active intrusion of the army into politics brought a period of almost uninterrupted civil war: by Sulla against the supporters of the dead Marius; by Julius Caesar against Pompey, who had overthrown the Sullan constitution; and by Octavian (Augustus), first against the conservative revolutionaries who had assassinated Caesar and then against his own associate in power Anthony. Military anarchy brought no noteworthy advances in war technique. It was, however, an age of brilliant generals, and from this galaxy it has been customary, with good reason, to single out for special consideration the generalship of Julius Caesar.

JULIUS CAESAR AND THE ARMY OF EMPIRE

When Caesar took command in Gaul, in 58 B.C., he was in his early forties and had had military experience in Asia and Spain. By 48 B.C., ten years later, Gaul (modern France and Belgium) had become a Roman province and Caesar a military dictator. The juxtaposition of the two facts is not coincidental. Caesar was a man of extraordinary versatility, of ability amounting to genius in many fields. The thread linking his talents was a driving political ambition; his reason for absenting himself from Rome for a decade was his recognition that the shortest road to high office lay in conquest of the provinces. Even his classic account of the subjugation of Gaul, *The Gallic War,* was in large part designed as propaganda to enhance his reputation in Rome. Many subsequent critics, including Napoleon, have slighted his military abilities, particularly at the tactical level, while conceding his unfailing aptitude to solve brilliantly embarrassing dilemmas caused by a slip in judgment. For example, when besieging a host of Gauls under Vercingetorix at Alesia, he allowed himself to be surrounded in turn by a relieving force. Responding to this grave challenge, he caused miles of fortifications to be strung around the city, facing inward and outward, and through tenacity and inspiring leadership secured the surrender of Vercingetorix and the dispersal of the relieving army. Though the Gauls had defeated lesser Roman armies in the past, they were no match for seasoned Roman troops led by an able commander. But, in 49 B.C., near Ilerda in Spain, when he was faced by a numerically superior army of Pompey's Roman legionaries, through a series of strategically memorable marches and countermarches coupled with adroit political propaganda, Caesar won its surrender without fighting a battle. As he himself put it, "It was no less worthy of a general to conquer by the wisdom of his decisions than by the force of his arms." Above all else stand Caesar's qualities as a leader of men. In an age when Roman soldiers felt no strong pull of allegiance to the state, Caesar won the loyalty of his troops through oratory and force of personality, a loyalty increased by victory and the rewards that went with it. This mixture of emotional attachment and self-interest enabled Caesar to ask a great deal of his troops; indeed, it was the devotion of the famed Tenth Legion which brought him to the political pinnacle of the Roman state.

The task of reorganizing the government was left to Caesar's great-nephew, Augustus. The monument to his work was the Pax Romana, the longest period of peace and stability that the Mediterranean world has experienced. Although Augustus preserved much of the old republican constitution, in essence the new imperial order hinged upon himself; its main prop, the imperial army of defense, was under his authority as *Princeps* and commander-in-chief.

The key elements in the Augustan military policy were consolidation of the frontiers and provision for their future security. It is not proper to say that the empire "went on the defensive." With the exception of the Parthians, Rome had no potential enemy of her stature. Moreover, Augustus himself annexed five new provinces and attempted to conquer Arabia and Ethiopia. He then abandoned an expansionist policy, partly because he required a period of peace in which to effect his imperial reconstruction, partly because of a deep war-weariness that had gripped the empire's population, but chiefly because a further expansion could bring no tangible benefits to Rome though it would subject the state to a greater military and administrative strain than it could bear. After the disaster of the Teutoburg Forest in A.D. 9, when three legions were ambushed and cut to pieces by German tribesmen, Augustus and his successors were content to regard the Rhine (and Britain) as the northern limit of the Roman *imperium.* The only substantial additions to the empire after the first century were those made by Trajan (A.D. 98–116) who annexed Dacia as a buffer for the Danubian frontier and conquered Armenia, Mesopotamia, and Assyria. The latter three provinces were relinquished by his successor, Hadrian, as too costly and extensive to defend.

The Augustan army reform followed the same trend, an emphasis upon stability and security. The legions continued to be recruited from Roman citizens, whether Italian or provincial, but non-Romans were also recruited and were given citizenship upon enlistment. All soldiers now swore to serve the full sixteen years (after A.D. 6, twenty years), thus making law what had been practice and ending turbulent agitations for discharge. On discharge, the soldier was granted a gratuity by the state, so ending his dependence on generals for land. The auxiliary units of cavalry and lightly armed troops were raised from the noncitizen residents of the provinces; they were paid less and served longer than the legionaries but received the prize of Roman citizenship upon discharge. The great bulk of the imperial army was based along the frontiers, the chief exception being the Praetorian Guard, the household troops of the emperor. The total strength of this frontier army was upward of 400,000, with a little more than half being auxiliary units; it was strung out, in numerous small garrisons and encampments, along the 10,000 miles of imperial boundary. Until the middle of the third century, this relatively small force, although a heavy financial drain on the state, efficiently performed its function of policing the boundary and maintaining peace in the provinces. Changes in its organization during this period were minor. The chief development was the large-scale application of Roman engineering techniques to strengthen the crust of the empire by permanent fortifications, like Hadrian's Wall on the Scottish border, and a tremendous system of roads in the rear of the army.

The imperial army was, however, more than a dependable outwork. Its permanent camps became centers of urban development and Roman civilization; the centuries of security it granted allowed Graeco-Roman culture to take root so deeply that the barbarian invaders were unable to dislodge it.

An element often overlooked in the Augustan defensive policy was the imperial Roman navy. Augustus, who owed his position to the decisive naval victory of Actium (31 B.C.), was the creator of the permanent imperial fleet. Its two major bases were at Misenum (near Naples) and Ravenna on the Adriatic. Together with subsidiary provincial squadrons, the fleets based at these points policed the Mediterranean so successfully that no other major naval battle was fought until the time of Constantine, although piracy, despite imperial propaganda, was by no means eliminated. Other squadrons were maintained on the Danube and the Rhine and in the English Channel and the Black Sea. The value of the latter squadrons, which were integrated with the army in border defense, was evident to the Romans; it was the misfortune of the Mediterranean fleet that once the initial task of clearing the sea had been performed, its worth was less apparent. Unlike the army, it had no battles to fight. The close connection which sea power always has with commercial prosperity meant little to a state in which legislation to promote private business was unusual. Lacking an interest in private commercial enterprise, Rome characteristically failed to appreciate the significance of sea power in relation to general prosperity. Yet the Roman Empire was, after all, a Mediterranean trading community. Augustus had designed the navy as insurance against emergency. But when that emergency came, the navy, which had survived for two centuries, had been dispensed with as a measure of economy.

Shortly after the middle of the second century, migrant peoples began to assault the frontiers of the empire; by the middle of the third century their numbers had multiplied vastly, and their attacks had become almost continuous. Such terrible external pressure inevitably brought revolutionary changes in a state already well advanced in decay. The great defensive structure of Augustus had been designed for a wealthy empire well able to bear the cost; but by the beginning of the barbarian onslaught, as part of that decline whose complexities have been probed by so many historians, the empire had passed through a period of economic stagnation into one of actual retrogression, accompanied by a decrease in population. This meant that the state, in order to obtain the revenues required to support the bureaucracy and army, had to resort to even harsher taxation policies. Successive emperors, caught between the protests of the civilian population and the urgent military needs of the state, came more and more to adopt the maxim given by Septimius Severus (193–211 A.D.) to his sons: "Be united, enrich the soldiers, and scorn the rest," until the armed forces, once the servant and protector of the government, engulfed it.

MILITARIZATION AND DECLINE

The third century, particularly the latter half of it, was a very dark period in the history of the empire, a period of anarchy during which the Praetorian Guard and the provincial armies contended to place their respective candidates upon the imperial chair, the office of emperor being auctioned off to the commander making the most bountiful promises. Behind the bitter conflict were irreconcilable social cleavages. The army, by this time consisting almost entirely of soldiers from the more remote provinces, was in violent opposition to the urban middle classes that had provided leadership in the Augustan empire. The struggle was one between a barely Romanized army and the civilian, more nearly Roman, elements of the population which were attempting to deny to the army the fruits of its enhanced status. The situation was further complicated by a military reform of Hadrian now making itself felt. He had made each frontier legion responsible for its own recruiting. Inevitably the legions drew their recruits from their immediate locale or from the plentiful barbarian sources across the frontier. This system, particularly in the Gallic and Illyrian armies, produced strong regionalist tendencies instead of a common imperial patriotism and accentuated the conflict for political power in which all the frontier armies were already engaged.

The second half of the third century witnessed a series of military reverses that threatened the security of the empire. Diocletian (284–305) restored stability by creating a co-emperor to rule east of Italy. Constantine (312–37) cemented the policy by creating a new capital in the East on the site of the Greek trading center of Byzantium. Though originally named "New Rome," the city soon came to be called Constantinople, after its founder. Each emperor had his own junior, a Caesar, to conduct military policy. The army, meanwhile, became more and more a cavalry force, both for mobility in moving to threatened parts of the frontier and for parity in dealing with the mounted enemies to the East.

Out of this century of chaos came a militarized empire, a despotism in which lip service was paid to the old Roman traditions but in which the government was actually carried on by members of a special caste. The emperor, himself a soldier, appointed only soldiers to positions of authority in the central and provincial administrations, thus creating a purely military aristocracy constantly being reinforced by barbarian recruits. The later empire was an oriental monarchy with the trappings of a police state, in which the citizen body no longer pursued its own ends but had become merely the exploited appendage of its defenders. There could be no turning back of the clock to the harmonized polity of Augustus: the greater the pressure exerted on the empire by the barbarian hordes, the more compelling became the justification for the extreme policies of social regimentation followed by the later emperors.

The reorganization of the empire, carried on chiefly by Diocletian and Constantine, was necessarily accompanied by a reform of the army, which had become undisciplined and corrupt during the military anarchy. Its numbers were substantially increased, mainly by the further recruiting of barbarian mercenaries (Roman citizenship was no longer a prize but a tax burden). The most sweeping reform was the creation of a central, mobile field force, largely Germanic in national origin, which could be shifted from one danger point to another along the frontier. This army was a mixture of horse and foot, with emphasis placed upon the cavalry. Its effectiveness was undeniable, since through most of the fourth century it managed to keep the bounds of the empire intact; but it bore little resemblance to the armies that had preceded it. Although the term "legion" was still in use late in the fifth century, the typical Roman soldier was no longer the heavy-armed infantryman but an armored mounted lancer, the cataphract. As for the imperial navy, it had disappeared during the third century. By A.D. 250 piracy had reappeared on a huge scale, and in 269 a war band of invading Goths sailed unopposed through the Hellespont. Diocletian could finance only the construction of local patrol squadrons on the Mediterranean; by the end of the fourth century the imperial navy was but a memory.

It is interesting to compare the differing approaches of two fourth-century military writers to the manifold problems of imperial defense. The unknown author of *De Rebus Bellicis* saw the cause of Roman decline in the grinding taxation policy. He therefore proposed to save money by reducing the size of the army through mechanization; to this end he submitted

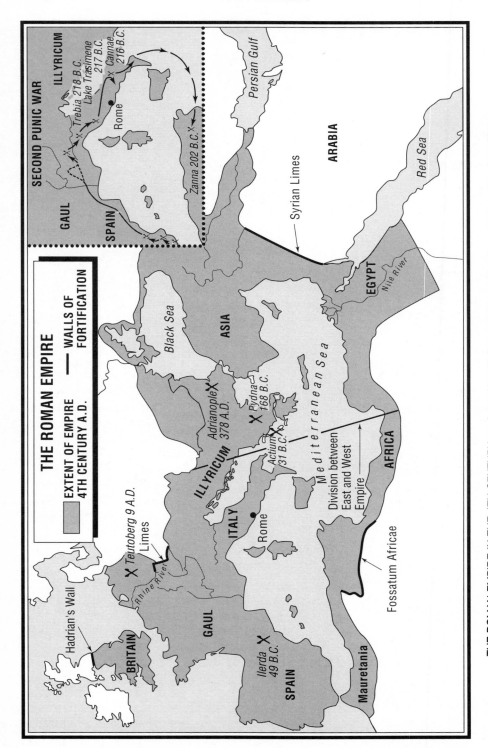

THE ROMAN EMPIRE IN THE 4TH CENTURY A.D. Inset: Hannibal's campaigns in The Second Punic War, 219–202 B.C.

plans and illustrations of a number of ingenious, if rather impracticable, military machines of various types. Vegetius, on the other hand, in his *De Re Militari,* found his solution in a return to ancient values and to the infantry-legion tactics which had won Rome an empire. "The reestablishment of ancient discipline is by no means impossible, although now so totally in disuse."

Discipline seems to have been at the heart of the problem, not the shift to cavalry that has been the primary focus of military historians in the twentieth century. Charles Oman's influential *The Art of War in the Middle Ages* identified the battle of Adrianople in 378 as decisive. A Roman army of almost 60,000 led by the emperor Valens fell upon a smaller Gothic force before either side was adequately deployed for battle. When mounted Goths returned from foraging to join the fray, they were able to outmaneuver the Roman cavalry and flank the legions. In the ensuing rout, as many as two-thirds of the Roman troops were killed or captured; Emperor Valens was one of the fatalities in the worst Roman defeat since Cannae. In spite of the critical role of the Gothic cavalry, however, this was essentially an infantry battle. The Romans lost primarily because of poor leadership and poor disposition on the battlefield. Oman nevertheless saw the loss as a victory of Gothic cavalry over the once-mighty legion. Adrianople, he said, marked the end of an infantry cycle dating from the triumph of the Greek phalanx over the Persian armies of Darius and Xerxes and the beginning of a cavalry cycle that would continue through the Middle Ages.

In truth, Oman's report of the death of the Roman legion was premature. The shift to a greater reliance on cavalry had begun well before Adrianople, partly in response to incursions by mounted barbarians, but also as part of the empire's greater reliance on a defensive system of fortifications on the frontier backed up by mobile reserves in the interior. The Roman army, still based primarily on the legion, remained militarily strong into the fifth century. Its decline and final collapse seem to have come at last from a breakdown in discipline and training that accompanied the increasing use of barbarians in the ranks. This dilution of the legion no doubt reflected larger currents in Roman society, including a decline in patriotism, political decay at the center, and a shortage of material resources. Whatever the reasons, the Roman empire and its style of warfare disappeared in the west by the end of the fifth century; they survived in the east in a significantly different form.

4

BYZANTIUM:
THE TECHNIQUE OF SURVIVAL

THE EASTERN ROMAN EMPIRE

When the Roman empire of the West split up into a number of Germanic kingdoms, its eastern counterpart lived on. The Eastern Roman Empire, or the Byzantine Empire, although the legatee of Roman civilization, was essentially a new departure, and one of astonishing vitality and persistence. Despite the fact that Constantine, when he founded Constantinople in A.D. 330 as the eastern capital of the empire, provided his new Rome with a senate and an exact duplicate of the imperial bureaucracy of the West, he made no effort to change its distinctive and deeply engrained features. The Eastern Roman Empire was Greek in culture, not Latin; it was the inheritor of the Hellenistic civilization diffused by Alexander's conquests; yet, the Byzantines thought of themselves as "Romans," and their chief tie to Greek culture was through the Christian church, to which Alexander was anathema.

It was the conjunction of its peculiar qualities with other, more fortuitous circumstances that enabled the Eastern Roman Empire to survive the passing of its western counterpart. Orthodox Christianity in the East eventually severed its ties with the Church at Rome. Eastern Orthodoxy was less universal in its appeal and much more closely allied with the state. This union of culture, politics, and religion accounts for the fervent nationalism of the East Romans, so strongly at variance with the passivity of the Romans in the West.

Less intangible, and no less important, was the economic health of the East. As the center of industrial production, the East had dominated the economy of the old Empire. Long before the collapse in the West, the traders of the East had obtained a near-monopoly of Mediterranean commerce by the export of textiles, glass, metalware, and other luxury manufactures, and they had also built up a flourishing trade with Persia. As a result the East had been by far the most populous area of the old Roman Empire. When the cities of Italy, Gaul, and Spain, which were generally administrative or military centers, were declining in size, the great cities of the East, like Constantinople, Alexandria, Tarsus, and Antioch, based upon thriving industries, were increasing. Citydwellers made up nearly half the population of the Byzantine Empire. Byzantine commercial dominance in the Mediterranean went almost unchallenged until the rise of Italian industrial competitors in the eleventh and twelfth centuries. At the same time the East was richly endowed agriculturally. The fertile provinces of Egypt, Syria, and Asia Minor fed its swarming cities. As in the cities, however, wealth in the countryside was distributed unequally. Most was in the hands of great landowners controlling large estates. The peasantry, while escaping the subjection and even servitude common in the West, was scarcely prosperous.

The Byzantine state never departed from the centralized, autocratic rule bequeathed to it by Diocletian and Constantine, and therefore was subject to the palace intrigue, official corruption, and naked struggles for power which an essentially nonconstitutional regime invites. Nevertheless the Empire was sound in its working parts; the civil service well organized if somewhat hidebound, the aristocracy well educated and politically responsible, the imperial office occupied by more than the usual number of original and statesmanlike emperors. The Empire had need of wise leadership. Its geographical position between East and West was an economic benefit. But, since it was placed squarely in the path of invaders from the east or the north who wished to occupy the derelict lands of the West, its situation was, at the same time, a military liability. Furthermore, its wealth was a permanent temptation to outsiders. Throughout its long history the Empire was called upon to withstand countless attacks from such disparate enemies as marauding Huns, migrant Slavs and Germans, fanatical Arabs, mercenary Persians, and later, from equally rapacious West European enemies, Normans and Crusaders.

For this reason, the military organization of the Byzantine Empire is of great significance. To defend itself from its multitude of enemies, the empire was compelled to devote a major part of its energies to the creation and maintenance of a military system superior to that of any of its opponents, and far superior to the methods employed in western Europe in the same period. During the fifth century, the rulers at Constantinople adopted the same dangerous expedient as the western emperors, hiring bands of barbarian mercenaries to defend their borders. Because however, of the patriotic, independent peasantry of Anatolia, Isauria, and Armenia, the proportion of native to barbarian troops never dropped as low as in the West, and it became the fixed policy of East Roman emperors to encourage the use of native-born as against mercenary troops, although the Empire was never able to dispense entirely with the services of the latter. The same century marked the construction of the first of a series of huge walls surrounding the strategic heart of the Empire, Constantinople, which rendered the city virtually impregnable to assault until the advent of gunpowder. These sane military policies, which testify to the inherent strength of the Empire, made it formidable to attack; and it required only the additional persuasion of Byzantine gold to deflect the Huns, and then the Ostrogoths, to the more vulnerable regions of southwestern Europe.

The recuperation of the Eastern Roman Empire under the careful husbanding of its fifth-century emperors convinced Justinian (527–65) of the possibility of reasserting the supremacy of Roman arms in the West and of reconquering areas lost to the Empire. A successful campaign was waged against the degenerate Vandal kingdom in North Africa and, while the Ostrogoths showed more spirit, substantial areas of southern and central Italy became Roman once more. These conquests were not the result of Byzantine superiority in numbers, for the imperial armies were invariably outnumbered, but reflected superior weapons, tactics, and generalship. The representative soldier of East Roman history is the horse archer, who, like the medieval knight, was clothed in mail. There the resemblance ceased. For the horse archer added to his mobility and shock power a skill with the bow which made him the most versatile and one of the most effective cavalry soldiers in the history of warfare. Belisarius, Justinian's great cavalry general, testified to the merit of the horse archer: "I found that the chief difference between the Goths and us was that our own Roman horse and our Hunnish *foederati* are all expert horse bowmen, while the enemy has scarcely any knowledge at all of archery. For the Gothic knights use sword and lance alone, while their bowmen on foot are always drawn up to the rear. So their horsemen are no good till the battle comes to close quarters, and can easily be shot down while standing in battle array before the moment of contact arrives." The horse archer was supported in the Byzantine array by heavy infantry and cavalry lancers; but it was the mounted bowmen who made the difference between the imperial army and other armies.

The small size of the expeditionary forces was one of the most notable features of the expansion under Justinian. North Africa was won with 15,000 men (6,000 horse archers), Italy with 11,000, plus 15,000 reinforcements. The excellence of Byzantine weapons and the professional approach of able generals like Belisarius and Narses are only part of the explanation. Under Justinian, Byzantium became the premier naval power of the Mediterranean. The Roman fleet could transport forces rapidly wherever they were required, and supply and reinforce them.

Justinian's imperial offensive was a departure from the broad sweep of Byzantine military policy. His successors had to cope with the problem of territorial overextension in both its financial and military aspects; and with the reforms of the Emperor Maurice (582–602) the Empire entered a defensive phase which was to be its military outlook for the next 500 years.

THE STRATEGIC DEFENSIVE

The system inaugurated by Maurice took many years to reach its mature state, but the organization here described may be considered as the completion of his framework. The aim of Maurice was to provide for maximum security at a minimum of expense and at the least possible danger to the state from the political aspirations of army commanders. Generals like Belisarius had surrounded themselves with large bodyguards of mercenaries who had sworn loyalty only to them; Maurice took the power to make appointments above the rank of centurion out of the patronage of his generals and placed it in the hands of the central government. The size of the professional standing army was drastically reduced and what remained was redeployed into a central strategic reserve and cadres for the frontier garrisons. The elimination of the distinction between mercenary and native units further strengthened the control of the central government over the army. Ultimately, by granting tax relief to serving soldiers and plots of land on discharge, the state succeeded in recruiting almost the whole of the army rank and file from the dependable inhabitants within its borders. The policy of land allotments apparently did not result in the kind of political turbulence it had produced in the days of the Roman Republic.

The small size of this standing army was made possible by the erection of massive defense works along the frontiers. On the line of the Danube, for instance, protection was given the empire by a cordon of fifty-two fortresses, strengthened in depth by a second line of twenty-seven further south. Constantinople was surrounded by a sixty-foot moat, guarding a triple ring of immense walls, each wall studded with towers at frequent intervals, and the innermost wall reaching a height of thirty feet. To man its system of frontier fortifications the government placed its reliance upon local militia forces stiffened by professional soldiery.

By the eighth century, Byzantine defense had been further refined, and somewhat decentralized, by a division of the empire into a number of self-contained military districts called themes, a policy quite similar to that earlier employed by the Roman Empire of the West in its frontier areas. Since the Empire was open to attack from any direction, often unexpectedly, a sudden thrust by an enemy force might, and occasionally did, penetrate its outer crust, bringing devastation to the defenseless interior provinces before the central army could deal with the danger.

The theme system was the response of a government acutely aware of the increasingly narrow margin between survival and extinction, and of the perpetual need for watchfulness and preparedness. Each theme contained a permanent army corps, bolstered by local militia and commanded by a *strategos* who was also the head of the area's civil government. Thus, just as in the Roman Empire of the third and fourth centuries, external pressure brought the militarization of political institutions. The process was carried further by the grant of land to peasants on the condition that the holdings would furnish soldiers in perpetuity. The themes were therefore firmly tied to the most stable and hardy class in Byzantine society. The size of each theme army varied according to its location. On the average, it probably numbered about 6,000, with frontier themes containing as many as 12,000 troops. The function of a theme army was to engage and blunt the momentum of an attacking force, and so give time for adjoining theme armies and the permanent central reserve to come to its assistance. The theme system had the merits of an elastic overall defense. It worked because the difficult communications problem was overcome by beacon networks, couriers, and great military roads.

The Byzantine Empire, much more than the Roman Empire, was a maritime state, deeply interested in matters of trade and industry. The military establishment made necessary by almost continuous defensive wars was very expensive; so were the requirements of the imperial court and bureaucracy; so, finally, was Byzantine diplomacy. It was a maxim of Byzantine diplomatists never to employ force, or the threat of it, if money would suffice, a policy which, because it was always backed by military power, was not so debilitating as is often the case with appeasement. Annual subsidies of 30,000 pieces of gold bought peace from the Sassanid rulers of Persia in the sixth and seventh centuries; at different points in her history the Empire pacified or purchased the support of Avars, Slavs, and Bulgars in the Balkans, Franks, Lombards, and Goths in Italy and Spain, Berbers in North Africa, and the rulers of Abyssinia. The very high level of public expenditure on war, administration, and diplomacy explains the existence of the paternalistic economic policies of the Byzantine emperors.

The importance of trade led to a preoccupation with naval affairs. It was the job of the imperial navy to control, not vast territories, but important Byzantine trading centers and the routes which connected them with foreign commercial areas. This meant controlling, either directly or indirectly, the Crimean coast, the vital water passage from the Black Sea to the Mediterranean, the port cities of the Adriatic, Syria, and North Africa, and the sea approaches to all these areas. As early as the seventh century, there were Byzantine naval bases and shipyards at Carthage, Acre, Alexandria, and Constantinople. By the eighth century, the extension of the themal organization to the navy gave the Byzantine naval system its final form. There were five fleets in the permanent navy: the imperial fleet stationed at Constantinople, and provincial fleets based on the southern coast of Asia Minor, Ravenna, Sicily, and the islands of the Aegean. As with the land themes, each naval district, commanded by a *strategos,* provided the major cost of the fleet assigned to it. The strength of these fleets varied over time, driven both by the naval threat they faced and by the ability and willingness of the central government and the themes to bear the cost of a standing navy. Byzantium's command of the sea was often compromised by lack of funds.

Other factors contributed to the naval supremacy that Byzantium long enjoyed over its competitors. The heart of its naval strength was Constantinople, with its harbor, dockyards, and arsenal. To reach the imperial capital a rival fleet had first to defeat the provincial fleets which protected the entrance to the Dardanelles, and then negotiate the narrow straits which were guarded by the imperial fleet. When the Arabs took to the sea by copying the *dromons,* or war galleys, of the Byzantines, they attempted to force the Straits on two occasions around the turn of the eighth century, but without success. The imperial navy possessed a decided advantage over the Arabs because it controlled plentiful supplies of naval stores and materials, particularly timber and iron, commodities which were available only in small amounts in the Arabic territories of Syria, Egypt, and North Africa. Finally, the Byzantine navy benefited from the monopoly of a secret weapon, the mysterious Greek fire, also known as "sea fire" or "wet fire." Even in Byzantine times its composition was kept a closely guarded secret, revealed only to the initiated. Today, the formula is unknown, though modern authorities agree that it was petroleum based. The Byzantines apparently preheated and pressurized the mixture before shooting it from movable copper tubes mounted on the bow of their *dromonds;* the mixture was ignited by a torch as it left the tube. Sticking to all it touched and burning even under water, this precursor of napalm was a formidable weapon. It was used with deadly effect against Arab fleets besieging Constantinople in 678 and again in 717–718. It was not a master weapon, but it gave a technological and psychological advantage to the superior Byzantine navy.

Enough has been said about the military methods of the Byzantine state to demonstrate that its approach to warfare was thorough and in many ways original. The disparity between the practice of war by the Byzantines and by the feudal society of the West has often been remarked. The decisive elements in Western war were courage, strength, and weight. The Byzantines surveyed war with the subtle and mature intelligence of an ancient civilization and applied to its study the same careful scholarship which distinguished their elaborate theology and public finance. But if war was not, for them, a heroic game, neither was it a merely academic exercise. It was precisely because the existence of the state hung so completely upon the capabilities of its armed forces that the study of war, so often ignored in more complacent and less threatened societies, attracted the best Byzantine minds. The Eastern Romans produced a number of military treatises for the direction of their generals, notably the *Ars Militaris* of the Emperor Maurice and the *Tactica* of Leo VI (c. 900), which were of a technical excellence unmatched in Europe until the sixteenth century. These works derived from Greek and Roman thought but nonetheless had a peculiarly Byzantine flavor.

The prevailing tone of these writings is one of caution. The underlying premises are that the most desirable victory is that won with the least expenditure in men, money, and effort and that fighting is something to be avoided altogether if other means will bring about a favorable result. Byzantine military thought was thus of a primarily defensive cast. War was regarded not as a temporary or unusual phenomenon but as the normal and inevitable outcome of the interaction of peoples. A state with tangible but finite resources could not permit itself the luxury of military adventures which would weaken it when the next war came along. To the Byzantine mind, war was not within the province of morality. To deceive the enemy, a general was encouraged to employ any method he thought expedient, including bribes to enemy officers, false surrenders to permit time for regrouping, and battlefield parleys as cover for ambushes. Leo VI had nothing but disdain for the chivalric code of the West, and for its exaltation of knightly courage and daring. To the Byzantine, courage was only one of the many attributes necessary to a good professional soldier, and daring usually resulted in pointless loss of life, the cardinal military sin in a society in which manpower was precious.

Military textbooks prepared a general for every conceivable situation. Explicit instructions based upon Roman military experience and, indeed, often going back to Greek manuals of the fourth century B.C. were laid down for the maintenance of discipline and training, for the entrenchment of camps, and for coping with supply problems in unfamiliar terrain. Byzantium was the only state of the period to give attention to the treatment of casualties; the textbooks reveal that the army had an organized ambulance corps of surgeons and bearers. The motivation was of course not humanitarian; the state was interested in restoring the wounded to battle fitness, and thus paid the bearers a bonus for every casualty brought in from the field.

The most impressive pieces of Byzantine military writing are those concerned with tactics and strategy. These were based on the principle that the methods to be employed must be varied according to the peoples to be fought. The Byzantines made it their business to learn everything of importance there was to be known about the methods of probable opponents, and they worked out thoroughly the means best suited to defeat each enemy. Leo VI, for example, observed that the most formidable feature of a western host was its initial charge, and therefore recommended that a pitched battle should be avoided. Instead, advantage should be taken of such western weaknesses as lack of discipline, careless entrenching procedure, uncertain morale, and absence of any supply organization. He recommended that western armies should be worn down by skirmishes and Fabian strategy, and by striking at their supplies. For Byzantium, this would be less expensive than great battles.

The military history of the Eastern Roman Empire covers almost a millennium. For the greater part of this immense period, its military institutions retained their original excellence and vigor, without any significant alteration in nature. During the first quarter of the seventh century the Empire fought almost continuously on two fronts against totally different peoples. In the north, migrating Slavs and Avars threatened to overrun the imperial provinces in the Balkans. Against them, the imperial forces fought an all-out war of annihilation. The danger was not from the military skill of these peoples, which was rudimentary, but from the vastness of their numbers. Therefore they pursued a ruthless policy of extermination. The struggle against the traditional Persian enemy on the east was of an entirely different character. The Sassanid rulers of Persia were in competition with Byzantium for control of the lucrative trade routes to the Far East; the struggle was therefore initially limited in aim and locality.

The strain of conducting two defensive wars simultaneously proved almost too much for the Empire. By 608 Egypt and Syria had fallen and Constantinople was menaced from the north and the east. Here Byzantine generalship and sea power took a hand, assisted by a great upsurge of religious feeling caused by the fall of Jerusalem. The Emperor Heraclius used the fleet to transport an army to the south coast of Asia Minor. Striking inland, he forced the Persians to withdraw from Asia Minor to protect their rear, and then, by marching into Persian territory, he compelled the withdrawal of Persian garrison troops from Egypt and Syria. Such brilliant maneuvering, pointless when employed against teeming hordes of migrating barbarians, was an apt strategy for use against a civilized enemy who placed great store in the maintenance of his communications and the protection of his homeland. Typically, when the war ended in 626 with a decisive victory at Nineveh, deep in Persian territory, Heraclius was content with a peace which restored the political situation as it had existed before the war. Byzantium had no appetite for indigestible territories.

THE ARAB THREAT

Barely had the long Persian war ended when a much greater danger threatened. Countless Arab horsemen, fired with religious zeal by their prophet Mohammed (d. 632), plunged out of the Arabian wastes intent upon plunder and the conversion of the unbeliever by the power of the sword. An exhausted Persia was overthrown almost immediately. Because of bitter doctrinal dispute with the church at Constantinople, the Christian inhabitants of Syria and Egypt welcomed as liberators the Arabs, who promised toleration. The loss of these rich provinces inflicted a heavy blow to Byzantine trade and revenue and gave the Arabs entrance to the Mediterranean. By 641 Syria and Egypt had again been detached from the Empire, this time permanently. Byzantine generals had encountered raiding parties of Arabian nomads in the past and had found that these lightly armored lancers could not stand up to their heavy cavalry and horse bowmen. But now the Arabs were no longer raiders and their fanaticism and numbers compensated for their faulty tactics. Moreover, the Empire always labored under financial difficulties and manpower shortages, although the latter seem also to have been connected with a decline in human fertility and with plagues.

It is a tribute to the high professional quality of Byzantine arms and to the latent reserves of stubornness in Byzantine society that the Arabs, in their first flush of fury, were fought to a standstill. They begged for a truce. When the struggle was resumed, each antagonist had learned much from the other. The religious fervor of the Arabs had been abated to some degree, but they had developed a political structure (the Caliphate) to embrace their conquests. They had also borrowed something of the Byzantine military organization: their lancers and horses were now as heavily armored as the imperial cataphract, and instead of throwing clouds of cavalry helter-skelter into battle, they had organized their horse into one deep, weighty line. The Byzantines, meanwhile, had countered the high mobility of the Saracen horsemen by perfecting the theme organization.

Despite the Arab switch to heavy cavalry, the Byzantines still possessed a preponderance of weight, but their leaders were not prepared to leave the issue of a battle to weight alone. During this period Byzantine cavalry tactics were developed, eventually attaining a level of performance beyond the reach of any other state during the entire Middle Ages. These tactics were strikingly similar to those employed by the Roman legion. A cavalry force was broken down into three lines, each composed of distinct units of about 450 troopers called *banda,* with other guarding units on the flanks. Such a loose grouping is as strong a testimony to the reliability and training of the Byzantine cavalryman as the maniple-legion was to the qualities of the Roman legionary. The single-line formation of the Saracens could not withstand the three successive waves of the East Roman horse. Thrown into a confusion which was exploited by the imperial horse- and foot-archers, the Arabs were usually driven from the field.

Thwarted on land, the Saracens challenged imperial supremacy on the Mediterranean. The reasons for the eventual triumph of the Byzantine fleets have already been touched on, but the contest was long, not ending until the middle of the eighth century. Because of its long preoccupation with Arabian sea power, the Empire was forced to give up most of the Italian lands won by Justinian, until by 754 only Venice and parts of southern Italy remained. At the same time the swaying fortunes of the naval war had allowed the Arabs to sweep from Egypt across North Africa and into Spain, unchallenged by the imperial navy. Spain was conquered by 717, and Arab raiding parties, filtering through the Pyrenees, probed into southern France. It was one of these parties which was defeated by a Frankish army in the desperate battle at Tours in 733. The fierce struggle of Charles Martel with one tiny tentacle of Arab power takes on a somewhat different significance from that usually attributed to it when it is remembered that what really stood between the Arabs and the more direct approach to western Europe was the double bastion of Byzantine land and naval forces. It need hardly be added that when the bulwark of Byzantium disappeared, Muslim armies striking through the Balkans were able to reach the gates of Vienna in 1529.

THE DECLINE OF NAVAL POWER

After the satisfactory close of the long conflict with the Arabs, the Eastern Roman Empire enjoyed nearly a century of stability and prosperity. There were, of course, numerous small wars on land and sea, but the tested formula of diplomacy, bribes, and military efficiency functioned smoothly and adequately. Naval supremacy gave the Empire a renewed mastery over Mediterranean trade. In order to retain this mastery, the Byzantine state instituted a number of regulations which have been compared to the British Navigation Acts. As with London, Constantinople was the metropolis of the commercial empire; through it, and it alone, were funneled goods from East and West. The strength of the Byzantine navy compelled Muslim traders of Egypt and Syria to divert their merchandise designated for Europe to the imperial capital for transshipment. At the same time the navy also controlled the termini of the Russian and Asian trade routes on the Black Sea. The final element in the imperial trading system was Byzantine control over the western ports for eastern goods; Byzantine merchants were forbidden to deal with any cities except those expressly enumerated in governmental regulations. These few favored cities were chiefly Italian, like Amalfi, Bari, and Venice; and all were in the possession of the Byzantine state.

This closed, rigid system undoubtedly brought immediate economic benefits; but in the end it defeated its original purpose by helping to weaken the Empire. The Italian cities, having the monopoly for reception and distribution of eastern goods, soon accumulated great capital surpluses for investment. In some centers, especially Venice, part of this capital was employed in the construction of merchant fleets which gradually encroached upon the Byzantine monopoly of the carrying trade to the West. By the beginning of the tenth century the Byzantine share of the carrying trade had very noticeably declined.

For the imperial navy this process had the most serious results. The navy depended for experienced manpower upon the maritime population of the empire; but, owing to the commercial shipping depression, the impressment organization of the naval themes enlisted fewer and fewer trained sailors. In consequence, the navy, upon which the whole regulatory system was based, diminished in the quality and quantity of its personnel, and therefore in its general effectiveness, and was unable to meet the rise of new Muslim sea power. The Aghlabid and Fatimid kingdoms of North Africa, shut out of the European market by Byzantium, had built their own fleets and by the early part of the tenth century had wrested naval control of the western Mediterranean from the Empire. But the Muslim victory cannot be wholly explained by Byzantine decline. There is some evidence that the Arab navies had narrowed the technological gap with their enemies by developing an incendiary weapon akin to Greek fire. Also, the Empire was in the same period beset by strong new enemies, in the Balkans by the Bulgars, and on the Black Sea by Kiev, an expanding Russian principality.

In the late tenth century the Empire underwent a spirited naval revival. The system of command was unified by placing all the fleets under a lord high admiral (*drungarius classis*); the speedy *dromonds* were replaced by vessels of an imposing size; and ingenious landing craft were constructed with sliding ramps, permitting the discharge of cavalry in full battle panoply. This resurgence regained much lost territory and gave at least a measure of sea control to the Empire; but the damage to her position during the naval lapse was irreparable. Although Constantinople was still the center of East-West trade, the control of trade had passed almost entirely out of the hands of imperial citizens and into those of foreign merchants, particularly Syrians and Venetians, who had established colonies in the imperial capital. Venice was now only formally a Byzantine possession; in actuality the city was a naval power in its own right interested in maintaining the political connection because of commercial advantages.

Although the empire continued to impress its many visitors with its wealth, its prosperity was derived chiefly from revenues incidental to the trade being carried on within its borders. Foreign control of East-West trade meant that opportunities for profitable Byzantine investment were shut off, and increasingly the wealth of the monied classes was diverted into land, with disastrous consequences. The assimilation of numerous peasant holdings into large estates meant, as it had in republican Italy, the displacement of the peasant class and the destruction of the social base of the army. The attempts of several emperors of the late tenth century to arrest this process by breaking up estates and redistributing the land touched off a crippling civil war. Order was restored long enough to allow the Emperor Basil II to wage a misguided and exhausting war of conquest against the Bulgars (1004–18). When it became necessary to fight a defensive war in Italy, social disruption had proceeded so far that the Empire had to turn for manpower to mercenary forces of Bulgars, Russians, and Scandinavians to fill out its depleted armies. By 1042 it found itself stripped of its last Italian lands.

Byzantine military power never fully recovered. Shortly afterward, in the midst of a civil war which had already spanned fifteen years and several emperors, there appeared on the eastern border an army of 100,000 Seljuk Turks, led by the Sultan Alp Arslan. The Seljuk Turks were typical nomadic horse archers from the Eurasian steppes and, despite their numbers, they should have been no more dangerous than other eastern invaders in the past. But at Manzikert (1071) a heterogeneous East Roman army violated one of its fundamental tactical canons: it failed to keep order during a pursuit of the enemy; and so was cut to pieces by the counterattacking Turks. Although the imperial fleet and the walls preserved Constantinople, the Asiatic themes and their peasant remnants were lost forever. By 1100, when the Crusaders had established kingdoms in Palestine, and incidentally had done much besides to weaken militarily the Christian bastion of Byzantium, Europe had made direct contact with the trade routes of the East, and the economic foundations of the Empire had disappeared. The Eastern Roman Empire lived on for three and a half centuries, surviving even the fall of Constantinople to Crusaders in 1204 and rule by Latin Christians until 1261. But only the shreds of Byzantine military organization remained; it was reduced to prolonging its precarious existence by the employment of mercenary troops unschooled in the traditions of the past.

5

WESTERN EUROPE: THE RISE OF THE FEUDAL ARRAY

THE DARK AGES

The inheritors of the Roman Imperium in the West were the Franks, a Germanic group of tribes. They moved slowly into Roman Gaul in the course of the late fifth and early sixth centuries from their original homeland in northwestern Europe. The German successor states which had been created by the Ostrogoths in Italy, the Visigoths in Spain, and the Vandals in North Africa had no firm base. They were governed by military elites which derived their authority from small armies of occupation. This Germanic upper layer was so rapidly enfeebled by its contact with Roman civilization that it could offer no great resistance to the Byzantines and the Muslims who established Eastern outposts in Western Europe. Medieval European civilization, essentially a union of Germanic and Roman-Christian culture, developed primarily in those regions where Germanic peoples settled in great numbers and merged with antecedent populations. In other words, the center of Western Europe shifted away from the Mediterranean to France and the Rhineland.

In the widespread devastation, uninhibited looting, and savage wars which accompanied the folk wanderings of the Germanic peoples through the rotten husk of the later Roman Empire, it was inevitable that not only would the rate of economic decline in these areas accelerate rapidly, but much of what was valuable in Roman civilization would be lost, some of it never to be recovered. Among the casualties of Western Europe's Dark Ages was the Roman military tradition. We have already seen that in order to counter the mounted mobility of the barbarians, the later Romans themselves had virtually abandoned the infantry legion.

There was, however, a brief period in postimperial military history when it appeared that the Franks might reconstitute the status of infantry. The peoples of Germany, forest fighters for the most part, had made little use of cavalry; the Goths of eastern Germany had become horsemen because, when deflected in their early migrations by the still-strong Roman frontiers, they had drifted into the south Russian steppes. The Franks, however, moved directly south into a shattered empire, and they came as infantry. They wore no body armor and carried bossed oval shields, javelins rather like the Roman *pila*, swords, and daggers. Their national weapon was the battle-ax (*francisca*), a heavy but well-balanced weapon which the Franks had learned to throw with great accuracy just before making contact with their enemy. Although little is known of their order on the battlefield, it is likely that it had progressed no further than a simple massing of warriors in heavy blocks. The primitive host of the Franks was a far cry from the supple and disciplined legion. The impotence of the Frankish infantry when faced with well-trained cavalry was demonstrated when a Frankish host invaded Italy in the middle of the sixth century. At Casilinum (554) the Franks were engaged by Byzantine infantry and cavalry and then encircled by horse archers, who shot down the nearly defenseless warriors almost at leisure.

It was only very slowly that the Franks, in whom, as in all noncivilized peoples, the rule of custom was strong, adapted themselves to cavalry warfare. Eventually, out of the shadowy interplay between military necessity and the facts of social and economic existence, the feudal system was to emerge in the ninth century with the mounted knight as its key figure. One of the most characteristic elements in the feudal system was specialization of function according to class, with the reservation of military functions to one section of the community. Some historians have held that the origin of the feudal relationship may be traced back to the *villa* of late Roman times, which was worked by unfree peasants.

Nothing could be more unlike feudalism than the social and military organization which the Franks and other Germanic tribes had brought with them in the fifth century. Their political institutions were simple. Folk assemblies, presided over by the tribal chief, dispensed customary justice. The only other function of the chief was to lead his tribe in war. There was no distinction between status as a member of the tribe and status as a warrior; when war came, all males automatically entered the tribal host. The *comitatus*, a military elite selected for bravery, formed the bodyguard of the chief. The Roman *villa* with its stratified society may have been an agent in the transformation of this barbarian social structure into medieval feudalism.

When, in 496, the war chief Clovis forced the various Frankish bands in Gaul to acknowledge him as their king, the tribal host was transformed into a national levy, which the king could raise whenever required. Moreover, the duty to serve in the host of the king was not confined to the Franks but was extended to the whole population enclosed in Frankish territory, the only classes exempt being the nonfree and the clergy. In practice, however, as Frankish society became stabilized and took on a predominantly agrarian character, the duty of military service became closely associated with the ownership of land. Each free household owed the service of one man with arms and equipment, an obligation that became hereditary.

The social origins of the feudal knight of later times are not, however, to be found in the mass levy of freemen but in the *comitatus,* the circle of paladins surrounding the king which became the core of the Frankish nobility. Increasingly, members of this small group, profiting by the lessons of battles like Casilinum, turned to the use of cavalry and body armor. Two developments in the early eighth century furthered the rise of cavalry: the introduction of the stirrup from the Orient via the Mediterranean world and the incursions of mounted Saracens into southern France. The stirrup greatly increased cavalry effectiveness, since it enabled a firmly seated rider to deliver a powerful thrust with his lance or to rise in the saddle to use his sword with greater leverage. The infantry of the Macedonian phalanx and the Roman legion had never had to face such firmly seated horsemen; and the raw Frankish foot soldiery was much less capable. Their weapons were ineffective against the armor of the horseman, and they could no longer topple him easily from his saddle. Not until the time of the Swiss halberdier of the fourteenth century did infantry again cope successfully with cavalry on its own terms.

Yet how could the Frankish state respond to the evident need for reshaping the nature and equipment of its armed forces? The horse, lance, sword, shield, and armor of the mounted soldier represented a costly investment which was far beyond the resources of the individual freeman. Moreover, the hard struggle for existence brought about by rude subsistence agriculture denied to the freeman the time necessary to master difficult cavalry tactics. The almost complete disappearance of trade and the decay of cities from the fifth to the tenth centuries (the Dark Ages) meant that there was little or no money available to the state, nor was there in any West European monarchy an organized, literate bureaucracy which could deal with fiscal matters on a national scale. The Frankish king was expected to "live of his own" and to provide for his family and attendants from the production of his own estates. He had no surplus revenue to devote to the establishment of a professional cavalry army.

THE HOLY ROMAN EMPIRE

A long step toward the solution of this dilemma was taken by Charles Martel (714–741), mayor of the palace for one of the weak Merovingian kings. In order to meet the Saracen menace, Charles widened in scope a practice that had probably already begun. Out of the royal estates and the holdings of the Church, he gave distinguished soldiers enough land for their support. From each vassal, in return for the estate granted, he extracted an oath of allegiance by which the vassal was bound to serve Charles as a soldier as long as he was physically able. At the vassal's death, or on his breaking his oath, his benefice was transferred to another, who assumed the military obligations which the land carried. In this way Charles secured a large body of mounted soldiers who could afford to purchase their equipment, had leisure to undertake military training, and were bound to their ruler by oath and interest. It was the mobility and power of the heavy Frankish cavalry, made possible by the innovations of Charles Martel, and not the tenacious stand of the Frankish infantry in 733 at the battle of Tours, that won victory over the Muslims in a long struggle that did not end until 752.

The many military successes and the widely flung conquests of Charles Martel's illustrious descendant Charlemagne (King of the Franks, 768–814) owe much to the constantly increasing proportion of cavalry in the Frankish host, although the mass levy of infantry continued to be employed. Charlemagne issued many edicts to assist the development of good cavalry. He forbade the export of armor, extended the system of vassalage to conquered areas, and defined in precise terms the equipment with which the mounted soldier was expected to appear when summoned to the host. A sweeping change was also made in the infantry levy in order to improve its quality and to ease the inordinate burden which Charlemagne's many wars placed upon the peasant. Freemen holding property below a prescribed minimum in size and value were no longer to be individually answerable for service but were to club together in small groups to send one of their number, suitably armed. Although this reform heightened the efficiency of the Frankish army, it further narrowed the social bounds within which the military art was practiced, and thus represented another stage in the slow development of the feudal state.

Charlemagne's military ability and fierce ambition resulted in the creation of an extensive empire which was bounded on the north and east by the Elbe and the Danube and included sections of modern Yugoslavia, all Italy (except Venice) to a point south of Rome, and part of northern Spain. Aside from his ambitions, Charlemagne sincerely felt it his mission to protect and expand Christianity and civilization; his empire was at once the legatee of Roman civilization and the political expression of the whole of Christendom, as was symbolized in his coronation as emperor by the pope in 800. Under his rule there was a great revival of learning; the last heathen Germanic tribes, the Saxons, were converted; there was even a momentary economic upturn.

The concepts of the twin unities of Christendom and the Holy Roman Empire remained strong and influential throughout medieval times, but Charlemagne's great political creation did not survive long. As a method of imperial government Germanic monarchy was too simple, too close to its tribal origins, to embrace the huge extent of the Frankish domain. The circumstances of imperfect communications, diverse populations at radically different levels of civilization, and the absence of a common citizenship and of relationships transcending the personal were compounded by civil strife resulting from the division of the empire among Charlemagne's three grandsons and by the terrible invasions of heathen marauders. By the ninth century these various factors brought a crystallization of West European society into the structure known as feudalism.

The reasons for the incessant raids of the Vikings in the ninth century and the Magyars in the late ninth and tenth centuries are obscure. It has been suggested that the Vikings represented a violent overflowing of areas of limited resources in which population had suddenly risen greatly; and also that the attempts of Scandinavian kings to bring order to their countries drove out the turbulent elements. The Magyars, on the other hand, were a people of Turanian origin from the eastern steppes; as with other nomadic peoples their movements were determined by the proximity of good grazing lands and the chance of easy plunder. Whatever the explanation of these sudden onslaughts, the military problems posed by the appearance of these twin scourges were the same. The Magyars, swift horsemen, were exponents of the hit-and-run attack; the Vikings had the mobility and freedom of action afforded by sea power. Wherever the Vikings landed, they scoured the countryside for horses to mount their infantry. Since their object was loot, not occupation, they avoided centers of possible resistance and, if cornered, slipped away to their ships. The slow-assembling, slow-moving Frankish infantry levy could not cope with either invader. Frankish heavy cavalry, however, was almost invariably successful if it caught up with a raiding party; and castles built by the lords provided secure bases for offensive and defensive operations.

FEUDALISM

The anarchic conditions of the ninth century, and the absence of organized central government, placed a premium upon the protective services of the cavalry soldier and also forced the people of every community to fly for help, not to the monarchy, but to the local lord in his castle. Everywhere landowners granted benefices, or fiefs, to soldiers in return for protection, and everywhere smaller landowners placed themselves and their feudal vassals under allegiance to a more powerful lord. In this way the great feudal ladder was formed, a progression from the knight below to the great feudatories and the king above. The core of the feudal system was a contractual relationship between lord and vassal at every rung of the ladder. The knight received his fee, comprising enough land and servile labor to support himself and his family, no longer as a revocable grant but in perpetuity, as long as he and his descendants fulfilled their contractual obligations to their lord. Petty barons yielded up their estates to great magnates and received them back under conditions which obligated both themselves and their vassals. The purpose of the contract, at whatever level, was essentially military. The lord was pledged to protect his vassals, while they were pledged to serve in his mounted array. In a larger sense, feudalism was not only a set of military relationships but the ordered response of a society seeking to avoid anarchy. However, no summary so brief as this can indicate either its growing complexity through time, its various forms in different regions, or its creativity in the realm of political, social, and economic relationships.

In this system, and under the conditions which brought it forth, there was little place for the freeman. Either he became a knight or, more often, he surrendered his holdings to a local lord powerful enough to give him protection and received the land back as a tenant owing dues and labor services to his lord. In time, as a result of such transactions, the status of the peasant was greatly lowered, and although many communities contained freemen who held their own land, the majority of European peasants were depressed to various conditions of servitude. The development of the seigneurial system accounts in part for the disappearance, except in England, of the national levy of freemen. We have already seen, however, that this military institution was inadequate to the demands made upon it and that the growth of the feudal system was brought about, in part at least, by the inefficiency of infantry.

The feudal system, then, embodied the supremacy of cavalry over infantry and the substitution of the castle for the infantry phalanx as the base for cavalry operations. In feudal eyes, no other event established this supremacy more convincingly than the conquest of England by the Normans.

English military institutions, although having the same roots as those of the Frankish kingdoms, had not developed at the same rate or in quite the same way. The Teutonic peoples who invaded Britain in the fifth and sixth centuries relied, like the Franks, upon the unarmored infantry supplied by their tribal levy, or *fyrd*, and it was upon this system that the military power of the several kingdoms of early Anglo-Saxon England depended. The ninth-century assaults of the Danish Vikings brought several modifications which owed as much to the pressure of circumstance as to the genius of King Alfred (871–99). Because of the constant fighting, the *fyrd* was divided, one part serving for a set period while the other continued in peaceful occupations. As on the continent, many stockaded forts, called *burhs,* were constructed to act as centers of protection for the inhabitants of the surrounding countryside and as points of resistance against the Vikings. Perhaps Alfred's most notable contribution was the creation of an English navy as an outer wall of defense which, if not maintained unimpaired by his successors, certainly established the English naval tradition. The Danish invasion in force in the eleventh century, which resulted in the elevation of Cnut (Canute) to the kingship, added to the English military structure the housecarls, a professional bodyguard for the king, equivalent to the Frankish *comitatus.* By 1066, all the great aristocratic houses had such bodies of retainers, who represented the peak of English military attainment. The *fyrd,* at least in theory, was still the core of the English army. In fact, however, between the time of Alfred and the coming of William, there was a tendency to reserve military functions to a particular class, a trend toward something not unlike continental feudalism. A member of this class, known as a *thegn* (thane) was above an ordinary freeman in status but below the small circle of aristocrats. He owed military service directly to the king, not in exchange for any land he held, but because of his social position.

The host that King Harold the Saxon marshaled atop a hill to meet the invading Norman army on the morning of Hastings in 1066 was drawn partly from the ill-armed ranks of the *fyrd,* but its nucleus was the body of housecarls and *thegns,* armed with the ponderous Danish battle-ax and protected by steel caps and chain mail. The most significant fact about the English army, that which caught the attention of the feudal world, was that it was wholly infantry. It trusted for protection against cavalry to its solidity behind a wall of kite-shaped shields.

The leading element of Duke William's Norman army, on the other hand, was its feudal cavalry. The Normans were the descendants of those Northmen to whom the French monarchy, convinced that the best defenders against Vikings were other Vikings, had granted the Duchy of Normandy in the tenth century. In time, they had been converted to Christianity and absorbed into feudal society; their abilities as cavalry soldiers were held in high esteem. The invading army also contained archers and infantry.

The victory of the Normans at Hastings was regarded in the Middle Ages as establishing beyond doubt the superiority of cavalry over infantry. Actually, the Saxon shield wall remained unbroken after repeated charges of the Norman horse. William won because the two wings of the Saxon army, the local levies, were drawn away from their strong hill position by cavalry feints and cut to pieces. Harold's heavily armed housecarls were not overcome until a concentration of missile fire aimed in high trajectory broke up their ranks and made them vulnerable to cavalry assault. The simplified medieval view of Hastings as a victory of cavalry over infantry is a demonstration of how military insight can be blunted by nonmilitary concepts. The mounted knight was no longer merely a soldier. By the eleventh century he was a member of an exclusive class which owed its social and political power to its control over the instruments and conduct of war and which could neither afford nor tolerate competition from below. Neither Norman England, nor any other European feudal state, utilized to any significant degree the real lessons of Hastings: the solidity of well-armed infantry and the decisive effect of missile weapons. Balanced armies, like those of the Eastern Roman Empire, were foreign to, and indeed incompatible with, feudal society.

Under the Norman monarchy, England became a feudal state similar in nature to other European countries, but one in which the crown retained a greater degree of power. As in continental Europe, the conversion to a feudal military system meant the construction of a great number of castles as bases for cavalry forces and as citadels of feudal power. Within a century after the coming of the Normans, there were about 1,200 castles in England, the earlier mounds and stockades soon giving way to great square stone keeps like the White Tower of London. Yet alongside the feudal cavalry army, the national military organization of Saxon England continued to exist. William retained the *fyrd* partly as a defense against his own barons: A national levy of English infantrymen was employed to quell the baronial rising in 1075. Although one of the results of the Conquest was to reduce many Anglo-Saxon freemen to the level of the unfree, and thus in turn reduce the number of those eligible for the national levy, this ancient system never entirely dropped out of sight. With some modification it was continued to become the militia in later times.

THE CAVALRY AGE

For a period of roughly three centuries, from the defeat of the Anglo-Saxon axmen in the eleventh century to the rise of effective Swiss and English infantry in the fourteenth, the cavalry soldier reigned as the dominant figure in warfare. The cavalry age coincides with the full development of feudalism as a mature form of political, economic, military, and social organization, common to much of Europe. It was therefore natural that the mounted knight, the individual representative of this system, should be regarded as the ideal soldier and the true backbone of an army. Conversely, the infantryman was held in contempt, partly because of his social inferiority, partly because the conditions for the development of efficient infantry forces did not exist.

However, infantry did not disappear completely from medieval battlefields. Levies of peasants or townsmen usually accompanied their mounted lords to battle, but the role assigned them was insignificant. Poorly and variously armed, with no internal or tactical organization, medieval infantry was incapable of withstanding cavalry assault, had little or no influence on the outcome of a battle, and was usually exposed to indiscriminate slaughter when its cavalry was routed. There are exceptions to this general picture: the infantry of Boulogne, for example, at Bouvines (1214) demonstrated great discipline and tenacity before succumbing eventually to repeated French cavalry attacks. Occasionally, too, as at Tinchebrai (1106) or the Battle of the Standard (1138), knights dismounted and fought on foot. This was an expedient growing out of the lack of versatility of cavalry; it was usually adopted either because the terrain was unsuitable to a cavalry engagement or because a weaker army, trapped and forced to give battle by a stronger foe, recognized the deficiency of cavalry when acting on the defensive and chose to fight as infantry.

The medieval period has often been described as one in which the development of European civilization was arrested, as a long interval of slumber between the collapse of Rome and the outpouring of energy with the Renaissance, as a time when society was stratified into a number of fixed castes. Similarly, military historians have viewed the art of war in the Middle Ages as stagnating because it was monopolized by a single class interested in preserving a mode of combat suited to its own abilities and position. There is much superficiality, and even gross inaccuracy, in these interpretations. The Dark Ages following the fall of Rome may well have been a period of decline, in culture, art, government, commerce, and technology—save for the brief

SUIT OF ARMOR OF CHARLES THE BOLD (1433–1477), last of the great dukes of Burgundy. (Charles Boutell, ed., *Arms and Arbor in Antiquity and the Middle Ages,* Appleton, New York, 1870, p. 149; copy in Special Collections, Perkins Library, Duke University.)

Carolingian revival. But even then, the seeds of change were fermenting. An agricultural revolution in the ninth and tenth centuries increased food production and made possible the commercial revolution of the eleventh century. Crusaders returning from the Middle East with a taste for spices, silks, and other Oriental luxuries found surplus wealth to indulge their appetites. The northern Italian city-states grew wealthy as transshippers of goods into central Europe, where cities revived or sprang up anew at the intersections of trading routes. Western scholarship, long reduced to arid copying of limited texts in monasteries, was invigorated by the rediscovery of classical learning transmitted through Arab sources; it soon spread to the universities springing up all over Europe. New technologies sparked an industrial revolution that might have taken off but for the economic and demographic scourge of the Black Death in the middle of the fourteenth century.

Characterizations of medieval warfare as stagnant, rigid, unimaginative, and devoid of tactics similarly ignore significant evidence to the contrary. The "feudal pyramid" was never the neat, ordered structure described by lawyers and theologians, but rather a crazy patchwork of conflicting loyalties and obligations, in which private war and struggle for power were endemic. Cavalry dominated the battlefield but never operated entirely free of infantry. The feudal array was often undisciplined, but it could nonetheless operate with cohesion and effectiveness when well led, and some commanders deployed their forces with strategic and tactical insight. The *De Re Militari* of Vegetius, written in the late fourth or early fifth century, was widely read in the Middle Ages. It served as the basis of an evolving body of knowledge on the art of war. Major developments were made in fortification, armor, and the techniques of siege warfare.

Nevertheless, medieval thought was not concerned with "progress," but with stability and order, and medieval institutions reflected this conservative resistance to change. Once established, the feudal system, with military power and gathering custom to support it, was difficult to shake off; peasant risings were invariably and ruthlessly crushed, the better to emphasize the permanent gap between those who ruled and those who served. In war, as in other departments of life, the feudal class refused to allow any encroachment by social inferiors upon what was regarded as its peculiar preserve or to tolerate any innovation which might threaten its monopoly. The castration of some South German peasants for taking up arms in support of the Emperor Henry IV in 1078, thus infringing on knightly prerogatives, is an illustration of the lengths to which the feudal warrior was prepared to go to defend his position.

The chief guarantor of stability in medieval Europe was the Church. Its influence, of preeminent importance in fashioning conduct and social attitudes in an age of almost universal faith, was placed upon the side of established order partly because the existing gradations appeared to reflect the differing capacities and purposes of man, partly because it taught the irrelevance of this life except as preparation for the next, and partly because, in every feudal state of Europe, the Church was the largest single landowner with the possible exception of the king. It was not surprising that the Church should come to the aid of the knightly class by banning the use of weapons inimical to cavalry warfare. An example of this attitude occurred when the eleventh century saw the replacement of the short bow, firing wooden shafts, by the crossbow. The latter weapon, consisting of a metal bow superimposed on a wooden stock, with the bowstring drawn back by a winch and released by a trigger, fired metal bolts which could penetrate the best chain mail. Missile weapons "which know not where they strike" were despised as unchivalrous by the mounted knight; the crossbow in particular was feared as a threat to the dominance of cavalry. In the eleventh and twelfth centuries a series of papal anathema proscribed the use of the crossbow, and its refinement, the arbalest, except against pagans and infidels.

The attempts by the Church to prohibit or restrict the use of such commoners' weapons as the crossbow were not very effective. By the thirteenth century most feudal armies had their quota of Genoese or other expert crossbowmen. The Church had rather more success in achieving its main objective, which was the limiting of warfare, both in frequency of occurrence and in the way it was conducted. In the absence of any centralized authority, feudalism in essence had meant the assumption, by local lords, of the functions of defense and order. Even before the threat of external attack declined in the eleventh century, it was inevitable that feudal magnates, their semi-independent status assured by strong castles and having at their disposal their own private armies, should turn to fighting among themselves. Throughout the later medieval period Germany was a jungle of feudal powers which the strongest emperor could not unite. In other parts of Europe conditions were scarcely much better. Even in England, where the king possessed a potential authority unique in the West, the combination of a weak monarch and a disputed succession threw the country into the disastrous anarchy of Stephen's reign (1135–54). In most of Europe, the sole authority overriding the crude appeal to arms was the Church. In the hands of vigorous popes like Gregory VII and Innocent III the threat of anathema, interdict, or excommunication could be an influential moral weapon.

It was not the intention of the medieval Church to eradicate war, which it regarded as an irrepressible evidence of man's sinfulness. Moreover, some wars were regarded as just, and their outcome divine verdict of where justice lay. The Church could lend itself to the encouragement of such wars, notably the Norman invasion of England to deal with Harold the oath breaker, or the holy wars of the crusading period. Religious sanction could also be given to the judicial trial by combat, until clerical revulsion against the sacrilege in so putting God to the test led the Fourth Lateran Council in 1215 to forbid the participation of priests in such trials (an event which, incidentally, furthered the system of trial by jury in England).

Generally, however, the Church tried to curb the prevalence of war and soften its nature. Beginning in the late tenth century, in the French provinces of Aquitaine and Burgundy, churchmen anathematized those who broke the "Peace of God," who despoiled church property and brought misery and death to noncombatants. Slightly later, excommunication was visited on those who broke the "Truce of God" by fighting at any time from Thursday night to Monday morning. Although these strictures of the Church may not have been given more than lip service, they did operate as a brake upon the more vicious and indiscriminate forms of feudal warfare; the right of sanctuary and the inviolability of church property were generally respected by the feudal conscience. When, in the twelfth and thirteenth centuries, the Church reached a peak of prestige and power, it was able to intervene in European diplomacy and war with considerable effect. The papal legate became a frequent arbiter in the settlement of international and internal disputes.

CHIVALRY

In a broader way, Christianity had much to do with the formulation of the code of honor known as chivalry. The ideal Christian knight, *sans peur et sans reproche,* was the pagan hero softened by Christian virtues, Grettir the Strong become Sir Galahad. The chivalrous knight was one who joined to bravery and the passion for adventure qualities of courtesy, truthfulness, loyalty, and mercy toward the weak and oppressed. Perhaps the individuals who met this lofty standard were few, and the number of those who aspired to it not much larger, but the mere existence of such a code raised the level of conduct in war, even if this meant only better treatment by the knightly class for other members of their international caste. In this behavior were the roots of modern international law, the codification of which began in the late Middle Ages.

The chivalric urge for adventure and glory, whenever it could not be satisfied by war, was met by the simulated warfare of the tournament. This lethal sport, which seems to have begun in France during the eleventh century, was immensely popular and undoubtedly was of some value in enhancing the skill of the feudal man-at-arms, but it was frowned on by the Church as a needless shedding of blood. In 1139 a decree of the Second Lateran Council forbade it, but without effect. In 1179, therefore, the Third Lateran Council denied Christian burial to those slain in such combat. This move did not halt tournaments, but it was sufficient to bring about the use of blunted weapons and eventually to reduce these events to opportunities for pageantry and relatively harmless display.

The clearest demonstration of the influence of the Church upon the military class is to be seen in the crusades. The promise of material reward and the hope of adventure found in the crusaders cannot be disentangled from spiritual motives. But the crusades are significant for another reason. Under feudalism, for the first time since the decay of Rome, Europe was powerful enough to take the offensive against the Muslim East, and in the process sack its former defender, Constantinople. The First Crusade (1095–99) saw the winning of the Holy Land and the establishment of the crusading kingdoms; subsequent crusades, which recurred until the late thirteenth century, were in the main directed toward maintaining the existence of these kingdoms.

The crusading armies were different from the usual feudal array in two respects: they were composed of volunteers, and they were even more incoherent because of their multinational origins and the rancor and jealously of their leaders. Yet, despite some memorable disasters caused by stupidity and recklessness, the success of the crusaders against the Muslims was on the whole pronounced, chiefly because the crusaders were compelled to employ tactics radically different from those of their customary cavalry melee. At Dorylaeum (1097) a Turkish army allowed itself to be trapped and crushed between two wings of heavy cavalry. After this, Muslim armies, composed of lightly armored mounted lancers and bowmen, never closed with crusading forces but only harassed them from a distance with missile fire. Finding their horsemen powerless to cope with a more mobile enemy, the crusaders soon realized the necessity for infantry support. Bowmen were required to counter the Saracen missiles (a bowman on foot could outrange a mounted archer); heavy infantry was needed to offer shelter for cavalry after a charge, and also to tempt the enemy cavalry to come within reach. Just such a combination of arms gave the crusaders victory at Antioch (1098) and won Jerusalem for them the next year. These balanced tactics, so unwelcome to feudal-class prejudice, had to be learned over and over again by successive crusading armies. Richard I of England applied them brilliantly against the able Saracen ruler Saladin at Arsuf and Jaffa (1191) in an unavailing effort to recover Jerusalem, which had been lost by extreme mismanagement in 1187.

The last European foothold in the Holy Land disappeared in the late thirteenth century. Military ability was not lacking; but the crusader principalities were without adequate permanent armies and were isolated from a Europe that was becoming increasingly concerned with more profitable projects. Warfare in the West drew little benefit from the experiences of the crusaders. The chief lesson, the advantage of a combination of arms which included infantry and bowmen, was too distasteful to be accepted by feudal chivalry. What developments there were tended only to strengthen certain aspects of feudal warfare. Quilted clothing, worn under armor, was a protection against missiles. Surcoats over it provided an added opportunity for heraldic display. The major result was to accentuate the already marked defensive nature of war, through developments in the art of fortification. Feudal magnates, impressed by the great castles of the East, returned to build similar structures, of which Richard I's Chateau Gaillard, built 1197–98 to guard the approaches to Normandy, was among the first and most notable. The new castle builders developed "concentric" fortifications, throwing out high curtain walls studded at intervals with round towers and eliminating square corners, which had proved easy to breach.

Despite the flowering of the chivalric ideal and the building of many bigger castles, by the thirteenth century feudalism as a political and military system was actually in decline. The justification for feudalism had been the order and security it brought to society during a period when central governments were weak, and in a broader sense, the preservation of the European religious and cultural heritage. This justification no longer had the same force. New factors were now uniting to dislodge the feudal aristocracy from the seat of power.

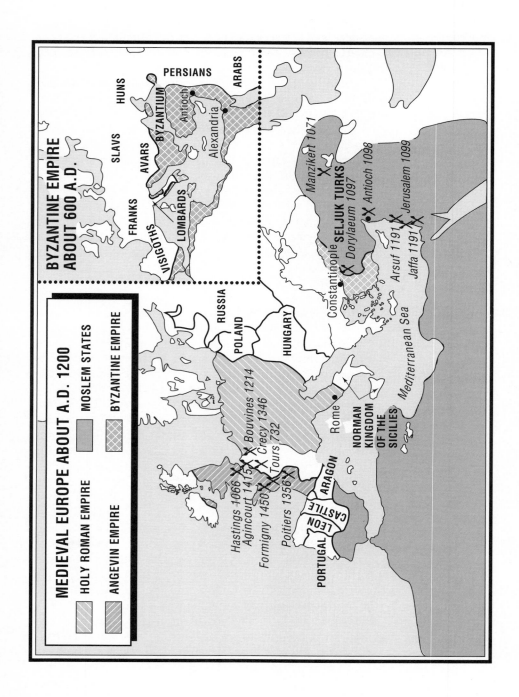

BYZANTINE EMPIRE
ABOUT 600 A.D.

PERSIANS

HUNS

ARABS

SLAVS

BYZANTIUM

Antioch

AVARS

FRANKS

Alexandria

LOMBARDS

VISIGOTHS

Manzikert 1071

SELJUK TURKS

Antioch 1098

Constantinople

Dorylaeum 1097

Jerusalem 1099

Arsuf 1191

Jaffa 1191

MEDIEVAL EUROPE ABOUT A.D. 1200

RUSSIA

POLAND

HUNGARY

Mediterranean Sea

HOLY ROMAN EMPIRE

MOSLEM STATES

ANGEVIN EMPIRE

BYZANTINE EMPIRE

Bouvines 1214

Crecy 1346

Tours 732

Hastings 1066

Agincourt 1415

Formigny 1450

Poitiers 1356

Rome

NORMAN
KINGDOM
OF THE
SICILIES

ARAGON

CASTILE

LEON

PORTUGAL

6

THE BREAKDOWN OF
FEUDAL WARFARE

WEALTH AND MONARCHY

Even when knighthood was in its full flower, there were basic weaknesses in feudal warfare which foreshadowed its replacement whenever a more efficient method should be evolved. Although the Middle Ages have an unmistakably martial flavor, they were a time of remarkably few battles. The raising of a feudal army by a king or magnate was an arduous process, lengthened by bad communications and fractious nobles. Once raised, its life was short: the usual term of service demanded of a knight was only forty days a year, and the campaigning season was sharply restricted. In England, the magnates objected to overseas service; on the continent, lords refused to serve outside the bounds of their particular state. In 1205, King John had to cancel a campaign to recover Normandy because his barons refused to cross the Channel.

A feudal army in the field was an indescribably undisciplined force. Many tenants-in-chief would take orders only from their immediate over-lord, the king; therefore an effective chain of command was impossible. There was a superabundance of courage, which tended to aggravate rather than to relieve the normal disorder. Long centuries of control of the art of war by one class, the exaggerated concentration upon cavalry warfare alone, and the absence of any provision for group training except in a restricted fashion in the reformed tournament meant that the study and practice of organized tactics had all but vanished. A vestigial remnant of tactical organization can be seen in the division of the feudal host into three "battles," or massed lines; but once the battle was joined, all semblance of order disappeared, and the struggle became nothing more than a confused melee of hundreds of individual encounters. The practice of taking prisoners for ransom instead of fighting to the death placed a positive pecuniary incentive on breaking ranks and making a fortune by private enterprise.

The same inability to achieve swift decision because tactical finesse was lacking is to be found in the broader aspects of war. So little attention was paid to strategy, or to such fundamental questions as supply and knowledge of the terrain and of the enemy's movements, that opposing armies often searched fruitlessly for each other without being able to come to grips, or used up the campaigning season in looting the countryside in order to maintain themselves. Moreover, although cavalry is the arm of the tactical offensive, it lends itself in a strategic sense to the defensive. No commander who felt himself weak need give battle; he had simply to retire with his force to the safety of the nearest stronghold. The art of fortification kept ahead of the revival of the Roman siege train and techniques. The primarily defensive nature of medieval war was preserved, as was the security of the petty baron against centralizing rulers.

Monarchical dissatisfaction with the limitations placed upon royal policy by the deficiencies of the feudal levy brought early attempts to break the military monopoly of the feudal classes, usually through the employment of mercenaries. Twelfth-century rulers did not have the financial resources to employ professional soldiers on any large scale, but by various expedients they could raise enough money to hire a few. Such an expedient was the English system of scutage, which was used by Henry I in 1125 and may have been even older. In its matured form it was a device through which the king, by assessing his tenants-in-chief a flat rate for each of the knights they had enfeoffed, was able to dispense with at least part of his feudal levy and buy the services of mercenaries who would fight as long as there was money to pay them. King John got into difficulties with his baronage because, among other reasons, he was more ingenious than his predecessors in using scutage and other feudal dues to squeeze money from them. Although the great age of mercenaries was yet to come, by the early thirteenth century they were a not insignificant part of every army. Richard I and John hired large numbers of Welsh foot, or reached across the channel for infantry from Brabant; the Genoese crossbowman was ubiquitous; and everywhere landless knights formed an obvious nucleus for a mercenary company.

The two great institutions of the Middle Ages were the feudal system and the Universal Church. The transformation of dissension-ridden feudal states into strong national monarchies and the breakdown in the temporal authority of the Church were due primarily not to military developments but to vast social and economic movements. At the same time, however, the art of war was divested of most of its medieval trappings and entered an age of rapid change. The new forms of war then became a vital element in the process which was altering the face of Europe.

One of the chief agencies in hastening medieval Europe into modern times was a great increase in wealth which began to be noticeable about the middle of the thirteenth century, primarily because of the revival of international trade. International commerce had never died but, until the eleventh century, apart from salt and iron, it had been largely concerned with such high-priced luxuries as spices, silks, furs, and especially slaves. The prosperity that came to Venice through its connection with imperial Constantinople and its naval power has been noted in a previous chapter; the crusades enhanced its wealth and importance and were instrumental in the rise of such mercantile centers as Genoa and Pisa. In the twelfth century the Low Countries became the most prominent cloth-manufacturing region of Europe and the Rhine a highway of trade; just as in northern Italy, trade and industry brought the rise of cities like Bruges, Ghent, and Ypres. England, as the principal source of raw wool for Flemish industry, gained greatly in wealth and also in urban development. In the same period, cities of North Germany like Lübeck, Hamburg, Magdeburg, and Danzig became the land carriers of Europe, uniting the economies of Scandinavia and the Slavic East with Italy. The weakness of the imperial government in Germany led these cities to form independent leagues, the most important of which, the Hansa, was founded in the late thirteenth century. By 1300 the Venetian galley-fleet had made the first of its annual trading voyages to England and Flanders, meeting in both countries Hansa merchants, thus signifying the completion of a trading economy that embraced the whole of Europe.

The new trade was not merely the exchange of the raw products of one region for those of another. It involved manufacturing, and the wealth stemming from it produced a demand for the importation or home production of luxury goods. Cities like Florence became famous for fine weaving and dyeing, Milan and Nuremberg for metalwork, Marseilles and Venice for metal and glass wares. Hand in hand with the flourishing of commerce went the emergence of an urban middle class engaged in business and manufacture and subdivided according to occupation in the specialized guild organizations. The appearance of a wealthy, numerous middle class had important consequences. The old simplicity of peasant, lord, clergy, and king was gone, disrupted by the addition of a dynamic class which was profoundly antifeudal in character. In regions like Italy and Germany, where the central government was weak, growing towns shook off the hold of adjacent lords and became in effect independent city-states. In England, however, and to a lesser degree in France, where the feudal duchies were strong, the towns became allies of the crown against the feudal aristocracy. Moreover, the growth of urban population and the spread of a money economy associated with trade had a strong impact upon the life of the countryside. Urban demands for food and raw materials made farming for profit possible; rising prices and a more plentiful money supply gave the peasant the means to purchase his freedom or commute his labor services for a money rent. In England, for example, enfranchisement and agricultural prosperity created a rural class of small farmers, the yeomen, who filled the ranks of English armies in the later Middle Ages.

One of the prime effects of the economic revolution was to enlarge the financial resources of the state, and therefore to expand its military power beyond the old limitations. Merchants with surplus capital soon became the bankers of royalty. During the Hundred Years' War, Italian banking houses lent money indiscriminately to both France and England; and the Hansa merchant, Tiedemann of Limburg, financed the early campaigns of Edward III. During the later years of this war, each country produced its own merchant-capitalists. Henry V reached the field of Agincourt on sums borrowed from men like Sir Richard Whittington, Lord Mayor of London; and the subsequent French recovery was in part made possible by immense loans from the famous merchant-banker Jacques Coeur. Increased revenue was also reaped directly from trade by customs duties; export duties on raw wool leaving England or wines leaving France swelled the royal treasuries. Of the utmost political consequence was the extended use of methods of general taxation, for in order to tax the new middle classes, European rulers found it wise to obtain their consent. At approximately the same time the characteristic representative institutions of late medieval Europe appeared: the Spanish Cortes and the German Reichstag in the late thirteenth century, the Estates-General of France in 1302, and the "Model Parliament" of Edward I in 1295. In all these assemblies the burgesses of the towns and the prosperous gentry of the countryside were represented, and it is interesting to note the close connection, particularly in England, between the monetary needs of the king in time of war and the growth in power of these middle-class institutions. The origins of the House of Commons as the central arch of the English constitution cannot be divorced from the history of the Hundred Years' War. Through its control of the vital power of taxation, the Commons was able to extend its competence to high matters of state and even, in 1399, to include the nomination of a king, Henry IV.

THE HUNDRED YEARS' WAR

By the late thirteenth century the monarchs of England and France had virtually extinguished private feudal war in their immediate domains. No longer was a castle an unconditional guarantee of baronial independence; the king, with his new-found wealth, could hire mercenaries for long sieges to overcome the wealthiest and most obstinate lord. In both countries, royal control was cemented by the building and maintenance of royal castles, commanded by a seneschal, a mercenary captain who was more likely of middle-class than aristocratic origin. Thus were established the two strongest monarchies of late medieval Europe.

There was, however, an important difference between the two. In France, the kings had used their wealth and power to harness their feudal nobility, but the feudal military system continued to exist intact. In England, not only was the baronage severely curbed, but the old military society was withering at its roots. Agricultural prosperity brought about subinfeudation, or the breaking up of the unit of military service, the knight's fee, into smaller parcels of land. It was obviously impossible to apportion military service fractionally, and so the many holders of a single knight's fee paid a

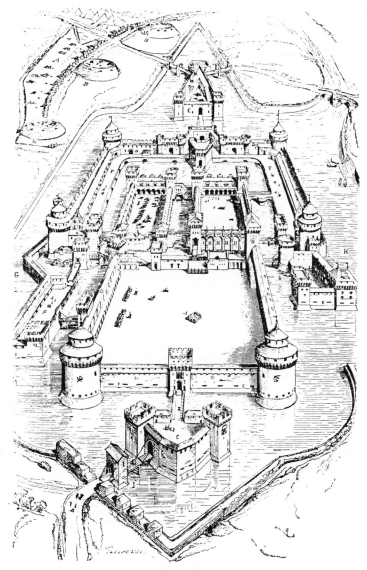

THE CASTLE OF MILAN at the beginning of the sixteenth century.
(Eugene-Emmanuel Viollet-le-Duc, *Military Architecture*, James Parker,
Oxford, 1879, p. 220)

fraction of its scutage instead. On the other hand, knights who continued to
hold land by virtue of military service found that while their services as soldiers were rarely demanded, they were being loaded by successive kings with
more and more onerous (and unpaid) duties of local government. In Henry
III's reign (1216–72) the English knight was required to serve on juries, conduct inquests, administer the king's forests, determine land boundaries, and

A TREBUCHET, throwing a dead horse into a beseiged town to spread disease. (Leonardo da Vinci, "Il Codice Atlantica," reproduced in Ralph Payne-Gallwey, *Projectile-Throwing Engines of the Ancients,* Longmans, Green, London, 1907, p. 30)

occasionally to sit in Parliament. For these reasons many men who held knight's fees refused to acknowledge themselves as knights; and at the same time that they rejected their civil functions, they ceased to provide themselves with armor and the other paraphernalia of knighthood. Edward I (1272–1307) attempted to shore up the feudal military system through a series of proclamations ordering those who held land above a certain value to assume the rank of knight and provide themselves with equipment. But no legislation of that kind could be effective. The last summons to the feudal levy of England was issued in 1385.

The Hundred Years' War (1337–1453) brought the complete discrediting of feudal methods of warfare and of the armored, mounted knight. Ever since the introduction of the crossbow, the feudal warrior had become increasingly concerned with strengthening his defensive equipment. Chain mail had been replaced early in the thirteenth century by much heavier and more expensive plate armor for horse and man. The only response the mounted knight could make to the new military techniques was to burden himself with heavy protective covering until, by the fifteenth century, he was virtually immobilized. Already, before the outbreak of the Hundred Years' War, French chivalry had received a damaging blow to its prestige on the field of Courtrai (1302). Count Robert of Artois, disdaining to wait until his Genoese crossbowmen had "softened up" an infantry army of humble Flemish journeymen, launched a cavalry charge across a bog. When the French were hopelessly ensnared in the wet ground, the Flemings, who were armed with pikes (a type of long infantry spear) and fired by a sturdy local patriotism, plunged to the attack. Seven hundred gold spurs were recovered as trophies from slain French knights.

That the French failed to learn from Courtrai (and other battles like it) brought calamity when they encountered the armies of the English. As we have seen, the place of the feudal levy in the English military scheme was one of steadily diminishing importance. Following the reign of Edward I, English armies were professional in nature, and troops were raised by indenture. The king contracted with well-known commanders to supply him with men, their number and time of service to be determined by the amount of money that changed hands. The type of troops required was always stated in the contract and usually consisted of a mixture of mounted men-at-arms, infantry, and archers. The introduction of the wage system and the dependence of the professional soldier upon his commander meant that the indentured companies had a degree of discipline unknown to previous, purely feudal armies. Another, though less important source of English manpower was the shire levy or militia, a lineal descendant of the ancient *fyrd*. The Assize of Arms of 1181 had laid down that every freeman between sixteen and sixty should provide himself with suitable weapons, according to the value of the land. In 1285 a statute of Edward I confirmed the Assize and also declared that the longbow should be the basic weapon of the shire levy. Occasionally during the Hundred Years' War part of this militia was employed on overseas service, its expense being shared among the remainder.

Mention of the longbow turns attention to the weapon which has become identified with the Hundred Years' War. The English seem to have encountered it in their wars with the South Welsh in the late twelfth and thirteenth centuries and were so favorably impressed that by the close of the latter century it had been adopted as their national weapon. The longbow, six feet in length, was pulled to the ear and was therefore very much more powerful than earlier bows. It was also superior to the crossbow because of its higher rate of fire, greater range, and the fact that in wet weather the bowstring could be more easily protected. Its power of penetration was great, although exaggerated by the legends which have since surrounded it. Its clothyard arrow could not pierce the best plate armor, but easily penetrated mail or the inevitable chinks in plate armor, and was deadly when employed against horses. The strength to draw the bow to the ear and the accurate use of the weapon required years of training. Its use in war was a virtual monopoly of the English yeoman, for whom the instrument was a mainstay of sport and recreation.

In wars against the Scots the use of the bow was perfected, and a genuine system of balanced tactics of horse and foot was built up around it. At Falkirk (1298) the archers riddled selected portions of the Scottish schiltrons (heavy infantry columns) and the English cavalry smashed into the gaps. With slight variations, these were the tactics which were to prevail in future

English battles with the Scots, from Halidon Hill (1333) to Flodden (1513). The English were defeated at Bannockburn (1314) because Edward II failed to make proper use of his archers. He opened the battle with a cavalry charge which degenerated into confusion. Then the archers, who were stationed in the rear, were ordered to fire, but probably hit as many English backs as Scottish breasts. Moreover, the bowmen had been left without any protection whatever, and were therefore easily dispersed by Scottish cavalry. By the time of the French wars, it had become customary to protect the archers from cavalry by placing stakes in front of their positions, and by intermingling their ranks with dismounted men-at-arms and foot soldiers armed with spears and bills. The bill was a six-foot cutting and thrusting weapon with a hook-shaped blade designed to unseat horsemen.

The causes of the Hundred Years' War, which were chiefly dynastic in nature, are not of great moment here, but the effects of the war upon military development and upon the history of the two nations which engaged in it were fundamental. The first and most apparent was the shock which French chivalry experienced at Crécy (1346) and Poitiers (1356). At Crécy the smaller army of Edward was divided into the usual three battles, each with a center of dismounted men-at-arms and wings of archers and spearmen. The French thrust forward their Genoese crossbowmen who, outranged and outshot by the English archers, were speedily routed or ridden down by their own cavalry. Throughout the rest of the day, charge after charge by the French knighthood withered against the solid phalanxes of dismounted English men-at-arms while from the flanks a hail of arrows took a heavy toll. When the defeated French retired at nightfall, more than a third of their number had been killed, the majority by missiles. At Poitiers the English were drawn up in much the same defensive formation, this time sheltered by a thick hedge. The French responded with tactics of such unbelievable fatuity that the only explanation can be their inability to credit the decisive contribution made at Crécy by plebeian archers. Believing that he was adopting the tactics which had given the English victory at Crécy, King John dismounted his knights and then launched them in three ponderous battles against the enemy position. Long before the struggling knights reached close quarters, they had been demoralized by English arrows. A flanking attack by men-at-arms led by the Black Prince, and another by a small cavalry force held in reserve, completed their rout. The French king and his son were among the many prisoners who fell into English hands.

The military disgrace of monarchy and feudal nobility at Poitiers was directly responsible for widespread political disorder in France. For a time Paris was virtually independent; the Estates-General demanded sweeping powers, while the provinces were convulsed by the bloody peasant rising of the *Jacquerie.* From Poitiers to the later years of the war the country was never to be free of the horrors inflicted by the "free companies," ravaging bands of mercenaries from all over Europe, whose contracts had expired and who owed no master.

Yet it is significant that the English were unable to capitalize greatly from either their victories or from French dissension and misery. Crécy and

SYMBOLIC REPRESENTATION OF THE BATTLE OF CRECY (1314), showing mounted knights, crossbow-men, and English longbow-men. One longbow-man in the lower front is preparing to fire; another is recocking his bow with a windlass. (John Froissart, *Chronicles of England, France, Spain, and the Adjoining Countries,* William Smith, London, 1839, Vol. I, p. 165; copy in Special Collections, Perkins Library, Duke University)

Poitiers were defensive battles in which the enemy had been beaten but not crushed; Edward III and the Black Prince were fine tacticians who did not rise above the general inadequacy of their times in strategic conceptions. Moreover, there were two insuperable obstacles in the way of large-scale conquests in France: the problem of supply and the unaltered dominance of the castle. The sole result of the victory of Crécy was the capture of a base, Calais, after a long siege; but then English generals had to live off the country by loot and pillage and could not undertake prolonged investments. Poitiers was fought only because the Black Prince could move no faster than his wagon train of loot and was unwilling to give it up.

After 1370, the tide of war turned for a space in favor of France. England was hampered by the passing of her outstanding leaders, the Black Prince in 1376 and the aged Edward III in 1377; by the succession of a minor, Richard II; by the rural unrest which culminated in the Peasants' Revolt; and by a political crisis which brought the deposition of Richard but did not end domestic troubles. In the Constable of France, Bertrand du Guesclin, King Charles V discovered a general with real ability, and not another rash feudal incompetent. Du Guesclin realized that it was not necessary to fight the English army in the field to regain French territory; the key to the retention of any area lay with the combatant who held its strong points, the castles. There was nothing of the chivalrous or the romantic in him. His refusal to consider the pleas of the feudal *noblesse* for yet another test of strength with the English removed them temporarily from the war and into their castles, out of harm's way. Using almost exclusively the professional soldiers of the free companies, du Guesclin and his successors fought a war of harassment, surprises, ambushes, sudden assaults, and slow sieges, which reduced piecemeal the English holdings in France.

Du Guesclin's brilliance was wasted on his contemporaries. On the field of Agincourt (1415), the chivalry of France showed that it had "learned nothing and forgotten nothing," as Talleyrand is reported to have said later of the Bourbon kings of France. The army of Henry V numbered only 6,000, of whom about 5,000 were archers; it was drawn up in the old way in three battles with cores of dismounted men-at-arms and wings of archers, and with the usual careful attention to terrain, its flanks being protected by woods and its front by almost a mile of sodden, freshly ploughed fields. The French constable, Albret, a pupil of du Guesclin, was unwilling to attack an English army so securely positioned, and wished to starve it out. But his slight authority could not withstand the exuberance of the nobility, confident in the knowledge that they outnumbered the English at least three to one. The rash formula of Poitiers was disinterred for the occasion. The French knights dismounted, sent their horses to the rear along with the crossbowmen and, in full panoply, began to plod through mud rising above their ankles toward the English lines. And the armor of 1415 was much heavier than that of 1356. The English archers, after easily repelling a minor cavalry assault, directed their fire against the slow-moving mass of French "infantry." By the time the French, greatly depleted in numbers, reached the English lines, they were so exhausted that they fell easy prey to the unarmored English yeomen, who abandoned bows for axes and swords. More than 4,000 French nobles and knights lost their lives in this crushing blow to feudal warfare; English losses have never been estimated at more than a few hundreds.

The ultimate French victory in the Hundred Years' War stemmed chiefly from two factors: the creation of a professional standing army and the rise of the imponderable spirit of French nationalism. The war has been called the first modern national war; from the start the royal armies of England, though professional in character, were almost wholly English in personnel. It was inconceivable, however, that a nation of two million, no matter how patriotic, should permanently triumph over one of sixteen; and when the mute antagonism of the mass of the French population was given expression in the inspiring figure of Joan of Arc, the days of English occupation became numbered. At the same time English enthusiasm for the war was being killed by heavier taxes, since it was becoming harder and harder for an English army to maintain itself in a hostile and thoroughly looted countryside. Moreover, livery and maintenance, an abuse of the indenture system of raising troops, had given rise to a bastard feudalism. The growth of private armies, unrestrained by an imbecilic king, weakened the national effort; and internal divisions culminated in the Wars of the Roses (1455–85) two years after the close of the French war.

In 1445 the first regular standing army since Roman times was organized in France by direction of Charles VII, partly to end the scourge of the free companies. Mercenary companies, now *compagnies d'ordonnance,* were taken into the royal service on a permanent basis, giving France a professional army of some 6,000 men. Each company, composed of men-at-arms, pikemen, and archers, was commanded by a noble who gave his name to it. The campaigns conducted by the French during the last years of the war are distinguished by a professional touch; in other words, by caution, good sense, and discipline. There were no more foolish attacks on a ready and waiting English army, but instead sudden assaults on marching English columns which allowed the enemy no chance to draw up his lethal defensive battle order.

When technical developments brought artillery to a state of reasonable effectiveness, even the defensive superiority of the English disappeared. From 1450 to 1453 an artillery train organized by the brothers Bureau methodically blasted the English out of their castles in Normandy and Guienne. At Formigny (1450) an English army, drawn up in traditional fashion, was enfiladed by two culverins placed on its flank. When the tormented archers broke ranks in order to seize the French guns, the *compagnies d'ordonnance,* no longer wary, closed in and cut up the English in a desperate hand-to-hand encounter. The English defeat at Formigny was as one-sided as their previous victories had been; and its verdict was confirmed at Castillon (1453) when an English army under the veteran John Talbot, Earl of Shrewsbury, relinquished the old tactics and attempted unsuccessfully to storm an entrenched line bristling with artillery.

The Hundred Years' War is in a sense an epitome of the revolutionary changes taking place in the art of war and the nature of society. Before its close the feudal principle had been replaced by the professional principle; aristocratic dominance had yielded to a democratization of the manpower and weapons of armies; the one-sided emphasis upon cavalry, and the tactical ignorance which accompanied it, had been gradually broadened until armies became combinations of all possible arms, tactics were restored, and new tactical departures encouraged; the new financial and political power of the state had widened the scope and lengthened the duration of wars and had corroded the moderating restrictions of the Church; and finally, the appearance of efficient artillery gave a hint of new revolutions to come.

THE SWISS PIKEMAN

In another part of Europe during the same period, the Swiss, although employing tactics utterly different from those of the English, produced a similar upheaval in the nature of warfare. Indeed, they went farther than the English, for they demonstrated conclusively the earlier lesson of Courtrai, that well-trained, well-armed, patriotic infantry could take the offensive against mailed cavalry and win. The Swiss Confederation, formed in 1291, was an alliance of the Forest Cantons of Uri, Schwyz, and Unterwalden, whose people were free peasants, directed against the feudal domination of the Austrian Habsburgs. As early victories solidified the virtual independence of the Confederation, the alliance was widened to include free towns like Lucerne, Zurich, and Berne. The underlying reasons for the emergence of Swiss military power were the compact, rugged, and defensible character of their country, the violent patriotism induced in a free people by the constant threat from a powerful external enemy, and in consequence so positive an accent upon military preparedness that training and service were compulsory for all men above the age of sixteen.

Swiss military development at an intermediate stage is illustrated by the battle of Morgarten (1315), at which the mounted array of Duke Leopold I of Austria was lured into a trap in a mountain pass. Powerless to maneuver, the Austrian cavalry was hewn to pieces by Swiss infantry armed with the halberd. The halberd, like the bill, united three weapons and as many functions on an eight-foot shaft: it was tipped with a spear point, had an ax blade that could shear through armor, and a hook below the tip to pull a rider from the saddle. The halberd, however, was primarily an offensive weapon; before the Swiss could deal with cavalry in country less favorable than their own mountain fastnesses, two problems had to be dealt with: how best to receive the shattering impact of a charge by heavy cavalry and how to prevent gaps from occurring in the ranks of their infantry which would permit cavalry to break up their formation.

The Swiss solution combined the methods of the two great infantry powers of antiquity. The standard formation was the phalanx; the standard weapon of defense the pike, an eighteen-foot shaft with a three-foot iron shank which prevented the spearhead from being lopped off by a sword stroke. To charging cavalry, the Swiss phalanx presented the same hedgehog appearance as had the Macedonian; the first four ranks leveled their pikes in an impenetrable barbed wall to the front, while those to the rear kept pikes upraised, ready to fill a gap left by a fallen comrade. Long training, discipline, and the patriotic feeling which made subjection to discipline possible gave the Swiss phalanx the steadiness required to maintain solidity in the face of cavalry attack. But, like the Romans, the Swiss were not

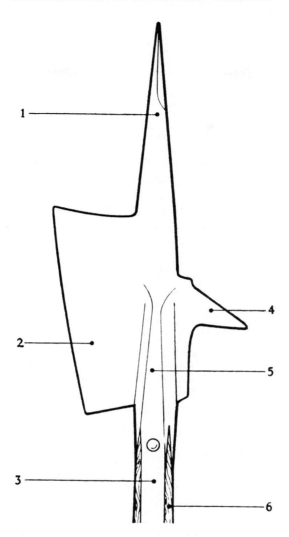

THE SWISS PIKE was a development from the halberd, which
traditionally combined a spear, an axe, and a hook to unseat
horsemen. The Swiss version, practical rather than ornamental,
could be used to pry open armor plate. 1, spike; 2, ax blade;
3, langet; 4, fluke; 5, socket; 6, pole.

content with mere defensive passivity: they thought in terms of attack, and
shaped their tactics to suit the offensive. Swiss troops wore no body armor
except steel cap and breastplate; the extraordinary swiftness this innovation gave them was schooled into order by tight discipline on the march.
(They were the first modern troops to march to the tap of a drum or to
music.) There was little time consumed in marshaling their forces for battle:

the Swiss marched in battle order of three columns in echelon. Their solidity and speed enabled them to attack bodies of horse; after the first push of
pike the way was paved for the halberdiers to do their deadly work.

The ability of the Swiss to meet and defeat cavalry on open and level
ground was demonstrated at Laupen (1339) when the feudal nobility of
Burgundy, unable to breach the pike wall or withstand the halberd, was
driven from the field. The supremacy of Swiss infantry over feudal cavalry
remained unbroken until the knight disappeared from the battlefields of
Europe. In the great campaign in 1476–77 Charles the Bold of Burgundy
challenged Swiss invincibility by adding to his feudal levy mercenaries from
many parts: archers from England, pikemen from Flanders, German harquebusiers, Italian men-at-arms. The Burgundian army, though formidable
in appearance, lacked the unity and confidence of the Swiss veterans. The
campaign brought defeat to Burgundy and death to Charles.

The reputation of the Swiss was so high that mercenary companies
of this warlike people fought for nearly every European state. (A notable
survival of the esteem in which they were held is the Swiss Guard at the
Vatican.) Their ultimate downfall was due in part to their unwillingness to
change their tactical formula when its victims, the feudal horse and ragged
medieval infantry, disappeared; from their weaknesses in the realm of strategy; from their over-confidence; and from their curious and unsatisfactory
custom of electing officers. It was also due to the improvement of field artillery. At Marignano (1515) the Swiss were halted by repeated cavalry
charges, and then smashed by concentrated French artillery fire. Even had
gunpowder not deprived the Swiss of their margin of superiority over their
opponents, the soldiers of other nations were pondering the ascendancy of
the Swiss phalanx and developing answers to it. The German *Landsknechte*
copies in every particular the pike and halberd tactics of the Swiss and at
La Bicocca (1522) in a furious push of pike managed to defeat them.

THE INTRODUCTION OF GUNPOWDER

Warfare in Italy during the later medieval period followed lines not dupli-
cated anywhere else in Europe. The citizen militia of the rich cities of the
north had thrown off the yoke of traditional feudal lords in the twelfth and
thirteenth centuries. Inevitably, wars against the feudality were replaced by
incessant wars amongst the newly independent city-states. Because it was
not possible to make perpetual demands for service upon citizens whose
industry created the economic well-being of the community, the employment
of mercenary forces became the primary method of making war. At first
mercenaries were hired from the free companies of Germans, Frenchmen,
and Englishmen who overflowed into Italy whenever there was a cessation
of martial activity during the Hundred Years' War. Soon, however, native
bands of *condottieri* were available in large numbers, attracted by the high
market value placed upon military services by centers like Venice, Florence,
and Milan.

The wars between Italian cities were almost purely economic in motive, and it is therefore not surprising that the mercenary captain should have governed his actions in accordance with strict business principles. It was found most economical, both for the *condottiere* and his employer, to restrict both the number and the type of troops hired. For this reason, infantry and bowmen were neglected, and the Italian wars of the fourteenth and fifteenth centuries were fought by armies of heavily armored horsemen, curious carry-overs from the feudal past. Moreover, since a victory could end the useful-ness of the *condottieri* as swiftly as a defeat, the mercenary captains (often by collusion) tended to avoid battle and drag out campaigns to the limits of their employers' patience and pockets. Innovations, particularly in firearms and artillery, were shunned as threats to military commerce; instead, sol-diers protected their personal investment by increasing the thickness and weight of their body armor so greatly that at Zagonara (1423), according to Machiavelli, "no deaths occurred, except those of Ludovici degli Obizi, and two of his people, who having fallen from their horses were smothered in the morass." The *condottieri* were military anachronisms, soon to be brushed aside by the modern army with which Charles VIII of France invaded Italy in 1494.

The English and the Swiss brought about the demise of the mounted knight as a power on the battlefield and thus assisted in the complex process which was ending his political and social dominance as well. Yet once the worth of their military systems had been established, both nations rested on their arms. It was left largely to other states to develop the weapons which administered the final *coup* to the inner defense of feudalism, the castle, and to the infantry of England and Switzerland as well.

The development of gunpowder weapons was exceedingly slow. Crude artillery made its appearance in the fourteenth century but, aside from the usual reluctance of soldiers committed to the prevailing weapons to accept a new one, it suffered from many technical defects. There was a distinct ele-ment of danger in the career of artillerymen, for the behavior of early pieces was at best unpredictable. Frequent confusion as to the mixture and handling of gunpowder, a very slow rate of fire, the lack of aiming devices, and the ineffectiveness of early solid shot against stone walls contributed to the mi-nor role which gunpowder played in the wars of the fourteenth century. Froissart noted that the three cannons displayed by Edward III at Crécy served only to frighten the horses; as late as 1418, Henry V, despite his pos-session of artillery, took six months to capture Rouen. The employment of field artillery was long hampered by its immobility; the early culverin, a field-piece with a one- to three-inch bore, was mounted on clumsy wooden sledges until about the middle of the fifteenth century, when wheels were substituted. The French were able to bring their culverins into play at Formigny only because of the stationary position taken up by the English. An interesting attempt to lend mobility to the new arms was conceived by the Hussite general, John Ziska, who mounted artillery in armored carts and employed these rudimentary tanks with considerable success against the armies of Catholic Germany (1420–30).

CANNONS OF THE FIFTEENTH CENTURY. *Top left:* Double cannon with the chamber for powder. C = A chamber with a handle. *Top right:* Double cannon with the wooden shield, or mantelet, to protect the gunners. *Bottom:* Cannon mounted on a carriage with a quadrant for ranging the gun. (Eugene-Emmanuel Viollet-le-Duc, *Military Architecture,* James Parker, Oxford, 1879, p. 172)

The most important single effect of the introduction of gunpowder during the later Middle Ages was in the sphere of siegecraft. Improvements in the casting of barrels, growing experience in the use of gunpowder, and the appearance of professional civilian artillerymen meant a higher level of gunnery and a heavier weight of shot. John Ziska's weapons included great bombards which threw stone or metal projectiles up to one hundred pounds in weight; by the end of the century more ponderous shot was not uncommon. The new artillery rendered obsolete the medieval stone castle, with its high, thin walls, which were vulnerable to cannon fire and unable to support defensive guns. With the castle went the last bastion of the medieval knight. In forty days in 1453, Turkish artillery destroyed the immense walls of Constantinople, walls that had remained impregnable for a millennium. Many lesser fortresses throughout Europe suffered the same fate in the same era. The costliness of the new weapons meant that of necessity they would find their widest employment in royal hands, and so gunpowder became yet another factor assisting the rise of centralized national monarchies. It was the alliance of gunpowder with the other resources of the national state that was to produce the recognizable beginnings of modern warfare in the sixteenth century.

7

THE BEGINNINGS OF MODERN WARFARE AND MODERN ARMIES

THE FRENCH INVASIONS OF ITALY

In 1494, King Charles VIII of France led an army of 30,000 men across the Alps into Italy. His professed aim was the capture of Naples in preparation for a crusade to the Holy Land. This wild venture marked the beginning of a new military era in which crusading, like all other medieval forms of war, had no place.

In the beginning, the march of the French down the peninsula was a military parade. Italy was hopelessly divided; her soldiers had no patriotic spirit; and the French had superior weapons. The Italians were stricken with awe by the great French train of horse-drawn bronze cannon which could keep up with the infantry on the march. Their own iron cannon, dragged laboriously by oxen, were much more cumbersome. But they were even more horrified by the military practices of the French. The "barbarians" from across the Alps fought to kill. The *condottieri,* and the cautious operations which they had substituted for real fighting, now seemed to be more fit for comic opera than for the drama of war. On their march south, the invaders cowed the Italian armies; and if the French campaign ended ingloriously a year later, the victor was disease rather than the Italians. But it is not proven, as has been alleged, that Charles's army became infected with, and spread, venereal diseases recently brought from America to Europe by Columbus's sailors. In 1495, at Fornovo when the French were fighting their way back, the Italian men-at-arms were routed by the French mounted *gendarmerie* and the French artillery. The next day, the Italian commander, the Marquis of Mantua, coming to ransom his friends and relatives, found to his horror that they had all been killed in battle. A new age had dawned.

Charles VIII's Italian escapade demonstrated clearly the power of new forces operating in society and in war. The French, through the use of improved artillery, had a decided technological advantage; but gunpowder had thus become a decisive element in warfare at just a time when a sociopolitical revolution was occurring. The Italian city-states, for all the wealth and culture they had garnered from their trade with the East, were no match for the new political organism, the nation-state, which had slowly evolved from feudal monarchies in England, France, Spain, and Portugal. In these new states, the centralization of power in the hands of the king had placed at his command the growing wealth of the new middle classes as well as of the old feudal landowners. In the course of the sixteenth and seventeenth centuries great merchants like the Fuggers in Germany, Burlamachi in England, and de Geer in Holland were prepared to manage royal financial transactions and to lend large sums on the credit of the state; men with financial experience in commerce served in royal treasuries and introduced new methods; new forms of taxation funneled wealth into the coffers of the state; monarchs like Henry VII and Elizabeth of England and Isabella of Spain and first ministers like the Duc de Sully in France showed themselves fully aware of the power of a full royal treasury and of the value of a busy commerce and a flourishing agriculture. Columbus's discovery of America had preceded Charles VIII's invasion of Italy by only two years; and Vasco da Gama brought the first spices from India by sea three years after it. The superiority of the modern nation-state over the late medieval city-state had been demonstrated even before the opening of oceanic sea routes transferred economic power from Italy to Portugal and Spain, and later to France, England, and Holland, and greatly increased the paramountcy of the rulers of those countries.

MACHIAVELLI

The real nature of the new forces operating in society and of their effect on warfare was clearly expressed for the first time in the writings of Niccolo Machiavelli (1469–1527), who became the Secretary of the Council of Ten of the city of Florence soon after the French invasions. Machiavelli's ideal form of government was that of the Roman Republic. But in his best-known work, *The Prince,* he analyzed the contemporary scene and flattered Lorenzo De Medici, who ruled Florence. In his book on the *Art of War,* and even more in *The Prince,* which has sometimes been described as the first book on modern political science, Machiavelli was the first man to write about the significance of modern warfare and to portray it accurately. Machiavelli discussed the new weapons and the problems arising out of their use. On the details of weapon development he was often wrong. Thus he believed that artillery would never be of more use than to scare peasants, that the day of cavalry was over, that infantry would always be used in large units, and that the sword would replace the pike. These mistaken opinions are explained partly by his lack of actual military experience; in his youth he had seen the clumsy fieldpieces of the *condottieri* and despised them. They are also partly explained by his passion for classical models and all things Roman; like many another modern writer on war, he was a disciple of Vegetius. But his errors in forecasting trends in tactics and weapon development merely show that even the most acute observer may be wrong in such details. They do not detract from his stature as the first interpreter of the nature of modern politics and war.

Machiavelli's aim in his writings was twofold. In the first place he wrote his book *The Prince* for the deliberate purpose of ingratiating himself with the rulers of Florence, the Medici, in order to regain his office, which he had lost in a revolution. Second, he wished to see Italy restored to independence and power, and he believed that this could be achieved only by the unification of the country under a monarch. Disgusted with the behavior of mercenaries, he advocated the recruiting of a national army of citizens, and he declared that only a powerful monarch, of the kind already found in other states, could revive Italy's glory.

It was this search for the secret of political power which led him to his revolutionary discoveries about politics, the nature of the state, and the significance of war. Whereas medieval writers had stressed that political affairs were subject to overriding moral considerations, Machiavelli divorced politics from morality altogether and claimed that success was based on force. Peace within the state was maintained by police power; and war was a natural condition in the relations between states. "A prince ought to have no other aim or thought than war." He comprehended the nature of the sovereign nation-state and argued that its existence must depend, in the last resort, on strength in war. He poured scorn on the limited warfare of the Italy of his day and showed that when states fight for their existence, there can be no limitation. Nor could the prince be limited by any moral consideration, either in his relation with other states or with his own people. Expediency must be his sole guide.

Machiavelli's ideas denied the existence of a universal Christian political society and hence were shocking to men whose whole intellectual background was based on the premise of such a society. Machiavelli, a product of Renaissance humanism, believed that humanistic scholarship based on the interpretation of history could shape policy. He correctly diagnosed that the national state was largely independent of any external control; and he also provided a textbook for princes, autocrats, and dictators which seemed to imply that success had no relation to morality. His direct influence on rulers in the following centuries—for instance, on Henry IV of France, who is said to have had a copy of *The Prince* on his person when he was assassinated—has often been noticed. His contribution to the growth of despotism is thus generally recognized, but although succeeding generations saw him as a wicked influence, today he is more often regarded as one of the first amoral realists among political thinkers.

Machiavelli drew his conclusions from a study of the political society to which he belonged. His realization of the strength of the nation-state came from actual contact with the new monarchies when on diplomatic missions, as his understanding of the fundamental principles underlying international relations was derived from his knowledge of the relations among the city-states of Italy. It was not without significance that his ideal prince was probably modeled on Cesare Borgia, one of the arch-scoundrels of history. Thus, Machiavelli was recording, as well as reflecting, the decline of public morality. Two years before his death there occurred a shocking event, the sack of Rome, the chief city of Christendom, by the troops of the Emperor Charles V, the nominal head of Christendom. Here was a dramatic symbol that the medieval ideal order had passed and that in this new day the old standards and the old authorities would be lost. War threatened the very foundations of western culture.

The imperial army that sacked Rome in 1527 is said to have included many Germans who had already accepted the doctrines of Martin Luther. The Protestant revolt that split medieval Christendom in the sixteenth century did not create the modern national state and modern international strife, but it undoubtedly expedited them. The assumption of a fuller control over religious and ecclesiastical affairs rounded out the omnipotence of national monarchs. In England, Henry VIII became supreme head of the church and an example for Protestant rulers everywhere; but even in countries which remained within the fold, monarchs began to assume a greater power to interfere in the affairs of the Church. The Reformation, directly or indirectly, ensured that the sovereignty and independence of national states was subject to no external or internal limitation.

Meanwhile, events had been drawing all parts of Europe into international conflict. The French king, Charles VIII, and his successors strove until 1559 to conquer parts of Italy, first Naples and later Milan. Their persistent determination inevitably brought them into conflict with the Austrian Hapsburgs who, as Holy Roman Emperors, had traditional claims to suzerainty over northern Italy, and also with the monarchs of the newly created kingdom of Spain, who had inherited interests in Naples. When all these dominions became united under Charles of Hapsburg, who was elected Holy Roman Emperor Charles V in 1519, the Franco-Hapsburg struggle enmeshed nearly all Europe. For Charles not only ruled lands which encircled France but also aspired to lead Christendom as his medieval predecessors once had done.

The French hunger for Italy, and Charles's dream of empire, had important consequences in the history of warfare. In the fifteenth century, wars had been fought in different parts of Europe, between the French and the English, among the various Iberian states, among the Italian city-states, among the Czechs, Germans, and Hungarians, between the Turks and their Christian neighbors, between the Danes and the Swedes, and between the Swiss and the emperor, but all these conflicts had been insulated from one another. In the sixteenth century, war became pan-European in its impact. Hence, the various types of troops which had developed in relative isolation in various parts of Europe—the Swiss pikemen, the German *Landsknechte* trained by the emperor on Swiss lines, the Italian *condottieri*, the light horse "genitors" and the sword and buckler infantry of Spain, and the heavy mounted *gendarmerie* of the *compagnies d'ordonnance* of

France—were thrown together into conflict just at the time when national antagonisms, made more bitter by religious differences, were ensuring the extension and prolongation of warfare. At the same time, the use of gunpowder in artillery and in small arms was creating for the tactician new problems which could be settled only in the crucible of war. Continual warfare made the sixteenth century a period of trial and experiment in which weapons developed rapidly and in which tactics changed with them, no one arm having primacy.

The basic problem to be settled was how gunpowder would affect the relative importance of the various military arms and how it could best be employed. During this century and the next, soldiers worked out methods of combining the firepower of shot with the defensive strength of the pike phalanx; cavalry tactics were adapted to the new warfare; artillery began to play a more important part in both the field and the siege; defensive works were adapted to meet new conditions; and the "new monarchies" experimented with professional standing armies which, by freeing them from the ponderous restrictions of the feudal array, or dependence on mercenaries, could serve them more efficiently than those older military organizations.

The battles in the first quarter of the sixteenth century proved that firearms would dominate future battlefields. Although Fornovo (1495) was won by Charles VIII's heavy cavalry and Novara (1513) was a victory for the old-style phalanx of pikes, at Marignano (1515) the Swiss echelon of pike columns was brought to a halt and to defeat by a combination of cavalry charges and artillery bombardment. It is probable that the artillery would have triumphed on the latter field even without the aid of cavalry charges. La Bicocca (1525) was won by harquebus fire from entrenchments. At Pavia (1525), after both sides had been dug in for three days, a surprise imperialist flank attack drew both armies out of their trenches, and Charles V's harquebusiers shot down the French cavalry as they came piecemeal to the attack.

THE ADOPTION OF FIREARMS

The lessons of war however, were as usual by no means so clear then as they are in retrospect. The adoption of firearms as the primary weapon was therefore fairly slow, partly because of cost, partly because of the conservatism of soldiers whose pride in traditional weapons is an important source of morale and therefore of fighting strength, and partly because new tactics still had to be worked out, but also because the older weapons still had a part to play. While the French had led in the development of artillery, they were strangely slow to adopt the harquebus. This was a handgun with a crook, or butt, for the shoulder. It was fired by a slow-burning match fixed to a cock and trigger and, when light enough to be easily handled, it fired a bullet of less than an ounce for about 200 yards; it was accurate, however, for less than half that distance. French kings preferred to recruit

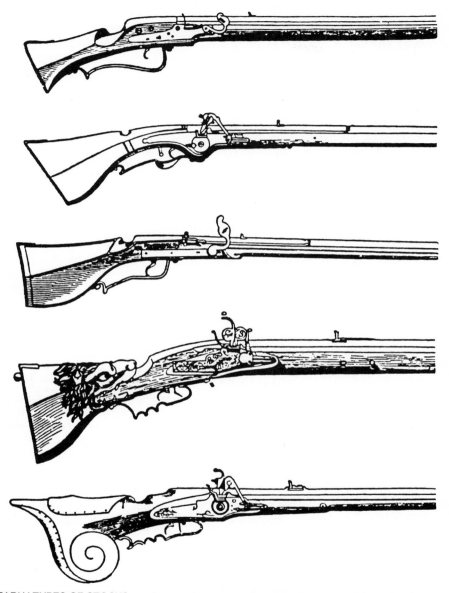

EARLY TYPES OF STOCKS used on harquebuses and muskets and including both wheel-lock and snaphance igniters. (Thomas H. McKee, *The Gun Book,* Henry Holt, New York, 1918, p. 34)

Swiss mercenaries trained for pike warfare. Only after Pavia did they begin systematically to increase the number of "shot" in their infantry.

The English, whose traditional weapons were the bill and the longbow, were even slower to turn to firearms and the pike. Henry VIII (1509–47) recruited Swiss pikemen for his continental wars, presumably because the eighteen-foot pike was better than the bill for holding off cavalry. But it was not until 1595 that a Privy Council ordinance finally declared that the longbow was not a suitable weapon for the English militia. This delay was not due merely to empty conservatism. When handled by an expert, the longbow was superior in range and penetrating power to the harquebus and much less dangerous to the user. But it could not penetrate the heavy plate armor of the late Middle Ages, which was vulnerable to the harquebus at short range.

It was the Spaniards who, at the beginning of the century, led the way in the use of the harquebus. Gonsalvo of Cordova, *El Gran Capitan,* first worked out tactics in which "shot" in field entrenchments played an important part against cavalry. However, influenced by the prevailing interest in classical studies, Gonsalvo also attempted to recreate the Roman legion by introducing sword and buckler men, a step which his contemporary, Machiavelli, heartily approved. But this development was short-lived. Only pikes could hold off cavalry charges. Although most men preferred the harquebus to the pike, because of the time taken to reload, the "shot" were defenseless in the open when enemy horsemen were at hand. The big problem for the tacticians, therefore, was to work out a combination of pikemen and shot which could operate together. Swiss experience had shown that the secret of using the pike was a tight, heavy formation and thorough drill. Introducing shot among the pikes weakened this formation. Tacticians therefore at first placed the harquebusiers on the wings or in front of the great phalanx of pikes. They could retire behind or beneath the pikes when threatened.

In the sixteenth century, mercenaries were usually recruited in companies which varied, according to time and place, from 100 men to 250. As early as 1505, the Spaniards began to group four or five companies together under a colonel, although at first only for organization and administration on the march. By 1534, they developed the *tercio,* a tactical unit of 3,000 men armed with pikes and harquebuses, which was a sixteenth-century Spanish variation of the fifteenth-century Swiss pike phalanx. About the same time, the French introduced a "regiment" of similar size. These massive formations came to dominate the battlefields of the mid-century.

The French continued to rely for their infantry chiefly upon hired Swiss mercenaries, while Spain, although following the sixteenth-century practice of building up armies by hiring men from different countries to use their traditional weapons, at the same time developed its own infantry. Recruited from the mass of the Spanish population, paid for by American gold, and trained and disciplined much like the Swiss, the Spanish tercio developed a morale and a prestige which gave it mastery. However, stories of the triumphs of small numbers of Spanish infantry over much larger numbers of Dutch during the revolt in the Netherlands (1568–1609) is a vindication of the combination of professional training and patriotic morale. Spain dominated the sixteenth century because she was enormously wealthy and could maintain a great professional army during wars.

THE MILITARY REVOLUTION

During the second quarter of the century, generals became more cautious and resorted more frequently to siege warfare. Thus there was ample opportunity for siege and defense methods to be adapted to the new weapons. In the previous century, the fall of Constantinople and the expulsion of the English from their castles in Guienne had seemed to spell the end of strongholds and walled towns. But in the sixteenth century the engineer got his revenge on the gunner. Fortification developed rapidly. Since high walls could be easily battered down by heavy artillery, fortifications went into the ground. Low walls, shielded by a ditch and a glacis and backed by earth, became the rule. Italian mathematicians took the lead in working out angles of fire; heavily gunned bastions gave a maximum cross fire. Medieval city walls were replaced by this type of defense. The castle gave way to the fort. For a time artillery was checked by the more rapid development of the art of fortification.

In the second half of the sixteenth century, society and warfare were alike dominated by the shattering impact of the Reformation. The dynastic struggles between the Kings of France and the Hapsburgs of Germany and Spain had hardly been brought to a halt in 1559 by the Treaty of Cateau-Cambresis when religious wars, which had already raged in Germany, broke out like a rash all over Europe. The intensity of the passions which religious fervor aroused embittered international antagonisms. Religion was often a basic cause of conflict and was as often a cloak for other motives. Ideological warfare meant that all restraints were cast off; in the name of sacred causes war became more ruthless and "unlimited." At the same time men who would not normally have been engaged in war took up arms in defense of their beliefs. The most important groups affected in this way were the middle classes in the towns whose commercial interests usually engrossed their attention. Religious strife, and to a lesser extent patriotic nationalism, swelled armies with elements very different from the normal type of mercenary soldier. When the religious wars ended and the fanatical citizen-soldiers returned to their homes, professional soldiers were left to fill the new armies which the religious wars had created.

The most important of these religious upheavals were the long civil wars in France brought on by the weakness of the sons of Henry II and the determination of the Mother-Regent, Catherine de Medici, to protect their patrimony against ambitious nobles who threatened it in the name of religion. A little later the patriotic and religious revolt of the Netherlands against the grim rule of Philip II of Spain began a long-drawn-out struggle that did not end until 1609, by which time geography and military engineering had decided that the northeastern portion, which could be defended by water, would become Protestant and independent Holland, while the southwestern half, now known as Belgium, would remain Catholic and be for two more centuries subject to Hapsburg rule.

Sixteenth-century armies continued to include large numbers of mercenaries, but there was a strong tendency toward the standardization of tactical methods and of army organization, with heavy borrowing from classical precedents. In place of massive formations like the Spanish *tercio* and the so-called regiment of Francis I, the true regiment began to appear, and within each of its four or five companies there were pikes and harquebuses. The shot were no longer attached as ancillaries to great phalanxes of pikes. Tactics were worked out by which pike and harquebus gave each other support; and the proportion of shot steadily increased. Gradually the harquebus was seconded and eventually replaced by the more efficient, but essentially similar, musket with a heavier barrel. Early models had to be fired on a rest but could propel a missile as heavy as two-and-a-half ounces. For a long time the musket was not popular because it was heavy to carry and took up to fifty-six drill movements to reload. After the piece had been fired, the musket rest was an embarrassment for which the musketeer really needed a third hand.

Maurice of Nassau, Prince of Orange (1561–1625), used pikemen and musketeers in formations of about 250 men, ten ranks deep, with musketeers on the flanks. After each man had discharged his piece, he retired through the ranks to reload. Maurice deserves special mention in the history of European warfare. Between 1590 and 1609, as one of the leaders of the rebellion in the Low Countries against Spanish domination, he pioneered what has been called "the military revolution" (Michael Roberts, *The Military Revolution, 1560–1660*). This meant not only smaller tactical units in battle, but, perhaps of even more importance, the introduction of parade-square drill. Drill not only prepared for more effective control during conflict, but also had a great effect on the discipline and morale of troops. It changed the feudal rabble, and the massed Swiss phalanxes and Spanish *tercios,* not only into smaller tactical units but also into modern armies. It gave the individual soldier a sense of belonging to a proud force; and incidentally, it came to occupy so much of his time that armies not engaged in operations were less likely to rampage off duty. Maurice also introduced marching in step, weapons firing drill, and firing by volleys; and after peace had been made with Spain he set up the first school for officers in 1619. His reforms spread across Europe and prepared the way for the emergence of standing armies as a new factor in the state.

As Machiavelli had noted at the beginning of the sixteenth century, cavalry was declining in importance. But before that century closed, by using heavier columns in place of a single line, it had regained some of its former importance. However, the growing use of muskets among the pikes made cavalry shock charges even more hazardous than they had been against the Swiss phalanx; so cavalry also began to use firearms. The firelock was difficult to use on horseback but the invention of the wheel lock, a gun with a revolving wheel which struck sparks against a piece of metal to explode the charge, gave the horseman a satisfactory weapon. The wheel lock was too delicate an apparatus to be used on the infantry musket, but in

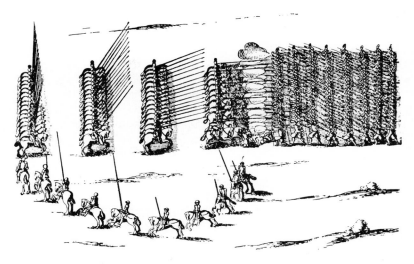

CAVALRY TACTICS AGAINST INFANTRY. (From *Instruction des Principes . . . de la Cavallerie,* Zutphen, 1621) (Massey Library, Royal Military College of Canada)

the sixteenth century it was widely adopted for cavalry pistols, two of which the horseman could carry in holsters with sometimes a third in his boot top. The result of the introduction of pistols was that cavalry began to avoid shock action and to rely on a tactical device known as the caracole, in which the front rank of the mounted column, when near enough to the enemy, discharged its pistols and retired by the flank to reload. Each rank in turn followed suit. Thus, cavalry declined in importance as a shock unit though it continued to be a valuable auxiliary when used for reconnaissance.

Perhaps the most significant development in armies in the latter part of the sixteenth century was the appearance of a modern rank structure and chain of command and of the ranks that have now become familiar. The basic unit, the company, was commanded by a captain. His lieutenant, or second-in-command, took his place when he was a casualty or absent. The command of several companies went to a colonel, who had a lieutenant colonel under him, usually one of the senior captains. The sixteenth-century "sergeant major" was a regimental officer, senior to the captain. He was responsible for the ordering of the regiment in battle array. As the title "sergeant major" was already associated with noncommissioned rank, the first part of the name was soon dropped, and the sergeant major became a major. Sergeants and corporals also appeared in the sixteenth century within the company. In charge of the army as a whole there was a lieutenant general deputized for the sovereign, who was, in theory, the commanding general. The lieutenant general had a deputy who was known by the peculiar title of "sergeant-major-general," later abbreviated to major general. He was responsible for drawing up the army in battle order and thus performed the same function as the "sergeant major," but at a higher level.

By the end of the sixteenth century, captains were usually "commissioned" by the sovereign and there thus appeared the nucleus of an officer corps. But the army was still far from being a "national" army in the fullest sense. Its pay was irregular and its food supply was at the mercy of greedy private contractors. Part of its clothing, powder, and rations were deducted from the soldier's pay. Often, as in Queen Elizabeth's expeditionary forces, the company was run by the captain as if it were a private business concern. If the captain were dishonest, he could make a fortune at the expense of the state and of his men by returning false musters or by trafficking in their necessaries. Armies were usually recruited for a single campaign; but disease and desertion, even more than casualties, often decimated them long before the campaigning season ended.

THE THIRTY YEARS' WAR

The Thirty Years' War (1618–48) saw further steps in the development of the art of war and in military organization. It began as a renewal of the religious warfare in Germany, but it rapidly drew in many of the great powers of Europe and soon became another round in the great struggle between the kings of France and the Hapsburg rulers of Spain and Germany. Thus, although this war had its roots deep in the bitter religious conflicts of the past, it was in fact a naked struggle for European empire and power. Catholic France, led by a Cardinal, Richelieu, was allied with the Protestant princes of Germany and the Protestant king of Sweden against the Holy Roman Emperor. The motives of the contestants rapidly became divorced from any real religious feeling, but the ruthlessness and ferocity that went with ideological warfare remained. The sack of Magdeburg in 1631, with the slaughter of 30,000 Saxons, was the most savage of many similar incidents. The combination of Machiavellian politics and religious fanaticism seemed likely to bring the destruction of society. After a generation of such conflict, Germany, which was the battlefield of Europe, was left devastated, with large areas completely depopulated. The Thirty Years' War, and the deep wounds it inflicted, increased German disunity. Germany, like Italy, was only a geographical expression until the nineteenth century.

Such was the nature of the war which saw the emergence of what is often claimed to be the first modern army. It came, not in one of the more advanced and wealthy states of the Atlantic seaboard, but in Sweden. National spirit, generated by the winning of freedom from the kings of Denmark in 1523, by the decline of the Hanseatic League's control of Baltic trade, and by Protestantism, was responsible for the rise of Protestant Sweden in the early seventeenth century. But the greatest factor in bringing Sweden to the fore and in maintaining her place as a great power was the ability of King Gustavus Adolphus and of some of his successors. Gustavus was a pioneer in military technological development and was at the same time able to take full advantage of the national and religious fervor of his subjects. He not only was a great tactician and a capable strategist; he also realized how to fashion a new army in keeping with the developments of his time. He built up a cohesive and fervent fighting machine of a modern type in place of the customary collection of companies of hired mercenaries and palace guards. He brought the military revolution to full flower.

Sweden had retained the medieval *levée en masse* which had virtually died out on the continent of Europe. Gustavus used this national obligation of military service for defense as a means of building up an army for overseas campaigns. He thus adopted a technique by which Elizabeth of England before him had used the militia to recruit ne'er-do-wells for overseas expeditions. But he did it much more thoroughly. Elizabeth's companies were collected by the lords-lieutenant and the justices of the peace in the various counties; but once transported overseas, the companies lost all identification and connection with the county of their origin and became attached to professional captains, who might well be from distant parts of England. Gustavus, similarly, recruited his armies for the invasion of Germany on a territorial basis, but he also arranged that they should be regularly reinforced from the same source. Thus, although lacking the deep purse which had enabled Philip of Spain a generation earlier to maintain his *tercios* for long periods in the field, Gustavus fashioned a fighting machine which was more permanent.

While the Swedes still made use of foreign mercenaries—for instance, Irish and Scots—these served as individuals. Foreigners were absorbed into Swedish units and were no longer hired in companies with professional mercenary captains. Foreign officers were, for convenience, usually employed in units that held a number of their compatriots. But the army which Gustavus led into Germany was a national professional standing army. However, as time went on, the proportion of foreigners in Gustavus's army greatly increased. Swedes were then assigned to garrison duties to hold captured towns securely while foreign mercenaries were kept busy in the field. This "Swedish" army's virtues were revealed in its discipline and in the proportion of fighting men which it contained. Whereas most armies of the day included hordes of camp followers who sometimes far outnumbered the soldiers, the Swedes marched without them; as a result, their efficiency on the march and in war greatly increased.

Gustavus's genius showed itself in every aspect of war. He further increased the mobility of his armies by reducing the weight of the weapons which the soldier carried. The heavy wooden rest, which had long encumbered the musketeer, was replaced by a lighter iron spike. The musketeer could then carry a sword. When attacked, he could now defend himself with the sword and also, if necessary, with his rest. Eventually a lighter musket was introduced and the rest was abolished altogether. A paper cartridge had been introduced earlier. Gustavus made it standard equipment. As the musketeer simply bit off the end of the cartridge, poured a little powder into the powder pan, and the rest into the barrel, dropped in a musket ball, and the powder cartridge case, and rammed it home, the movements necessary for reloading were greatly reduced in number and fire power was proportionally increased. As a result, battle drill was altered. Instead of the old ten-rank formation, the Swedes introduced a three-rank formation in which all three ranks fired together. The front rank knelt, and the others stood upright. Sometimes a second set of three ranks was stationed immediately behind in close support. Thus the Swedes were responsible for perfecting the technique of those deadly vollys which were to advance infantry to primacy on the field of battle by the eighteenth century. While the smoothbore musket was accurate only at the short range of about fifty yards, the bloody carnage caused by the volley at close quarters was frightful.

One significant result of the increase of infantry firepower was the decline of the pike. The weapon itself was shortened and the proportion of pikemen among the infantry was greatly reduced. Hence the great pike phalanxes which had been a feature of all armies since the great days of the Swiss pikemen in the fifteenth century began to disappear from the order of battle. But until the end of the seventeenth century, groups of pikemen were still diffused among the musketeers.

Another significant change was the abandonment of armor. Iron corselets gave too little protection against the increasing effectiveness of muskets. They also increased the extent of wounds. It is said that the Swedish musketeers began to go into battle without their armor because Gustavus himself, finding that his armor hurt an old wound, had abandoned it. Only the pot helmet, the predecessor of the modern steel helmet, was retained. As a result, the musketeers gained greatly in mobility. Nevertheless, pikemen and cavalrymen continued for a long time to wear the cuirass or breastplate, partly through conservatism and partly because their close tactics gave them more need of such protection.

Gustavus also made reforms in his battle array. The small company of a hundred to two hundred men had long existed as an administrative unit. In battle, companies had usually been joined together into an inflexible phalanx like the Spanish *tercio* of three thousand men. Following Maurice of Nassau's precedent, Gustavus split his forces up into "divisions" (more properly battalions or battle groups) of four or five hundred men. The battle groups were units enlisted on a regional basis and then deployed as tactical units. These smaller formations could be maneuvered more freely; and they were less vulnerable to artillery fire. Some carried pikes; the "shot" had muskets. Skillful combinations of the different arms was the key to success. They could stand up to cavalry attack.

At the same time, Gustavus revitalized the cavalry so that it became once more a decisive force on the battlefield. He insisted on resort to shock attacks with the sword and on the use of the pistol only at very close quarters. Thus he revived the use of cavalry as shock troops and checked the prevailing tendency to resort to the caracole, which had proved to be spectacular, rather than useful. Furthermore, he introduced the first dragoons, mounted infantry who served as light cavalry in the attack but reverted to infantry in defense.

Gustavus greatly simplified the artillery, which at that time included a confusing number of types. He adopted three standard guns: siege, field, and regimental. The weight of the guns was reduced. The regimental piece, a four-pounder four feet long, was a thousand pounds lighter than those used in other armies of that day. Gustavus also introduced the first artillery cartridge and thus not only made the handling of ammunition much safer but actually made the artillery rate of fire higher than that of the infantry. His regimental gun was mounted on a gun carriage and could be dragged by a single horse. The artillery could then maneuver in action. Although the gunners were still civilians, they were placed under army officers. Last, the proportion of artillery in the Swedish armies was greatly increased.

Gustavus's reforms were justified at the battle of Breitenfeld (1631) when he defeated the imperial general, Tilly, largely because of the superior mobility of his new battle groups of musketeers and pikemen, which he coordinated very successfully with artillery and cavalry. But his greatness as a tactician must not be allowed to obscure his ability as a strategist. His reforms made his army more strategically mobile. Greater mobility gave him an advantage in this sphere also. Before embarking on his great campaign to destroy Catholic power in Germany, Gustavus had first secured a firm base in Pomerania on the coast of the Baltic, opposite Sweden. Once that was done he marched rapidly, sought battle eagerly, planned his line of march and of attack with a view to future operations, and endeavored to coordinate the operations of armies in different parts of the country. However, the imperial army's mercenary commander, Albrecht Wallenstein, outmaneuvered him strategically in the 1632 campaign. It is possible that if he had not been killed in the ensuing battle of Lützen (1632), Gustavus's military genius, and his new type of army, might have enabled him to conquer for Sweden the greater part of Germany. As it was, the Baltic became a Swedish lake for a century. His methods, and especially his army organization, were admired and promptly imitated throughout Europe.

THE ENGLISH CIVIL WAR

When civil war in England broke out in 1642, after the constitutional and religious dispute between the king and Parliament, English and Scots soldiers of fortune carried Gustavus's ideas to Britain. In the first year of the war, both sides relied on calling out county militias that were little better than rabbles. King Charles I had an initial advantage because the landowning nobility provided him with the nucleus of a mounted force. The Royalist cavalry was entrusted to a German nephew, Prince Rupert, a young man with experience in the Thirty Years' War, who trained his men to break into the opposing forces before using their pistols.

But Parliament held London and the southeast, the richest part of the country and the most predominantly Puritan. The train bands of the City were the most nearly efficient militia forces in England at the outset of the war; and the citizens were prepared to fight what they believed were Romish tendencies in royal ecclesiastical policy. The flourishing gentry of the Eastern Counties had likewise been strongly influenced by continental Protestantism.

One of these was Oliver Cromwell, who trained a parliamentary cavalry force, the Ironsides, to overcome the hard-riding cavalry of Rupert. He was not a professional soldier and had nothing like Rupert's experience in war; but at Edgehill (1642), where he had seen the Royalist cavalry drive the Parliamentary horse from the field, he realized that courage and the aggressive spirit were not enough to ensure victory. Discipline and tactical training were needed as well. He therefore insisted upon rigorous training for the horse which he raised for Parliament in the Eastern Counties.

The wealth of London merchants and of the sheep-farming gentry provided Parliament with funds to raise a long-service professional army called, significantly, the New Model. This, like Cromwell's Ironsides, was driven by religious enthusiasm which gave it a morale superior to that of the Royalists; and yet it was at the same time a highly disciplined force. All the Civil War armies adopted foreign, and especially Swedish, battle formations; they all employed officers who had served as mercenaries on the continent; but the vital principle of long-service enlistment which was adopted by Parliament may have been an innovation caused by the pressure of circumstances. Whatever its source, the effect was important. Precisely how far the New Model drew upon Swedish precedents is not clear. But its greater military efficiency defeated both the Royalists and the Scots. Cromwell, as Protector, was the first ruler of England to conquer the whole of the British isles. His new armies laid the foundations of the British army of the future. Indeed, the Coldstream Guards traces its history directly back to the Puritan armies of the Commonwealth. The traditional scarlet of the British soldier was first worn by the armies of Cromwell. The foundations of a permanent standing army had thus been laid in England by the circumstances of the Civil War.

The victorious Puritans had taken up arms against Charles I's pretensions to rule by "divine right," a political theory derived from medieval precedents but which had been used to bolster a most unmedieval absolutism. One of the irritants which had led to the Civil War was the quartering of royal soldiers in civilian billets, a practice intended to relieve the king of financial worries by avoiding the necessity of resorting to Parliament for funds. However, the New Model and other Puritan armies, raised to defend constitutional liberty, soon in fact destroyed it. They were the armies of a fanatical minority which used them to overthrow Parliament and to force its religious and moral ideals upon a largely unwilling community. When Cromwell divided the country into districts ruled by major generals, the perils of stark military dictatorship were brought home. The country soon yearned for the return of the son of the king whom it had so recently executed. The restoration of Charles II in 1660 was, however, possible only because the Puritan army had its own unsettled grievances—namely, arrears of pay. The military organization which Cromwell had fashioned had not been fiscally sound. Certificates given to the rankers in lieu of arrears of pay had been bought up at a discount by some of their officers, who became war profiteers and parvenu landlords. Since the end of active warfare, the citizen soldiers had gone back to their shops and their farms, leaving the armies of the Protectorate to purely professional soldiers. Pay was the bond which tied these to the state. When they were not paid, the regime fell. When the restored monarchy paid them off, they went peacefully to seek civilian pursuits.

LOUIS XIV

On the continent the growing military strength of monarchies based on permanent royal armies had already shown itself able to withstand any such debacle as the Stuarts experienced. For instance, although weakened by minorities and regencies, the French house of Bourbon had gained in strength and had been able to crush a dangerous revolt, the Frondes (1648–53). Nevertheless, the army was still in need of reform. Its system of supply was riddled with corruption, its regimental organization was chaotic, its pay was irregular, its discipline harsh and capricious.

During the minority of Louis XIV, the secretary of state for war, Michel le Tellier, set about the reform of the army. Later his son, the Marquis de Louvois, Louis's war minister, created a civil administration for the affairs of the army and thus gave it a firmer and sounder place within the framework of the state. Through the minister of war the king now exercised as firm a control over the army in peacetime and in winter quarters as he had formerly exercised through his generals while it was on campaign. These developments were in line with the contemporary growth of bureaucracy in other spheres. The seventeenth century saw much expansion of national civil services. The intendancies and the *noblesse de la robe* in France, the Admiralty, customs, and colonial civil services in England, and the civil service in Prussia were all expanding in this period.

At the same time Louvois introduced important reforms within the army itself. Battalions were integrated into permanent regiments and brigades; the grosser swindles in supply were checked; a quartermaster general's department was set up; white uniforms were made standard; the troops were taught to take up stations by word of command instead of having to be laboriously put in place by the sergeant major; a system of inspection established uniformity and obedience throughout the army; and the name of one of the royal inspectors, Martinet, became a byword for rigid discipline. Louvois cut the pattern for the regular army of the future and thus provided the support for the despotic monarchies of the next century.

But Louis could never have fashioned his great armies or fought his many wars had he not been served faithfully by another great minister, Jean-Baptiste Colbert, who put his finances on a sounder footing and, following mercantilistic theory, fostered trade, industry, agriculture, internal communications, colonies, and shipping to increase the wealth of the state. Mercantilism had had its origins in the sixteenth century, when the acquisition of gold- and silver-producing American colonies had brought great wealth and strength to Spain. Money was said to be "the sinews of war." The first mercantilists were "bullionists" who taught that power flowed to princes who discovered mines of precious metals in their own territory or in overseas colonies. Later arrivals in the competition for overseas empire sought always for gold. When they could not find it, they often resorted to thinly disguised piracy, as had the Elizabethan sea dogs who preyed upon the galleons of King Philip II of Spain. But economic thinkers soon realized that gold could be accumulated within a country by a favorable balance of trade and by the acquisition of colonies which could produce the exotic staple products now in demand throughout Europe. Hence the power of the state was used to protect industry and agriculture, to control the flow of trade, to acquire and govern colonies, and to foster merchant shipping. The latter policy, of which the English Navigation Acts are the best example, had a closer connection with military strength than had other mercantilist regulations. It provided ships and sailors for the navy. But all mercantilist policies were designed deliberately to build up the wealth of a country and pay for standing armies and navies.

AN ECONOMIC REVOLUTION

The German economic historian Werner Sombart (1863–1941), argued that the growth of armies and of military organization in this period had a very special bearing on the fashioning of western society. He ascribed the beginning of large-scale metal industries, without which all material and technical development would have been impossible, to a large extent to the demand for cannon; and the growth of financial enterprises like that of the Fuggers in the sixteenth century and of the great Jewish banking houses of the seventeenth century, he attributed to the demand for loans for military purposes. He also suggested that the large-scale military undertakings of the seventeenth century were the first examples of the kind of organization later to be followed in commerce and industry; that the qualities of the military adventurer were those of the "entrepreneur"; and that the "military virtues" of order and discipline were akin to the spirit of the new capitalist world.

Many of these generalizations were later challenged by J. U. Nef, a Harvard historian, in his *War and Human Progress* (1950). Nef said, for instance, that it was the medieval demand for church bells, instruments of peace, and not for bronze cannon, that stimulated mining and metal working. It must nevertheless be accepted that war, and the preparation for war, was a powerful incentive, even though perhaps not the only one, for the growth of modern industrial, commercial, and financial organization. Gustavus, Louis XIV, and Peter the Great of Russia all founded arsenals which were among the first examples of the factory system in modern times. Adopted for economies of scale and to enhance royal power, seventeenth-century arsenals anticipated, and blossomed, more readily than the mills and factories being built to utilize water power and, later, steam. The funding of the national debt and the establishment of the Bank of England, fundamental innovations in state finance of the most profound importance, were direct results of the need for providing capital to fight Louis XIV.

By the second half of the seventeenth century, the military systems of the new national states of Western Europe were being fashioned into efficient machines. Long before this time European military superiority had expanded European influence and power into the four corners of the globe. The geographical discoveries of the fifteenth and sixteenth centuries had been made possible by inventions like the mariner's compass, the astrolabe, and ocean-going caravels; but the conquest of overseas territory was only possible because the Europeans had superior weapons and superior military and political organization. The power of the national state, exercised either through viceroys or through chartered trading companies, governed large empires in Asia teeming with native races only a little less easily than it penetrated empty forests and plains in America. In turn, the possession of colonies brought wealth to Western Europe, and to the colonial powers in particular, wealth which could be converted into military strength. Hence, at the time when the states of Europe had been unified, usually by a monarchy, and when their military weapons had been tempered for use, colonial and commercial rivalry was pushing nations into conflict and taking the place of religion as one of the greatest causes of international strife.

8

THE BEGINNINGS OF
WESTERN SEA POWER

THE VIKINGS

The story of the Viking raids provides a prologue, or false dawn, for the history of sea power in the West. In 734 Charles Martel, better known for his triumph over the Moorish vanguard at Tours, crushed the Frisians, who had been the leading maritime people of northern Europe for the previous two centuries. Thus the door was opened to the pagan Scandinavian seafarers at a time when social and population pressures among them encouraged the undertaking of what was to be the last wave of the great Germanic invasions.

One branch of the Vikings drove southeast from Sweden to settle Russia, to lay siege to Constantinople (865), and to penetrate even as far as the shores of the Caspian. But our concern is with the Norwegian and Danish Vikings who concentrated their seaborne raids on the British Isles and the littoral of Western Europe. During the century following the first mention of them in the Anglo-Saxon Chronicle (787), their expeditions increased in size, frequency, and range. Utrecht and Antwerp were burned, Cadiz and Seville sacked, Paris besieged, Ireland overrun, and the centuries-long conflict on English soil launched. At least two major raids extended into the Mediterranean, where cities of southern France, northern Italy (Pisa and Luna), and Morocco were captured. Before the invading Northmen were finally absorbed by the Christian peoples of Europe in the eleventh century, they had wandered far afield to colonize Iceland, Greenland, and Newfoundland, and to fight the Red Indian in the New World.

From such sources as the epic *Beowulf,* the Norse sagas, and archeological findings, scholars have been able to reconstruct the warrior civilization of these people. Though notoriously fierce in battle (they contributed the word "berserk" to the English language), the Vikings contributed nothing to the development of land warfare. Their unique place in history rests on the special advantages that they gained through the use of the sea. Although their numbers were small, their ships and their seamanship made them the scourge of the West. Only at the siege of Paris is a force as large as 40,000 men and 700 ships mentioned, and those numbers are now generally believed to be grossly exaggerated.

The typical Viking warship was an open boat of lapstreak construction about eighty feet long, propelled by sixteen pairs of oars and a square sail on a single mast amidship. In 1893 a replica of such a ship made speed up to 11 knots under sail while crossing the Atlantic. With her high dragon bow and fine lines, she proved an excellent sea boat. According to the sagas, the largest Viking ship of them all, the *Long Serpent*, was a 165-foot vessel capable of carrying about 400 men. That was a great king's ship; but smaller vessels, assembled in fleets of a hundred or more, transported the major Viking expeditions.

Plunder was the immediate incentive for the original Viking raids, probably single-ship expeditions conducted by exiled petty nobles and their thanes. One successful raid led to others; richer goals, better protected, required the joining together of raiding groups; blood feuds led to foreign colonization by the more discontented or the more ambitious. Their only major sea fights were among themselves. These battles were primitive infantry engagements afloat. Their naval tactics had not developed to the skillful maneuvering and ramming practiced by the Greeks and Romans. For defense, the Vikings simply lashed their ships together, side by side, in one large platform. When on the offensive, they relied on boarding tactics against the flanks of a formation similarly constructed. They met little opposition at sea until, according to early English accounts, King Alfred had larger and better ships built and beat back the invader. The subsequent history of Viking successes makes clear, however, that Alfred enjoyed no more than a temporary naval success which the English lacked the power to follow up. After Alfred's death the Viking raids continued unabated.

Viking navies represent only a false dawn of sea power nonetheless, for the Northman's ship, though a good sea boat and resolutely manned, was primarily a vehicle for the transportation of land forces and an adjunct of land warfare. The Vikings did not control the sea in the modern sense of that phrase; rather, by means of their ships, they enjoyed the advantages of a cavalry of the sea, tactical surprise and rapid movement of a limited raiding force. At length, both at home and in the lands they had plundered, the feudal system enveloped the Christianized Vikings in the mainstream of European civilization. The progressive development of medieval sea power in northern waters derives not from the Vikings but from the conversion of the seaworthy, decked merchantman sailer into a man-of-war. For example, and by contrast with the long, fast, open galley–type Viking ship, the fifty-seven vessels which the English Cinque Ports furnished the king under their feudal agreement in the thirteenth century were "round-ships" with a length-to-beam ratio of about 2:1. They were slow, difficult to maneuver, driven by sail, and built and maintained as merchantmen, except as the king might modify them while they were in his service. At such times, when they had lightweight sterncastles and forecastles added to make them men-of-war, these ships provided superior fighting platforms for the king's crossbowmen and archers. They were, in fact, floating castles, stoutly built for good sea-keeping characteristics and able to withstand the weapons in use in that day.

MEDIEVAL SEA FIGHTS

The Battle of Dover (1217) illustrates the major features of the medieval sea battle. In this encounter a force of sixteen large and twenty lesser English warships—i.e., temporarily converted merchantmen—under Hubert de Burgh, governor of Dover, fell upon a force of seventy small craft and ten warships carrying 900 troops and supplies to reinforce the French in London. The English came downwind, throwing lime to blind their opponents and raining crossbow bolts. Thus gaining the upper hand, they boarded, cutting down the French sails and trapping the crews under them. The contest was then decided by hand-to-hand fighting. Fewer than a score of the French vessels escaped. Besides being the first naval victory of the English over the French and introducing 600 years of intermittent war at sea between them, the battle provides noteworthy examples of basic principles of naval tactics and strategy. Tactically it was a melee, but so conducted by the English as to wrest victory by the exploitation of an upwind position. Strategically, it suggests an awareness of the potential of sea power. By assailing the enemy's seaborne communications, de Burgh forced the French to abandon their holdings in England, thus accomplishing with ships what English land power had failed to do.

This pattern of naval warfare continued through the Hundred Years' War with certain refinements, as in the Battle of Sluys (1340), on the Flemish coast. There, Edward III with a force of about 20,000 men in 250 ships attacked a French force of equal size. The English enjoyed a slight superiority in the number of heavily armed soldiers and crossbowmen; otherwise the two forces were evenly matched. Against a triple French line, the English king advanced with his best ships in the van, so arranged that those carrying men-at-arms for boarding could forge ahead between those carrying the crossbowmen after the latter had engaged from a distance. This combined use of long-range weapons and heavy shock troops exactly paralleled the king's land tactics in method, and with equal success. Edward's strategy in using sea power to take the battle to the enemy revealed the offensive aspect of the naval component in the overall defensive policy of a maritime nation. Englishmen then, like Americans later, learned that the sea could provide the advantage of doing their fighting on the other fellow's shores.

Medieval navies in northern European waters departed somewhat from typical feudal organization in two particulars. The first concerned the ships, and the second their crews. Arms could be stored for future use, but a ship cost so much and when laid up deteriorated so rapidly, that it had to serve a continuing purpose in peace. Thus, as long as warships were converted merchantmen, a country's normal activity determined the size and quality of ships available for naval service in time of war. A king would maintain only a few ships for his own transportation and diplomatic errands; they were his personal property; otherwise he was dependent on levies from his merchant marine. Hence the king granted rights and privileges in order to encourage the merchant class and its trade so that he could obtain more and better ships for his occasional use. As England was dependent for economic well-being on the wool trade with the Low Countries, she enjoyed an important advantage in comparative sea power. The second difference arose from the warship's need for experienced shiphandlers as well as fighting men. Unlike the peasant who could drop his plow, obtain his familiar arms from his lord's arsenal, and be off to the war, the ship's officers and the sailors were professionals performing the same duties in peace as in war. Thus it came about that the medieval warship was commanded by the king's representative, a noble, and fought by his men-at-arms, but was sailed by the professional merchantman, a "boatswain," and his nonmilitary crew. The weapons were those of land fighting, and so were the tactics. As we have seen, battles were won on sea as on land, by crossbow artillery and the soldier's spear and sword. The warship was a floating, miniature castle, and siege weapons and techniques were adapted for naval use. However, of many ingenious contrivances, none was sufficiently successful to challenge boarding as the decisive tactic in sea battles.

NAVAL CONFLICTS IN THE MEDITERRANEAN

Concurrently in the Mediterranean the distinction between sail-propelled merchant ships and oar-propelled warships, dating back to the ancient Greeks, had led to the evolution of the dromond as the standard fighting ship of the Byzantine navy. A typical dromond was about 100 feet overall, with fifty oars to a side arranged in two banks, and with a single large sail amidships. There were a sterncastle and a forecastle. On the latter was a bronze tube for throwing Greek fire. These ships, sometimes built much larger, were successful in turning back the Saracen naval attacks on Constantinople, but by the time of the Crusades they were being replaced by the galley, a type which continued in limited use as late as the eighteenth century.

Galley warfare constitutes the peculiarly naval part of the story of the Italian Renaissance. The galley type, with oars all on one level, was developed by the Italians and was used by them to convey the treacherous

Fourth Crusade, which sacked Constantinople (1204); to protect their expanding commerce; to conduct their internecine wars; and finally, in the great battle of Lepanto (1571), to smash the Turkish bid for mastery of the whole Mediterranean.

From its inception as a major fighting ship, the galley, 130 to 160 feet in length, was equipped with a great "spur" extending forward like a bowsprit. This was intended to break the enemy's *aposti*, or outriggers for the oars, rather than to ram the hull in the fashion of the ancients. At first the rowers were freemen or mercenaries; later they were more frequently criminals and slaves, chained for life to their benches. With 100 men or more at the oars, the galley could make a top speed of about 7 knots for a short time; with her great lateen sails, usually two on the larger ships, she might make 12 knots under ideal conditions.

The battle of Lepanto reveals galley warfare at its ultimate stage, influenced by the introduction of gunpowder, but otherwise little changed from the land-fighting-at-sea of earlier naval engagements. The opposing forces were almost equal, the Turks being superior in number of ships and men, the Christians in the weight of armor and armament.

The Christian fleet consisted of about 200 galleys, half of them Venetian with freemen at the oars and the others Spanish ships rowed by criminals and slaves. Counted as Spanish were the Sicilian, Neapolitan, and Spanish royal galleys as well as 25 mercenaries hired from *assentiste*, the naval equivalent of the *condottieri*. Galley armament included five bow guns, a 36-pounder flanked by 9-pounders and 4½-pounders, as well as three 4½-pounders on each broadside. For hand-to-hand fighting the galleys carried large numbers of heavily armed and armored infantrymen, many of them armed with harquebuses. While going into battle, the rowers were protected by light wooden sideworks (*pavesades*) and, especially if freemen (as on the Venetian galleys that carried fewer troops), were also supplied with weapons. There were also six Venetian galleasses, which deserve special mention. This was a new type of warship built to carry guns. It differed from a galley in being wider, heavier, and higher of freeboard, with a gun deck over the rowers. The armament consisted of 6 guns firing ahead, 6 astern, and 18 on each side, for a total weight of shot of 326 pounds. These ships had fifty oars and required seven men on an oar. At Lepanto, with only four men to an oar, they had to be towed into position. Under the command of Don John of Austria, brother of the king of Spain, the combined Christian fleet carried a total of approximately 70,000 men.

The Turkish fleet numbered 210 galleys, many of them of the largest size, and forty galliots, a somewhat smaller galley type, rowed by Christian slaves. They were armed with only 3 bow guns, did not use *pavesades*, and carried few harquebusiers, for the Turks had supreme confidence in the more rapid fire of their archers. Their entire force totalled 75,000 men, of whom 25,000 were soldiers. The Christian slaves at the oars were a potential liability. They welcomed an opportunity to revolt when a Turkish galley was boarded.

A VENETIAN GALLEY. Galleys combined oars and sails; they lacked heavy ordnance. (R. A. Fletcher, *Warships and Their Story,* Cassell, New York, 1911, p. 42)

With the two great fleets drawn up in line abreast, the encounter developed much like a land battle of center to center and wing to wing. Ali Pasha, the Turkish commander, took the offensive, sending his ship directly against that of Don John. The force of the Turkish charge was less effective than usual, however, because the Christian leader had placed the galleasses well forward between the opposed battle lines, two each at the center and at the left and right wings, like forts around which the Turks had to detour. He had also cut the spur from his ship, as had many of the other Christian leaders, so that the bow guns could be depressed to fire directly into the oncoming enemy. Although the overwhelming Christian triumph was finally won by hand-to-hand fighting, it is reasonable to conclude that Christian superiority in adapting weapons to sea warfare supplied the margin of victory.

The course of events immediately after Lepanto illustrates two complementary principles of naval warfare: like fights like, and the proper object of a fleet is the enemy's fleet. The first was illustrated in 1572 by the attempt of the Christians to use twenty-four sailing ships (nefs) to supplement their galleys in an action with the Turkish fleet. The sailing ships with their guns were too formidable for the Turkish galleys to attack and, in turn, too slow to support an attack by the Christian galleys. Future naval fighting in the Mediterranean proved that the galley could usually escape the nef by rowing upwind or by venturing into shallow waters, whereas the armed nef's superiority in fire power and greater weight protected her from the galley. As a consequence, the two were not likely to fight. The second principle was illustrated in a negative fashion by naval policies until 1573, when the desultory war came to an end. The Christians, more interested in immediate personal profit than in the destruction of their enemy after Lepanto, turned aside to territorial expansion. The Venetians signed a secret "free trade" treaty with the Turks; and the Spanish, now that the threat to their own Mediterranean possessions was removed, concentrated their efforts on bringing home great riches in gold and silver across the Western Ocean. The Turks, quickly making good most of their losses and also avoiding naval action, contented themselves with the defense of their landholdings. As a result the Turkish menace remained.

Lepanto marked the end of an age; within seventeen years an entirely new concept of naval warfare was demonstrated in the defeat of the Spanish Armada (1588). Three-masted sailing ships had come into use in the early fifteenth century; such vessels were employed by the Spaniards and Portuguese in the great voyages of discovery. The peoples of northern Europe, quickened by the Renaissance, were the more adept in marrying ships of this type to guns in a union that was to give a new meaning to the term "sea power." The English were the leaders in this naval revolution. They had to share with the Dutch the claim for first place in seamanship and maritime experience, but their island sanctuary permitted them, and at times compelled them, to give priority to naval, rather than to military, development. Under Henry VIII, who followed a course of action his father had already indicated, were born both the Royal Navy as a permanent institution, and the prototype ship-of-the-line.

ENGLISH SEA DOGS AND THE ARMADA

Any history of the organization, administration, personnel, and equipment of the Royal Navy must begin in the sixteenth century. What was unique and also important was the evolution of the sailing warship. There had been little modification of the design of ships from the time of the battle of Sluys until the introduction of naval artillery. The first naval guns were breech-loading mankillers for use in repelling boarders. More and more were added, castles were built higher and higher on stern and bow to make more fighting platforms for guns, and the result was an unhandy, top-heavy vessel with possibly more than two hundred pieces of light ordnance. In actual tests Henry VIII proved the value of the cast, muzzle-loading "Great Gun" as a shipkilling weapon; but its installation required revolutionary changes. To use it, the English invented gun ports and a gun deck, for it was much too heavy to be installed in the castles. Once it had intruded in the cargo space, the differentiation between the functions of warships and merchant ships had an important effect on naval architecture. The length of the ship determined the number of guns that could be installed in a broadside. When gunfire was more important than cargo capacity, ships became longer in relation to their beams, and the cumbersome castles could be reduced in size because their functions had become secondary. Thus in 1513 the *Mary Rose,* with her increased length-breadth ratio and "Great Guns" on the main deck (which caused her to capsize), when remodeled with a gun deck and gun ports, became the prototype of the ship-of-the-line. England's new sailing warships, of which Drake's *Revenge* is a good example and Nelson's *Victory* only an improved descendant, combined the three great naval virtues of speed, steadiness as a gun platform, and maneuverability, because of greater length-to-beam, low center of gravity, and decreased windage.

The merits of this new ship as a weapon, and its influence on the tactics of naval warfare, were only partially revealed in the defeat of the Spanish Armada. One question, its superiority to the galley, had been answered by Drake in the spring of 1587. Instructed to "impeach the provisions of Spain," he was using his fleet in that classic role of naval strategy, striking the enemy's fleet and communications far from home. With twenty-five ships, some of them converted merchantmen, he sailed right into Cadiz harbor, spent two days destroying eighteen Spanish ships and taking six prizes, and sailed right out again. Twelve Spanish galleys protecting the port were beaten off by day and by night, whether the English ships were at anchor or under way. "We have now tried by experience these galleys' fight and I assure you," wrote Drake, "that these her Majesty's four ships will make no account of twenty of them in case they might act alone."

King Philip had had his beard singed too often to be ignorant of what the English could do at sea. The activities of the English sea dogs on the Spanish Main, a nursery of fighting seamen in which Queen Elizabeth had financial investment, was a threat to his treasure supply that he could not ignore. When he determined at last to attack England, he planned what was not a naval encounter, but a major amphibious operation. Sound in concept, except that it called for a degree of cooperation between his land and sea forces and for a resolution that were lacking, Philip's plan called for the use of the famous Spanish infantryman as his main weapon. His soldiers were

VESSELS OF THE ARMADA CAMPAIGN

(figures for the 45 most effective ships in each fleet)

	English	Spanish
Tonnage	17,110	35,508
Men	8,171	15,235
Guns	1,600	1,350
Broadsides	7,000 pounds	4,500 pounds
Ammunition	50 rounds per gun	50 rounds per gun

Note: These figures have been compiled from data in W. L. Rodgers, *Naval Warfare under Oars* (Annapolis: Naval Institute Press, 1939).

to board and capture any English ships that interfered with their passage through the narrow waters of the Channel. They were then to join Parma's veterans in the Low Countries, cross to the Thames estuary under protection of the fleet, and march triumphantly upon sparsely defended London. For this task Philip collected sixty-four large ships, four galleasses, four galleys, twenty-three supply ships, and thirty-three small craft for dispatch and reconnaissance. Although manned by some 30,000, including more than 8,000 sailors and about 19,000 soldiers, the Armada was shorthanded, the seamen of poor quality, and the infantry diluted by raw recruits who had never before been to sea.

Although the largest English ships were about the same size individually as any of the 14 Spanish flagships, the advantage in tonnage and manpower was with the Spanish; but the advantage in armament was with the English, especially in the number of larger guns and the skill of the gunners.

The Armada entered the Channel in a multiple line-abreast crescent formation reminiscent of galley warfare. It reached Calais after a series of what the Spaniards called "skirmishes." Driven from an offshore anchorage by fire-ships, the Armada had a final skirmish off Gravelines and was then forced to try returning to Spain by going north about Scotland. Of a total of 250 fighting ships on both sides, only thirty English and twenty-five Spanish had been engaged. Total English losses in battle were 60 killed; Spanish, 600 killed and 800 wounded.

The defeat of the Armada was a complete strategic victory. Further attempts to attack England in the following years never amounted to much. In spite of the very limited action, the Armada was a landmark in naval warfare. The Spanish desired to board. The English, with inferiority in size of ships and numbers of men, kept the windward (i.e., westward) position. Taking advantage of their guns and their faster, better-handled ships, they were able to do so. Time and again they sailed in against the most windward Spanish ships, discharged a broadside when well beyond grappling range, wore about, came back to discharge their other broadside, and then retreated (ran away, said the Spaniards) to reload their guns.

Shortage of ammunition, the fact that most of the heavier English guns were 18-pounders (too light to inflict fatal damage), the reluctance of some English ships to follow the leader in attack, and finally, the slow rate of fire (probably not more than two broadsides an hour) prevented a clear-cut tactical victory. During the entire ten-day engagement only three Spanish ships were actually sunk, but the Parthian methods of the English succeeded in demoralizing the Spanish leaders and crews. Unable to close for boarding, the Spaniards were frustrated in every effort to conduct the battle to their advantage. By the time the Armada reached Calais, it was not bringing aid to Parma but was asking for it. The Dutch fleet kept Parma's transports blockaded, and the two Spanish forces never joined. When fireships drove the Armada from Calais, many ships cut their cables and thereafter lacked ground tackle to keep them off a lee shore. Finally, at Gravelines, where three ships were sunk, two others were disabled and were later captured by the Dutch. Damage to hulls and rigging contributed to the loss of many others in the hard weather off the Scottish and Irish coasts. Only about half of the men and ships of the original Armada returned to Spain.

THE RISE OF THE ROYAL NAVY

The English had achieved their tactical victory by making the outcome hinge upon the weapon in which they were superior. Henceforward, great-gun fire, not hand-to-hand fighting, would constitute the primary effort in naval warfare in the age of sail. During the seventeenth century English naval power capitalized on the strength, and corrected the weaknesses, that had characterized Lord Howard's fleet in 1588 and in so doing led the way for all other European navies. The weapon had been discovered, and it was soon perfected. Drake's *Revenge* had a single gun deck, by 1610 the *Prince Royal* appeared with two complete gun decks, and in 1637 the *Sovereign of the Seas* was launched, a true "battleship" of three gun decks and one hundred great guns. As time passed, the three-deckers were customarily employed as flagships and, with the two-deckers of seventy-four guns, formed the line of battle. Single-deckers of twenty-four or thirty-six guns were known as frigates and became the eyes of the fleet, supplemented by sloops which were smaller vessels with no gun decks but with all their guns on the weather deck. Vessels became ship-rigged, with three masts carrying square sails, and a fore-and-aft sail on the mizzen (aft) mast. The structural limitations of wood set the top limit of size at about 200 feet, 2,000 tons displacement, and three gun decks. Ordnance was standardized and the rate of fire improved, with the shot-weight of the most common ship's gun increased to thirty-two pounds.

    The hard-fought Dutch Wars of the century resulted in the development of tactical doctrine published in the form of "fighting instructions" and culminating in the Permanent Fighting Instructions of 1691. The follow-the-

A SHIP OF THE LINE, *H.M.S. Sovereign of the Seas,* in the age of sail. Such ships had two, three, or occasionally, four gun-decks. The average man could not stand fully upright between the decks. (Edward K. Chatterton, *Sailing Ships,* Sidgwick & Jackson, London, 1914, p. 229)

leader tactics of 1588 had constituted an almost accidental line-ahead formation which brought the broadside power of the ships against their opponents. In a line-abreast formation, or galley-style warfare, sailing ships would obviously mask their own most effective fire power. The line-ahead formation with ships at 100-yard (half cable length) intervals assured that all would engage, particularly if the opposing lines were conterminous, i.e., van to van, center to center, and rear to rear. This necessity was driven home by repeated instances in the Dutch Wars, when the lead ships did the fighting and the others hung back. Two schools of tactics developed, the melee-ist and the formalist. To the former belonged aggressive leaders and innovators of tactics, often former Cromwellians, such as Robert Blake, England's greatest admiral of that century, who desired the minimum of doctrine and the maximum of freedom in making an unforeseen advantage the basis of a great victory. The formalists, primarily seamen, seemed to be more concerned with fighting in the correct way than with victory. As Charles II's brother, James, duke of York, lord high admiral and later King James II, was a formalist, that school had the greater political power. Formalism was not used in the seventeenth century to avoid conflict. But when the Permanent Fighting Instructions were published, they introduced a rigidity that resulted in ninety years of frustration. The essence of the Instructions was their prejudice in favor of maintaining a strict line-of-battle formation which, in effect, penalized initiative and placed a premium upon playing safe.

The professionalizing of the naval establishment of England under the Protector, Oliver Cromwell, had been a natural outgrowth of a permanent navy of specialized ships. Merchant ships might be converted into privateers, but not even a great East Indiaman was a match for a ship-of-the-line. The warship, her officers and crew, her maintenance, and her employment in forwarding the national policy, could no longer be left to chance. A period of half-solutions was brought to an end in the reign of the restored King Charles II (1660–1685) by that remarkable diarist, and even more remarkable administrator, Samuel Pepys. The Admiralty, a political body, controlled naval policy; the Navy Board, a professional body that dated from 1546, was assigned to supply and maintain naval material. As a secretary to the former and clerk to the latter, Pepys contributed more to the establishment of proper methods for the successful direction, administration, and organization of the modern Royal Navy than any other man.

Pepy's work came at a time when it was most needed. Through the long and uncertain course of the Dutch Wars, England ultimately defeated her most powerful maritime rival. She was strong in her insularity and now had a navy which could make sea power serve national policy on a world-wide scale. From the time of the Tudors, English statemen had realized that the fisheries were a school for seamen and had encouraged them by legislation. In the Navigation Acts, deliberately directed against Dutch sea power, she had found a means of stimulating her mercantile marine and therefore her naval strength, and incidentally exploiting her colonies. In the Navy Board she had the means of maintaining and supplying a navy in peace and of commanding it in the emergency of war. The Board exercised authority over what was by far the largest industrial and commercial organization of the time. It controlled the Royal Dockyards and also the contracts for building and repairing in private yards. From 1683 supply was undertaken by a separate Victualling Board. The Master of the Ordnance, an office dating from 1546, was responsible for guns and powder. These boards and officers, only loosely supervised by the Admiralty, which was concerned with personnel and operations, lasted until 1834. They were the basis of England's sea strength over two and a half centuries. That sea strength contributed to the development of a vast overseas empire, to domestic wealth, and to political power for a commercial middle class which had been partly spawned by the colonial corporations set up for trading and colonizing.

9

THE LIMITED WARFARE OF
THE EIGHTEENTH CENTURY

AN AGE OF REASON

By the eighteenth century a moderating trend in European warfare had come about as a result of a moral revulsion from the atrocities of the Thirty Years' War and the accompanying realization that war had so ruinous an effect upon the human and material resources of the state that, unless it were curbed, it would cease to be a worthwhile means of achieving political ends. The intellectual climate of the time favored such a moral and rational approach to the problem of war. The fierce religious partisanship which had embittered the wars of the recent past was ebbing away; the old certainties were receding before the onslaught of new, and less immediately inflammatory, ideas. The mathematical genius of Sir Isaac Newton (1642–1727) demonstrated the relationship between the behavior of the planets and the fall of an apple; and it seemed to his contemporaries that through the application of reason to man himself eternal, immutable laws governing human society could be laid bare. Rationalism was ultimately to be a revolutionary force but the early rationalist reformers were moderates who put their trust in the conversion of established rulers to a philosophy of "enlightenment." The rationalist mistrust of extreme solutions and the emphasis upon practicable and "natural" behavior was an intrinsic element in eighteenth-century life, and it left its mark on warfare as on other activities.

The eighteenth century was an age which protested its devotion to the proprieties and its dislike of excess, but the term "limited warfare," usually applied to the conflicts of the period from the latter half of the seventeenth century to the outbreak of the French Revolution, conveys a false picture

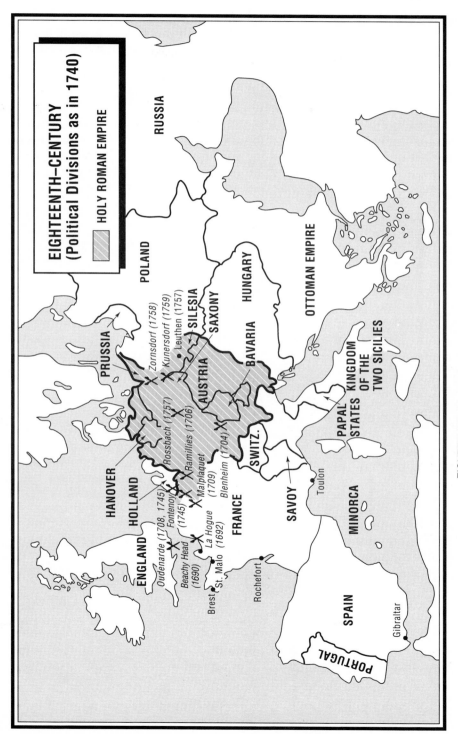

EIGHTEENTH-CENTURY EUROPE as of 1740

unless correctly interpreted. It has nothing to do with the frequency of wars, which were more numerous than in the nineteenth century; nor does it refer to the size of field armies or national armies, which were larger and of a more permanent type than in previous centuries. Eighteenth-century warfare, however, was severely limited in the way in which it was fought, and in the objectives that policymakers sought to gain by its use. In the rationalist era, the belief that all men are entitled by natural law to equal rights took firm root, at least in theory. The best society would be that in which man's natural rights are least interfered with. Hence, one of the most influential by-products of rationalist thought was the humanitarian ideal. Cruelties visited upon civilians in time of war were barbarous infringements upon the rights of humanity. At the same time, the intellectual habit of viewing mankind as a whole was deeply opposed to strident, divisive nationalism. "Patriotism," said Dr. Johnson, "is the last refuge of a scoundrel."

A combination of humanitarianism and internationalism had already led to the first modern attempt to formulate the laws which should govern the conduct of nations toward one another. The Dutch jurist Hugo Grotius (1593–1645), in his *Rights of War and Peace* (1625), treated states as individuals existing in a broader society, the society of nations. The law of nations (*jus gentium*) was equivalent to that regulating relations among individuals; that is, respect for the rights of others and of obligations contracted with them. Jean Bodin (1530–96) and Thomas Hobbes (1588–1679) put forward the theory of the absolute sovereignty of the state; Grotius insisted that such sovereignty must, in practice, be qualified. If the only law of conduct between nations were the Machiavellian law that might makes right, then international society would inevitably be destroyed. Hence relations among states, in war as well as in peace, must be subject to forms of law.

The legists of the eighteenth century added little to the work of Grotius. The Swiss jurist Emmerinck von Vattel emphasized in his *Droit des Gens* (1758) that since war was unfortunately the only means available of obtaining justice, then all wars must be assumed to be just. He warned that the justice of war did not permit the use of any conceivable means to bring victory, for the essential object of war was to gain an equitable and durable peace. For this reason, noncombatants should be protected, not injured, and treaties should be moderate rather than severe. Jurists agreed, however, that heads of state could ignore the laws of war in the face of "necessity."

It is obvious that since no international authority existed to enforce them, the humane and honorable precepts of international law were more a reflection of the spirit of the age than an actual code of conduct for European states. Yet the influence of the philosophers of the Enlightenment on the reigning monarchs of Europe was considerable: Voltaire's friendly relations with Frederick the Great of Prussia and Catherine the Great of Russia are cases in point. To have the ear of a king was of the utmost importance, for with the exception of the parliamentary monarchy of Great Britain, the leading states of Europe had become absolute monarchies conforming to the model of the France of Louis XIV. The word of the king was law; the interests of the state were identified with those of the ruler. Royal authority was clothed in the theories of the divine right of kings to rule, and of the indivisibility of sovereignty. But the real basis of power was military. The security of a dynasty was closely related to the size and strength of its standing army. The growth of continental standing armies had brought about the suppression of religious and aristocratic opposition. In the eighteenth century the "benevolent despots" ruled absolutely through control of standing armies officered by members of the aristocratic classes who had formerly led opposition to monarchical power. Everywhere, except in Holland and England, this combination of royal and aristocratic military power had ensured the suppression of representative and constitutional government.

The situation in England is illuminating because of the contrast with the growth of royal military power elsewhere. From the Royalist and Roundhead armies, regiments of foot and horse had been established after the Restoration to protect the person of Charles II; but Parliament, reflecting the country's fear of a standing army, was suspicious of even these small forces. James II brought battalions of the Irish establishment to England in an attempt to bolster royal power and his Catholic policy. In the resulting crisis he was deserted by his army; and his daughter Mary and her Protestant Dutch husband, William of Orange, ruled in his stead. The terms of the "Glorious Revolution" established the principle of parliamentary control of military power. Standing armies could not be maintained in time of peace without the consent of Parliament. In time of war the machinery by which this principle was guaranteed was the passage of the Mutiny Act, which, for one year at a time, legalized the disciplinary system of the army. The annual grant of revenues to pay the army was, however, the chief source of parliamentary authority. Thus a method of providing the king with an army for defense without endangering the liberty of the subject had been worked out in England. Elsewhere royal control of the defensive forces of the realm left the crown with unchallenged power within the state.

DYNASTIC WARFARE

The fundamental drive behind most of the frequent wars of the eighteenth century was the desire of absolute monarchs to strengthen their position at the expense of other states through the acquisition of territories and population. A second factor leading to wars, particularly those which involved the maritime states with colonial empires, was competition for trade to augment the wealth of the state. According to the mercantilist view, the amount of wealth in the world was constant; the only way, therefore, to increase one's share of it, other than by a favorable import-export balance, was to deprive another state of its colonies or its trade by force of arms. In Austria, Brandenburg-Prussia, and Russia, the power motive was uppermost. In England and Holland, strongholds of the commercial classes, the mercantilist drive was a major element in foreign policy. France found herself torn between the desire for power on the continent and the necessity of protecting her commercial and imperial interests abroad; while Spain vainly endeavored to maintain her position in Europe and overseas.

The early wars of Louis XIV (the War of Devolution, 1667–68; the Dutch War, 1672–78; the War of the League of Augsburg, 1689–97) have been seen as attempts to expand France to her "natural frontiers," or as part of a policy of bringing all French-speaking areas under French rule. Neither interpretation is supported by the documents or by events. Throughout his long reign, Louis XIV sought prestige and glory. Both for him and for his nobles, whose chief occupation was war, real glory was to be found only in military victories. Louis's bellicosity brought into being a succession of anti-French alliances. The states of continental Europe feared the rising power of France, and England could not tolerate the control of the trade of the Low Countries and of the shipping of the English Channel by an unfriendly power. The threat to the European balance and to world trade posed by the prospect of a union of the Spanish and French thrones brought the Grand Alliance and the War of the Spanish Succession (1702–13). The terms of the Treaty of Utrecht disclose the motives of the contending states: the thrones of France and Spain were never to be joined, Austria and Brandenburg-Prussia received Spanish territories in Europe, Holland was rewarded with a trade monopoly of the River Scheldt, and England won certain French colonies in North America, Gibraltar, and Minorca from Spain, and the exclusive right to trade in slaves and other merchandise on a limited scale with the Spanish colonies. The later wars of the period show the same interplay of motives and interests. The War of Jenkins' Ear (1739) was a commercial struggle waged between England and Spain over the trading rights conceded in the Treaty of Utrecht. It was absorbed into the larger conflict of the War of the Austrian Succession (1740–48). This war was precipitated by the rapacity of Frederick II of Prussia, who seized the province of Silesia when Maria Theresa ascended the throne of Austria. France and Spain joined Prussia in the attack on Austria, and England, while supporting Austria with cash subsidies, took the opportunity to challenge French colonial rule in India and America. The fear of the new Prussian power drove Austria and France, traditional enemies, into alliance, a coalition joined by Russia. As usual, Great Britain benefited from French preoccupation in Europe by making gains abroad, and the result of the Seven Years' War (1756–63) was the loss of most of French India and Canada to England. Meanwhile Prussia, aided by English money, not only survived the war but received permanent title to Silesia.

The essential feature of all these wars was the relatively limited nature of the objective involved, whether dynastic or commercial. Wars were fought to achieve tangible ends which required, not the absolute destruction of an enemy, but rather a military decision favorably affecting the diplomatic bargaining to follow. Moreover, the personal ambitions of a king or the economic interests of a class did not carry the intense emotional appeal necessary to inflame whole populations. Wars in the Age of Reason, while hardly conducted in private, scarcely affected normal peacetime conditions and were participated in by only a very small percentage of the total population.

SOLDIERS

Limited participation was due partly to the social structure of European armies, which in turn reproduced the social structure of the dynastic state. Throughout Europe, society was headed by a privileged aristocracy exempt from taxation and given the chief offices of state and church in exchange for support of the monarchy. The function of the middle class in the despotic state was to produce wealth and endure taxation, and to bear without complaint its exclusion from political power. Only in England was there little distinction in function between the nobility and the middle class, because only there were the two classes nearly united in the pursuit of the same interests. Nowhere in Europe did the lower classes of town and country have political rights; indeed, in most of continental Europe, the status of the peasant was still serfdom.

The sharp distinctions in society were carried over to the army. The officer corps was reserved to the nobility, while the mass of the soldiery was recruited by voluntary enlistment from the peasantry and urban unemployed. The aristocratic nature of the officer corps was the result of deliberate royal policy, for the army provided a convenient means of buying aristocratic support through patronage. This meant, however, that positions were often created to satisfy the clamoring of the nobility; and so the armies of Europe were overloaded with superfluous officers. In the army of Louis XV there were so many that commands were rotated on a day-to-day basis in order to give everyone something to do. In Prussia, Frederick William favored the land-owning nobility (*Junkers*) to such a degree that at the end of his reign all officers above the rank of major were noble except three. The exceptions to this class monopoly were necessitated by the fact that a certain amount of technical knowledge was required of artillery and engineering officers. It was this necessity which led to the founding of the first engineering schools of the modern world. (Woolwich founded 1721, reestablished 1741; Ecole Militaire, 1751; Academy of Engineering, Potsdam, 1768.) The Ecole Militaire was at first established for sons of impecunious nobles who would submit to the discipline and hard work which a professional career entailed. It is significant, however, that under the Ancien Régime there was a strong tendency for the wealthy nobles to push their way into the college, to shoulder aside their harder-working fellows, and to introduce the standards of idleness and display for which, on the whole, their class was noted. Incidentally, the fact that the officer corps of Europe all belonged to the same caste made for the observance of an aristocratic code of honor which had a direct line of descent from feudal knightly chivalry.

Military tradition and class prejudice were not alone responsible for the exclusion of the middle class from positions of command. From the mercantilist point of view, an army was necessary to protect the wealth of the state without making inroads upon it. Therefore the composition of the army was restricted to the "unproductive" classes of the state. For the same mercantilist reasons the rank and file consisted of the lower orders of society. Soldiers were enrolled by voluntary enlistment from the peasant classes and from the urban unemployed. Since enlistment could fill only part of the required quota, every army in Europe employed large numbers of mercenaries. Squads of energetic recruiting officers scoured the countryside, and even foreign countries, seeking to entice the unwary into the service of the king. One-third of the French army prior to the Revolution was German. Seldom has the military career been held in less repute. The lowly origins of the soldiery earned them the contempt of civilians and the distrust of their officers; their heterogeneous national background prevented the development of patriotic feeling. The immense social gulf between the men and their officers made impossible the group solidarity essential to high morale.

The only state to employ a draft for peace-time recruiting was Prussia because, while pursuing the foreign policy of a great power, she ranked only twelfth in population in Europe. Two-thirds of the Prussian army was recruited from the peasant class, artisans being exempted because their preeminent value to the state lay in their productivity capacity. By the time of the Seven Years' War, however, other states were also beginning to find it difficult to maintain their armed forces by voluntary enlistment alone. Russia and France resorted to drafting, but only of peasants. Austria and Spain followed suit shortly after.

The harsh discipline for which eighteenth-century armies were notorious was due in part to the untrustworthiness of the men in the ranks. Corporal punishment was resorted to as the best method of obtaining unqualified obedience. In the Prussian army, the most methodically ordered in Europe, an officer had no hesitation in striking a common soldier for the merest infraction of dress or parade regulations. It is not surprising that desertion was endemic in the armies of the period. The provision of barracks, which became almost universal, was done not because of a solicitous regard for the well-being of soldiers, but to lessen the opportunities for escape from service. Almost everywhere he went, the soldier was subjected to the strict supervision of officers and sergeants.

TACTICS OF THE LINE

The nature of battlefield tactics also demanded a close attention to matters of discipline. The bayonet, which first appeared when men plugged knives into the muzzles of their muskets at Ypres in 1647, was given a permanent place in modern armies by the invention of the "ring" or "socket" bayonet in 1678. The flintlock musket, introduced at the end of the seventeenth century, sharply reduced the number of misfires to which the old matchlock had been prone, and this made possible a far greater fire power. The musketeer could now defend himself against cavalry; he no longer had need of an ancillary body of pikemen. Within a fairly short time the pike became obsolete. With the passing of the pike went the deep phalanx or column formation. The three-line formation invented by Gustavus, usually known simply as "the line," was found to be adequate even when infantry had to face cavalry attack, although it should be noted that the infantrymen of Gustavus protected themselves with the "Swedish feather" (a very long pike). For all-round defense against waves of horsemen it was quickly discovered that the "square" of "lines" was the ideal formation. Deployment and maneuver of "the line" for attack and defense necessitated rigorous training, much of which took place on the parade square.

There was no room in linear tactics for the exercise of individual initiative. Linear tactics were designed to exploit the new fire power to its fullest extent through the simultaneous volley upon word of command. A meticulous order upon the field of battle was of the highest importance, for if there was the least deviation from the mathematically precise line arrangement, the musketeers would damage their fellows more than the enemy. Prussian officers were reputed to "dress" their companies with surveying instruments; certainly it was not uncommon for troops advancing under fire to halt in order to re-form. Moreover, it was a leading principle of this kind of war, with weapons of very limited range, that to obtain the maximum moral and physical effect upon the enemy, the volley should be withheld as long as possible, and then followed up by a bayonet charge. For troops to expose themselves unflinchingly in the face of enemy muskets and artillery while advancing at the agonizingly slow cadence of eighty paces to the minute required discipline of the very highest kind. Because of tactical necessity and the nature of their troops, the military leaders of the eighteenth century uniformly subscribed to the maxim of Frederick the Great that the purpose of discipline was to teach the soldier to fear his officers more than he did the enemy.

The formal tactics of the line decidedly restricted the potentialities of an army in the field. To marshal an army into battle array was a slow business, and the huge frontage of the linear army required a broad and reasonably level plain. The joining of battle, therefore, could occur only by the tacit consent of two equally confident commanders, since to refuse an engagement merely involved withdrawing to wooded or rough country. Even when battle was given, it was seldom that victory was complete, since the line could not be adapted to a relentless pursuit. Moreover, so great was the fear of desertion during the confusion of a rapid advance that pursuit was officially forbidden in many armies.

Many other factors operated to reduce the frequency of battles. One of these was the heavy casualties that always resulted from the murderous exchange of volleys. In proportion to the numbers taking part, the losses in killed and wounded were much higher in this period than in other eras. At Malplaquet (1709) the victorious Marlborough lost 33 percent of his effectives; at Zorndorf (1758) the Russians lost 50 percent and even the Prussian victors lost 38 percent; the next year, when defeated at Kunersdorf, the army of Frederick the Great lost 48 percent of the men committed. It was natural, therefore, that generals should refuse battle unless circumstances favored them. In war of this sort, battle was the last resort. That this was so was not because heavy losses imposed an unbearable strain upon the nation itself. They did not. The army formed a very small proportion of the total population and was divorced from it. But to train a soldier to meet the exactions of linear warfare represented the work of at least two years; to find a replacement for him was a difficult task in an age when recruiting officers competed fiercely for likely candidates in every country. The professional soldier, though held in contempt as a human being, had cost his king a sizable amount in hard cash to be fashioned into a martial robot.

Not without reason were soldiers called the "toys of kings." Frederick William I of Prussia (1713–40) labored through his entire reign to fashion his army to a state of perfection, yet never once committed it to battle. Even more solicitous was the treatment accorded the "guards" regiments which adorned every European monarchy. These units, purportedly the crack regiments of the army, were actually designed primarily for show; dressed in the most brilliant uniforms, they played an integral part in the elaborate ceremonies which surrounded the despot. The most useless of these military showpieces was the regiment of "Potsdam Giants," the beloved plaything of Frederick William I. Each man in this unit was over six feet in height; Frederick William's agents toured Europe for lofty prospects; and persons seeking his favor sent him batches of outsize soldiers. Peter the Great alone dispatched over two hundred tall Russians as gifts. Frederick William drilled his "Blue Boys" personally; but he could not bear to think of them in action. One of the first acts of Frederick II was to disband the 3,000 giants bequeathed him; but the guards tradition itself remained unbroken.

SUPPLYING WAR

A further check upon the intensity of war was the system of supply in general use. Every command in Europe followed the practice of establishing magazines for food, clothing, ammunition, and other supplies throughout the home territory, and additional depots upon invading another country. The aim was to make the army self-sufficient, not only because the barbarities visited by plundering armies upon civilian populations in past wars offended the humanitarian spirit of the age, but because it was feared that soldiers foraging for themselves would desert in droves. The result of the system was greatly to reduce the mobility of armies. A field force was tied to the distance over which it could carry its bread. Magazines were three days' march apart. "Ovens," supplementary magazines, were set up at one-day intervals. The formidable baggage that accompanied an eighteenth-century army materially reduced the speed of march and range of action. The soldiery, largely unaffected by ideals, had to be courted by creature comfort; their officers had no intention of denying themselves their peace-time indulgences. When, in 1707, Lord Peterborough lost his personal baggage in Spain, it included sixteen wagons, over fifty mules, and several valuable horses, while the Duke of Northumberland proceeded to active service in Flanders with a retinue of three gentlemen attendants, one page, two footmen, a wagoner, a sumpterman, and three grooms.

Not only did logistics hamper the movement of armies, but general economic conditions checked the size of a force in the field. Field armies were bigger than in the preceding century but circumstances limited further growth. Marshal Hermann Maurice de Saxe (1696–1758) was of the opinion that a field army should not number more than 45,000 men; greater numbers would only be an embarrassment to the general. The difficulty of maneuvering mass armies in line was only one reason for this generally held principle. The size of an army was rigidly governed by the low state of agricultural productivity and the cumbersome distribution system of the time, particularly since soldiers were no longer permitted freely to prey upon the countryside. Similarly, strategy had to conform to the dictates of the seasons. No campaign could begin until there was plenty of green forage available for the horses and the immense number of draft animals, nor could it continue in any particular district beyond the time when such forage was exhausted. Roads were so bad that, in autumn, operations had to be suspended until the late spring. The unreliability of roads placed a relatively high value upon water communications.

Almost inevitably, the Low Countries, an area of high agricultural productivity, well furnished with a network of waterways, became a favorite theater of war. The Low Countries, however, were also thickly studded with towns, fortresses, and other easily defended points. Here, and to some extent in most other regions, the necessity for securing lines of communications made inevitable the slow business of siege warfare. A general, no matter how impetuous and desirous of battle, could not ignore the need for systematic clearance of his lines of supply.

This tendency toward the defensive was assisted by the state of weapon development and military engineering. It was not until the time of Frederick the Great that any appreciable advance was made in siege artillery. Before his day, European armies had at their disposal guns which were deficient in range, accuracy, penetrating power, and durability. Meanwhile, Louis XIV's great engineer, Marshal Sebastian de Vauban (1633–1707), and his many students had gone far to check the advantage that artillery had originally given the offensive. With emphasis upon one or another of Vauban's "three systems," engineers constructed complex defensive works designed to impede the progress of an invading army and to exhaust it in unrewarding and time-consuming sieges. Although the eighteenth century saw no radical improvements upon Vauban's techniques, forts were built to give the widest play to enfilading fire against attackers, and to provide opportunities for sudden sallies by defenders. The offensive, on the other hand, benefited from Coehorn's invention of the trench mortar in 1673, and from the perfection of Vauban's methods of attack by digging approaches and parallels and by the siting of batteries to enfilade enemy defenses with ricochet fire. The refinement of fortification and of siege methods transformed this branch of warfare into a geometric exercise, and a spectacle to delight the ladies of the

VIEW OF TRENCHES DURING A SIEGE. "In the final stages of a seige, the attacking gunners would protect themselves from enemy fire by placing their cannons behind gabions, wicker baskets usually open at both ends and filled with dirt." (Eugene-Emmanuel Viollet-le-Duc, *Military Architecture,* James Parker, Oxford, 1879, p. 213)

court. The defense was too formidable to allow frontal assaults by irreplaceable soldiers, while unrestricted artillery bombardment of civilian houses was not indulged in, since it did nothing to assist the prosecution of the attack and was wasteful if and when the town fell. Instead, the attack was placed in the hands of engineers, who through the precise application of

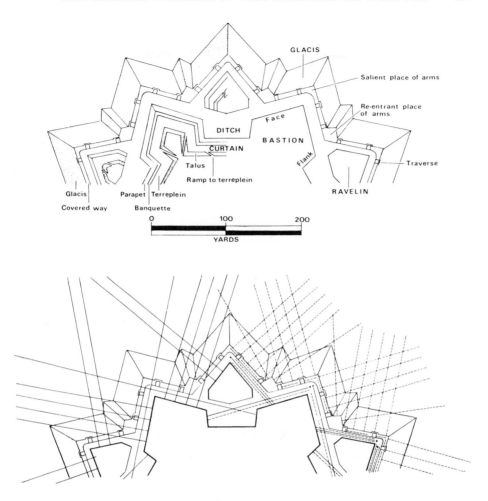

Top: PLANS OF EARLY MODERN DEFENSIVE WORKS. *Bottom:* Diagram of fields of fire from the same defensive works. With the development of seige artillery, fortress walls were, unlike medieval castles, sunk into the ground and protected by a glacis and ditch. (Christopher Duffy, *The Fortress in the Age of Vauban and Frederick the Great, 1660–1789.* Routledge & Kegan Paul, London, 1979, p. 3)

mathematics brought their trench network and batteries to such a position that the defending commander, caught in the toils of Euclid, could honorably yield up his fortress.

The importance of supply and the costliness of battle imposed upon the whole conduct of warfare the attributes of a complicated game. The essence of generalship was not to force battle, for the effect of heavy casualties upon small armies would be catastrophic. It concentrated on destroying the enemy lines of supply. The eighteenth century was the age of maneuver, of march and counter-march, of diversions and deceptions, as rival commanders attempted to menace the communications and supply areas of their opponents. The aim of the general was not necessarily to bring his enemy to battle, but to make his position so untenable through adroit movements that he would be compelled to fight at a pronounced disadvantage or else concede defeat.

GREAT CAPTAINS

The great exemplar of the war of maneuver was the French marshal Henri de la Tour Turenne (1611–75), whose career bridged the dissimilar eras of the Thirty Years' War and the wars of Louis XIV. Turenne's victories, and there were many of them, were drab compared to the bloody triumphs of Gustavus or Wallenstein, but they were won with a minimum expenditure of manpower. Turenne and most of his successors regarded battle as a last resort, to be accepted with caution and then only when conditions seemed favorable. He demonstrated many times his ability to maneuver his opponent into areas of meager supply while keeping control of his own lines of communication, and to mask his movements by sudden feints which, by deceiving the enemy, permitted the seizure of desirable positions.

The generals of this age have been accused of a positive antipathy toward fighting and have often been contrasted, to their disadvantage, with the vigorous Revolutionary commanders of a later era who fought war to the hilt and who sought battles rather than avoided them. It is true that generals like Turenne and his brilliant eighteenth-century counterpart the Marshal of France, Count Maurice de Saxe, believed that warfare, like other human pursuits, had its laws which could be discovered. Saxe, in his posthumously published *Mes Rêveries* (1756), even declared that it was not only possible, but reasonable, that a successful general might wage war throughout his career without resorting to battle. Skillful movement and close attention to logistics should bring a commander the desired results without bloodshed. Yet it is noteworthy that Saxe himself commanded the victorious French army in one of the most decisive, and bloody, actions of the age, the battle of Fontenoy (1745).

Moreover, despite the many limitations upon war, few periods have witnessed so many battles, or a more illustrious array of outstanding generals. The number of battles was a result of the many wars, which in turn were possible only because war, involving almost autonomous armies, placed little strain, other than financial, upon society as a whole. The nature of eighteenth-century warfare put a premium upon fine generalship. European armies were nearly identical in weapons, tactics, and systems of supply; military genius, therefore, became one of the chief factors in determining the outcome of war, particularly because the strategy of the indirect approach demanded a leader of high intellectual qualities, though he might lack the dash of a Condé or a Rupert. Two generals alone transcended the limitations of their age. The Duke of Marlborough (1650–1722) and

Frederick the Great combined dexterity in maneuver with a strong belief in battle as the decisive element in war, an attitude not in keeping with prevailing military spirit.

One of Marlborough's major contributions to the eventual defeat of Louis XIV's bid for hegemony in Europe was the fashioning of an overall plan of campaign in cooperation with the Austrian Prince Eugene of Savoy (1663–1736)—despite the cautious parsimony of his Dutch and English employers. Of even greater weight was Marlborough's ability to come to grips with his foe on his own terms and to capitalize on the ensuing victory. He was aided by the fact that the early advances of the French had penetrated beyond the heavily fortified border areas in the Netherlands and thus allowed him to indulge in warfare of swift movement. Thus, in the Blenheim campaign (1704), he abandoned the profitless safety of the Netherlands and, to the horror of Dutch and English politicians and to the surprise of the French generals, plunged across Europe to the Danube to deal with the French threat against Austria. Joining forces with the armies of Prince Eugene and Margrave Louis of Baden, Marlborough lured the French to attack him at Blenheim and won a hard-fought victory. As a result of this battle, Austria was saved from occupation, Bavaria ceased to be of any assistance to France, the immense prestige of French arms was destroyed, and the armies of Louis XIV were forced back on the defensive.

Ramillies (1706) was won because Marlborough was also a great tactician. During the battle, English troops were withdrawn from the allied right, passed behind the lines, and swelled a massive Dutch and Danish assault that shattered the French right. Oudenarde (1708) was an "encounter battle" in which the allied troops were committed piecemeal as they reached the field. Here Marlborough retained a steady grasp of a very confused situation and finished off the battle by taking the French in the rear with a cavalry envelopment. Yet these victories of generalship, decisive though they were, could not overcome the limitations imposed by contemporary politics. Marlborough could not carry the cautious Dutch with him in a direct offensive against Paris, nor could the Grand Alliance compel Louis to accept its exorbitant terms. When political power at home shifted away from Marlborough's Tory party, the great commander was dismissed (1711), never to fight again.

FREDERICK THE GREAT (1712–1786)

The stature of Frederick the Great as a general is closely linked to the rise of the military power of Prussia, and also to the decline of French arms. The army of Louis XIV had become the model for the rest of Europe; that of Louis XV reflected the dry rot that was attacking the foundations of the Ancien Régime. The provision of commissions for the nobility had reached such an extent that the upper echelons were overstaffed, and organization and discipline were breaking down. The growing practice of giving commissions by purchase added another element of disorganization by carrying over into the army the civil discord between the aristocracy and the middle class. The maintenance of too many elite regiments sapped the fighting strength of the army; even the rank and file were losing precision as the old drill regulations went unenforced.

The rise of Prussia was due largely to her army and to the fact that her rulers felt that the chief duty of a king was to be a soldier. Military ends so shaped policy that, in 1752 some 90 percent of the budget was devoted to expenditures on the armed forces. Prussia's only claim to the rank of a great power lay in military strength, for in population and economic resources she was well down in the scale of European states. Under Frederick William I, the energies of the state were directed into military channels to a degree unsurpassed in Europe. The royal bureaucracy was expanded and made more efficient, paving the way for the tripling of the revenue of the crown during Frederick William's reign. This increased revenue went to foot the bill for an expansion of the army from 38,000 men in 1714 to 80,000 by the end of the reign in 1740. The formerly hostile Junker class was firmly welded to the monarchy by reserving to its members positions of high rank in the army. In this respect, Prussian militarism was more exaggerated than that found elsewhere, since the officer caste was accorded officially a political and social precedence second only to that of the king. By intensifying the recruiting methods of his father, and by the acquisition of populous Silesia, Frederick the Great entered the Seven Years' War with an army of more than 150,000, exceeding that of the much larger Austria.

A few months after Frederick succeeded his father, in 1740, he embarked upon a policy of undisguised aggression, seizing Silesia, which he successfully defended against Austria (1740–48), and moving then against the armies of Austria, Russia, and France in the Seven Years' War. Militarily he was aided only by English gold and English-subsidized German forces.

Frederick's achievements can be partially explained by his variations from the norm of eighteenth-century strategy. His *Instructions for His Generals* paid due reverence to the strategic orthodoxies of the time, e.g., that "hunger exhausts men more surely than courage," and he stressed the importance of starving the enemy (troops, not civilians) by maneuvering them away from their sources of supply. Yet he saw also that maneuver alone could not bring a decisive verdict. "War is decided only by battles and it is not decided except by them"; Frederick's willingness to give battle, savoring strongly of the Napoleonic approach to war yet to come, was what set him off from other leaders of his age. In a time of limited war, he gained early success because he was different: he was prepared to take risks, to commit to battle, whenever it was advantageous to him.

A thirst for battle would have been suicidal if Frederick's men and armament had been inferior to, or even only equal to, those of his antagonists. But the Prussian emphasis upon drill and discipline had produced soldiers who could march faster, change from column to line more rapidly, and load and fire more swiftly and effectively than any other troops in Europe. The mobility and precision of Prussian troops meant greater scope for generalship, since Frederick could make unprecedented demands upon his armies with a reasonable assurance that his orders would be fulfilled. It was mobility alone which made possible Frederick's frantic struggle on several fronts during the Seven Years' War. Ringed around by enemies and deficient in numbers, Frederick taxed the endurance of his troops to the utmost by a series of speedy marches unequaled until the Napoleonic Wars. To fight Rossbach (5 November 1757), Frederick marched 170 miles in less than two weeks, a feat beyond the capacity of his baggage-laden enemies; after defeating the French, he immediately had to retrace his steps to crush the Austrians at Leuthen (5 December 1757).

Frederick's superiority in armament was of crucial importance in beating the Austrians. The Prussians employed the iron ramrod (and had done so for over forty years); the Austrians, in company with other European armies, still relied upon a wooden ramrod, liable to warp. The combination of a better ramrod and practice at loading and firing drill meant greater Prussian infantry fire power; it was the steady volleys of the Prussians that won the decisive battle of Möllwitz (1741). In the Seven Years' War, Frederick's opponents sought to counter Prussian mobility and fire power by the use of more big guns, and Frederick, although this meant placing a grave strain upon the treasury, was compelled to match them. Characteristically, however, he made original departures in this phase of military technique. In keeping with his attention to mobility, Frederick had his guns drawn by four horses in file, so that artillery could shift position during battle. He was also the first commander fully to exploit high-angle fire; one third of his artillery were howitzers.

There have been wide differences of opinion respecting Frederick's abilities as a tactician, now and during his own lifetime. Most arguments center on his perfection of the tactical device known as the "oblique order," which involved a flanking thrust by one wing of his army while refusing the other. This stratagem arose because the Prussians found themselves chronically outnumbered. The oblique order gave local superiority at one point, and as fresh battalions arrived on the scene in column of march and then wheeled meticulously into line, the enemy was rolled up from the flank. Although the oblique order was employed at Leuthen, Zorndorf, Torgau, and, disastrously, at Kunersdorf, Frederick himself (and Napoleon, a student of his campaigns), gave the credit to his big guns, "the most to-be-respected arguments of the rights of kings."

Is Frederick, because of his strategy of the offensive, to be regarded as the precursor of Napoleon? Napoleon was to disregard entirely the complicated sparring over supply depots and communications which retarded the pace of warfare before his day. For him, maneuver was a necessary prelude to battle, and battle meant the total concentration of force to obtain, not merely the defeat, but the annihilation of the enemy. Napoleonic strategy was feasible because of the vast resources the emperor could command, but Frederick had no such blank check on human lives. He took the offensive for precisely the opposite reason. Forced to fight a war on several fronts, yet being hardly equal to any one of his three opponents in money and manpower, he had to shuttle from point to point of his kingdom, dealing a series of offensive blows to prevent a fatal junction of the armies opposing him. That he succeeded is a tribute to his energy and strength of mind, and to the capacities of his men and arms. He was not fighting a Napoleonic war of annihilation, however. Annihilation or total victory was beyond his resources. (It is noteworthy that he lost half of his sixteen battles.) He fought a defensive war within the guise of an offensive. It is significant that after the glorious year of 1757, Frederick resorted increasingly to the war of maneuver and to the use of fortifications to ward off his enemies. He gave battle only when other alternatives were exhausted. In later life, rendered cynical and prematurely broken by his tremendous exertions in the field, Frederick not only discarded his youthful doctrine of the efficacy of battle but also began to echo Voltaire's bitter attacks on the barrenness of war.

Just as the Prussian state was an extreme version of eighteenth-century absolute monarchy, Frederick the Great was the supreme example of eighteenth-century generalship. Wider scope was given to his talents in the field chiefly because he, as king and general in one person, could exercise an unparalleled authority over the lives of his subjects.

10

THE GREAT AGE OF SAIL, 1689–1815

THE ENGLISH STRATEGY

The great age of sail, covering a century and a quarter of almost constant conflict between France and England, illustrates the profound effect of the new, permanent, professional navy with its fleets of stately ships-of-the-line. Study of the naval component of this long struggle is particularly worthwhile for the light it throws on the role of sea power in determining French and British foreign policy, on the way that sea power expanded warfare between European powers into global, or worldwide, conflict, on the means by which England achieved tactical superiority in naval warfare, and, finally, on that supreme test of sea power versus land power in the age of sail, the British struggle with Napoleon.

The wars between England and Spain in the sixteenth century, and between England and the Netherlands in the seventeenth, had been essentially naval, sea power versus sea power. Queen Elizabeth's navy had demonstrated its worth in the defense of the island state. The English depended absolutely upon it. Probably neither then nor at any time since would it have been possible for England's army alone to repel such an invasion as her enemies had the military, but not the naval, power to stage. The English had found the weakness in Spain's national policy. Spain's commitments in the Mediterranean and in the Low Countries made her dependent upon wealth garnered in her colonies. Therefore, the English ships that threatened the galleons bringing treasure from the New World endangered Spanish operations and interests everywhere, not merely those directed against England. So was it also in the Dutch Wars. The Netherlands was dependent upon her fisheries and her revenues as a commercial carrier; without those resources the Dutch could not protect their land frontiers. Thus the Netherlands proved as vulnerable as Spain to the kind of attack that could be made by the ships of the Royal Navy.

But the emergence of France under Louis XIV presented a much more difficult problem. A relatively self-sufficient land power, France was not so susceptible to naval attack as Spain and the Netherlands had been, and when France became a sea power as well, she threatened England's maritime supremacy. Under Jean-Baptiste Colbert, Louis XIV's great controller of finances (1661–83), the French pursued a full-scale mercantilist policy. Colbert encouraged the growth of all forms of business activity, especially those centered in French trade and possessions overseas. French naval architects soon proved themselves the finest in Europe, surpassing the Dutch, who had led in most of the seventeenth century. Colbert was able to build a navy adequate for his projected commercial program. The interdependence of commerce and naval power, long recognized in England, justified Colbert's naval expenditures and also a system of *inscription* to man the warships. Before long, a superior French fleet lay at anchor in its newly developed base at Brest, where it outflanked the English fleet in the Channel and challenged for control of the North Atlantic.

France did not have to wait long to test her new sea power. The Glorious Revolution brought William of Orange to the throne of England in 1689 and involved him in a double struggle with the French; in Ireland, where he had to fight for his crown against James II, whom Louis supported, and then on the continent, where French armies had invaded the Netherlands. Here was an opportunity for decisive French naval action; none was forthcoming. Louis himself had allowed William to sail unopposed to England because he thought that James II had grown too independent and needed a scare. The next year the French admiral, Tourville, did nothing effective to interfere with William's campaign in Ireland, where James was defeated in the Battle of the Boyne (12 July 1690). One day earlier, Tourville had routed an Anglo-Dutch fleet off Beachy Head in the Channel and had gained supremacy at sea for the time being; but it was too late to influence the decision in Ireland. But the French made little effort to use their control of the sea, largely because Louis's attention was directed to his continental campaigns. On the English side, after Beachy Head, Admiral Torrington had acted upon the policy of maintaining a "fleet-in-being" to keep the superior French fleet in check. Two years later, a superior Anglo-Dutch fleet defeated Tourville off La Hogue and regained control of the Channel. The naval war then developed into a stalemate, after certain inconclusive efforts by the English fleet in the Mediterranean. The main campaigns were fought on land, and neither side aggressively employed naval power.

Thus it was that England's minister of finance, Lord Godolphin, asking King William for "speedy directions" in the employment of "our great useless fleet" against the French in 1696, posed a question that struck at the heart of the matter: of what use is a navy? Or, more inclusively, of what use is sea power against land power? The answer to the question, of course, varied with the particular circumstances of time and place, with the resources available to each of the opponents, and above all, with their national policy. Those elements which are the constant determinants of sea power were, however, empirically arrived at by the opening years of the eighteenth century. At the same time, from the experience which France and England gained from their first major struggle in modern times there were evolved the broad policies which were to govern them in a hundred-odd years of conflict; they are applicable to all warfare between a maritime nation and a land power with maritime ambitions.

England found in an alliance with continental powers—in this instance the Grand Alliance, of which, among others, Sweden, the Netherlands, and Spain were members—a possible answer to Godolphin's question about sea power versus land power. King William in his conduct of the war pursued a "continental policy," making England's primary effort the support of her allies in a land campaign on the continent, while the fleet played a secondary, almost incidental role. English statesmen who objected to the continental policy were aided by the skilled pen of Jonathan Swift in advancing a "maritime policy" as a more effective method of employing the island's resources in the war. In its extreme form, Swift's argument was for maritime isolation, cynically leaving the ally to do the land fighting, while England enriched herself with conquests overseas and the ruination of the enemy's commerce.

In practice, a middle course became the guiding principle of British policy in the eighteenth century. If such a compromise policy had been pursued against Louis XIV, King William's "useless fleet" would have been directed to undertake the screening and support of amphibious operations against French naval and privateering bases, since the French fleet refused to come out and fight. Such vigorous action, however, was beyond British resources, as long as William's primary commitment was to a continental land campaign; further, it necessitated wholehearted cooperation between army and navy, for which neither one was at that time adequately prepared, mentally or physically. Amphibious operations in the West Indies were notoriously productive of disputes between land and sea commanders.

As time passed, England's maritime policy became in essence a proposal for answering the sea power versus land power question by the indirect approach, concentration on economic (or "strategic") warfare directed against the enemy's sea communications and overseas resources, with only secondary expenditures of British resources to support their ally's land campaign on the continent. A modification of the maritime policy, important because it proposed consignment of a major British military force to continental fighting, is frequently referred to as the "Low Countries policy." It was an exception which arose from the fact that England had to prevent the seizure of Antwerp and the mouths of the Scheldt by any major maritime power. That strategic area lay athwart British communications with the Baltic, could serve ideally as a base from which to launch an invasion of the British east coast and up the Thames, and was a serious rival of London's position as the European depot of worldwide trade.

England's first participation in a continental alliance had demonstrated both the need for allies when she was at war with a power of the magnitude of France and the necessity for a national policy realistically based on the best employment of the means available. As a maritime nation, England could not afford to isolate herself and risk the domination of Europe by a single major state. She had to follow a balance-of-power policy, taking pains that her primary contribution in an alliance would be sea power supplemented by land power, not vice versa. The strength of that sea power would be limited by her national wealth, in large part directly attributable to her merchant marine, by the size, condition, and tactical skill of her navy, by access to strategic materials (which in that day were primarily Baltic naval stores) for the upkeep of the fleet, and by the possession of bases from which the fleet could operate in strategic areas. Whether she possessed these four "elements" of sea power in sufficient magnitude was dependent upon the wisdom of the British government.

England's merchant marine, the first element, flourished despite great losses to French privateers and commerce raiders. During the long struggle against Louis XIV the French merchant marine was swept from the oceans. Mercantile interests were generally well represented in Parliament. The Navigation Acts were in effect a subsidy which encouraged the building of sound types, as well as large numbers, of merchantmen. Trade with the colonies was monopolistic, of course, and conducted so as to produce the maximum possible benefit. For example, tobacco from the American colonies was imported to England exclusively in British bottoms, and then 90 percent of it was exported to the European countries, again in British bottoms. Britain's wealth, her merchant marine, and her colonial empire grew simultaneously.

England did less well by her navy, the second element, which had seemed an onerous financial burden ever since the "ship money" days of Charles I. The situation in 1701, when only half of England's 130 capital ships were seaworthy and not even those could be manned, may be considered typical at the outbreak of a war. Manning the fleet became increasingly difficult throughout the eighteenth century. About half of the sailors were "pressed" into service from merchantmen, a quarter might be volunteers (including foreigners), and the final quarter might be undesirables, unemployed, debtors, the dregs of foreign waterfronts and riffraff generally. Decay of the fleet in peacetime became, indirectly, a cause of the next war, for English unpreparedness unquestionably influenced French decisions to resort to war. Shortage of ships tended to discourage aggressive naval action, a tendency fortified by the formalist tactical straitjacket of the Permanent Fighting Instructions, with the resultant danger, often apparent during the first half of the eighteenth century but never quite realized, that the fleet might be considered an end in itself and not an instrument of war.

Access to strategic materials, the third element, was clearly comprehensible to an island people. Even while the English let their fleet rot "in ordinary" (laid up in reserve), they would become highly incensed by the possibility of being shut off from overseas markets. Dependence on Baltic naval stores was a problem as acute then as dependence on oil from overseas is today. Foresight had so far assured a supply of English oak for hulls; spars, pitch, and tar had to be obtained from the Baltic. Special legislation was passed as early as 1704 to develop North American naval stores; but Sweden remained England's principal source for years to come. To safeguard itself, the Royal Navy sent fleets to the Baltic for seven consecutive years beginning in 1715, and subsequently as necessary.

Even in sailing-ship days, the ability of a fleet to maintain itself in any given area arose from the propinquity of good bases, the fourth element, which were more likely to be obtained by war than by peaceful diplomacy. On the continent, Lisbon alone was usually a friendly port adequate and available to meet the needs of the fleet. William III put a fleet into the Mediterranean on a year-round basis, but England did not have a base there until 1704, when Rooke captured Gibraltar. Four years later, urged on by Marlborough's strategic genius, the British also occupied Port Mahon in Minorca. The Royal Navy could then challenge the Mediterranean nations and exercise a remote but powerful influence on European strategy generally.

THE FRENCH CHALLENGE

By comparison with the British, French naval power was less an absolute necessity than a luxury. When the expensive fleet inspired by Colbert's mercantilism failed to achieve a decisive victory against England, some French statesmen concluded that theirs was the truly "useless" fleet, a drain on resources that they could ill afford because of the demands of continental land campaigns. For a while they hoped that the use of privateers and naval commerce raiders in *guerre de course* would be cheaper and more effective than a fleet. Their one truly aggressive naval officer of that day, Jean Bart, advocated *guerre de course* as valuable in itself and as a means by which Britain's naval strength could be dissipated. Bart's many successes led such a military authority as Vauban to urge the entire abandonment of the ships-of-the-line navy and the concentration of all maritime effort against British commerce. Before peace was negotiated in 1697, some 4,000 British merchantmen had been captured. Even so, the French policy was only superficially successful. Britain was not defeated, and her commerce expanded even during periods of war.

The principles actually determining French naval policy before the Revolution were those of any ambitious nation which possesses valuable overseas colonies but is not dependent on sea power for survival. Colonies became hostages to sea power. Time and again France's continental victories would be forfeited at the peace table in payment for colonial possessions that had been lost to the British fleet. Even when France conquered Flanders itself, she had to relinquish that prize in order to retrieve Louisbourg and other

New World colonies. French pride and ambitions, if not survival, were vulnerable to British sea power. *Guerre de course* had to be supplemented by as large a battle fleet as French finances could afford. When that fleet was too weak to challenge the British directly, then at least as a fleet-in-being it might hold enough of the British navy in home waters to keep French colonies safe while French commerce raiders drove British merchantmen from the high seas.

The Seven Years' War (1756–63), the climactic conflict of the age of mercantilism, proved how false French hopes were and how much British sea power could accomplish when directed with understanding and intelligence. Hostilities, begun as a North American border incident, had become a true world war. Seriously concerned with the financial drain of another war with Britain, some French statesmen had advocated maintaining peace on the continent and concentrating their undivided effort against Britain in a maritime war; but the possibility of defeating Britain in Germany, by making George II's Hanover a hostage to French continental power, was too tempting to let pass. British statesmen, on the other hand, hurriedly found a continental ally in the king of Prussia. Nonetheless, the first round went to the French. They diverted British attention by massing barges on the Channel, as though for an invasion, and with that feint screened a successful campaign for the capture of Minorca. The British fleet was forced to withdraw from the Mediterranean; and to direct the war the British government had to bring in the leader of the opposition, that staunch advocate of a maritime policy, William Pitt, later Earl of Chatham, who tailored his aims to the available resources.

"Pitt's system," as it is frequently called, was based on the cardinal principle of making the war turn on the weapon, sea power, in which the British were superior. Pitt waged war against continental France by subsidizing his German allies and by amphibious raids on the coast of France to interfere with the proper concentration of French armies; he waged war against French resources by means of the "Rule of 1756," which prohibited neutrals from taking over French colonial trade in time of war from which they were restricted in peace (i.e., before 1756), by direct attacks on French colonial possessions in America and in India, and by blockading the French ports. Except for the German subsidies, Pitt's system was an offensive extension of sea power; even the subsidies were possible only because of Britain's maritime sources of wealth.

The successful implementation of Pitt's policy saw the development of a new force in modern warfare, what the eighteenth century called a "conjunct expedition." Joint operations by military and naval forces, most brilliantly displayed in Wolfe's campaign against Quebec, made possible the conquest of Canada and of the choice French sugar islands of the Caribbean, contributed to Clive's victories in India, and led to the capture of Havana and Manila (after Spain entered the war). The experience for these successful amphibious operations was first gained on the coast of France where, however, the raids were tactically bungled. In September

1757, Pitt had the first conjunct expedition organized. This was a joint force, with a battle fleet covering 9,000 troops in transports, directed against Rochefort, a naval and shipbuilding center on the Bay of Biscay. After exasperating delays and a demonstration of the effectiveness of naval gunfire against land positions, the expedition returned to England without attempting to assault its major objective. Lack of operational planning, indifference to achieving tactical surprise, and absence of an aggressive spirit had resulted in humiliating failure. The next June, 13,000 troops were skillfully landed on the flank of the Channel port and privateering center of St. Malo. After eight days ashore it was decided that an assault on St. Malo's prepared positions would be too costly and the force withdrew, but not until 100 privateers had been burned and the entire countryside alarmed. In August 1758 came the third and most successful of the conjunct expeditions, which captured Cherbourg by means of a well-coordinated amphibious assault aided by the use of special landing craft that might have served as prototypes of some developed in World War II. Unfortunately for the British, their successful troops got entirely out of hand and were badly cut up before they could withdraw, their failure arising from the violent French reaction to their initial victory at Cherbourg.

The surprising fact is that these raids were strategically successful. Not only did they destroy French shipping and divert troops needed against the Prussians, but they proved such a nuisance that the French resolved on a counterattack, if they could concentrate their fleets in the Channel. France always suffered in its naval conflicts with Britain because French naval power was normally divided between a Mediterranean fleet based on Toulon and an Atlantic fleet based on Brest. A British fleet based on Gibraltar overlooked the only route by which the two French fleets could concentrate. The Toulon fleet escaped from the Mediterranean but was caught off the coast of Portugal and destroyed as a fighting force. Shortly afterward, the British admiral, Edward Hawke, caught the Brest fleet in the Bay of Biscay and smashed it in the overwhelming victory at Quiberon Bay.

The Peace of Paris, 1763, found Britain with more extensive domains and greater power than ever before. The entire east coast of North America from Key West north beyond the St. Lawrence was hers; and increasingly the great wealth of India poured into her coffers. Her navy numbered almost 150 ships-of-the-line and more than 100 frigates. The key to her success had been the exploitation of control of the sea through amphibious operations; and an amphibious doctrine was evolved. But what had been painfully learned was now quickly forgotten. Possibly the stultifying effect of the Fighting Instructions on so many naval encounters had made British officers wary of doctrine; but more likely there was a general failure to recognize that amphibious operations are a special art of war and not amenable to casual improvisation.

The results of Pitt's brilliance were soon jeopardized by subsequent British naval policy. As usual the Royal Navy fell into disrepair, but the French, under the energetic direction of Choiseul, spent large sums on their fleet and welcomed the opportunity that the American Revolution gave them to challenge British sea power again. Stretched thin, the Royal Navy temporarily lost control of the seas. Perhaps we should say "forfeited control," for De Grasse's victory off the Chesapeake Capes, which resulted in the surrender of Cornwallis at Yorktown and all its consequences, was a tame affair as battles go.

TACTICS AT SEA

Throughout the eighteenth century naval tactics had followed a formalist pattern, perhaps appropriate to the age, but not the begetter of decisive victories. The French, usually on the defensive and unable to outmaneuver the British, preferred to fight from the leeward where the elevation and range of their guns were increased by the heel of the ship. Their objective was to disable the enemy by shooting down his spars and rigging; then they could escape from a superior force or eventually overtake and destroy an inferior one. The British preferred the weather gauge. Being upwind, they could control the battle, attacking when they chose or retiring if outnumbered. Gunfire was so inaccurate that they strove for a large volume of fire at short range and found that decisive results might be attained by bearing down to within pistol range and pouring their broadsides into the enemy's hull. In practice, however, the British suffered from the doctrine of the conterminous line and the limitations placed on command decisions, both by the Fighting Instructions and by the inadequate communications system. Not until 1776 was there an official Signal Book, which arranged the signals logically, for an entire fleet; and not until 1790 was there a single Signal Book for the entire navy which made possible an unlimited flexibility in the orders issued by fleet commanders. Naval encounters between the French and the British, though often strategically advantageous, were tactically indecisive until a means was found for regaining tactical concentration by breaking the enemy's line and by thus restoring the melee as the crisis of the battle. Theoretically, a ship passing between two ships of the enemy's line could rake each with a broadside fire, but throughout the approach the enemy himself could deliver a raking fire; and if he fell off to leeward somewhat, he could draw out the attacker's approach and defeat his purpose.

The battle which probably did most to break the stranglehold of conterminous-line fighting was Rodney's celebrated victory over De Grasse in the Caribbean at the Battle of the Saints (1782). The French admiral was engaged in the preliminary stages of a Franco-Spanish campaign against Jamaica when the British fleet caught up with him and, after several days of maneuvering, forced the battle. Rodney had an advantage in the number, size, and speed of his ships; he had more three-deckers and more ships that were copper-sheathed to prevent fouling. The French sailed down to windward from the north on a light easterly wind, their line of battle rather ragged as a result of maneuvering for position. Rodney's fleet sailed on a parallel and opposite course, to leeward. Just as the two fleets were approximately conterminous, the wind veered to the southeast, forcing the French ships to alter course toward the British line and tending to carry the British ships into the French line unless positive action were taken to avoid it. And so, largely by chance, Admiral George Rodney broke the line, captured Admiral Comte de Grasse in his great three-decker, the *Ville de Paris,* and decisively defeated the French fleet.

Much of the credit for the victory must be given to Sir Charles Douglas, Rodney's captain and a gunnery expert. He had worked effectively on improving both accuracy and rate of fire. His improved methods enabled gunners to train their pieces as much as 45° right or left for "oblique fire" and, under ideal conditions, actually pound the enemy with three broadsides in two minutes, as compared with Douglas's own hopes for two rounds in three minutes.

The enemy's rate of fire rarely exceeded 50 percent of the standard British performance. These improvements were achieved through training and through certain technical devices. Douglas replaced the old linstock with a flintlock which ignited a goose-quill of powder specially prepared for firing the gun; he introduced flannel powder cases which, unlike the silk used earlier, did not leave a smoldering residue that had to be wormed from the barrel; and he controlled recoil by forcing the guns to roll up an inclined plane and fastening them by steel springs as large as ten inches in diameter. Thus gunnery became safer and faster, and the general adoption of Douglas's technique made the sustained rate of broadside fire in the battle fleet of the Royal Navy vastly superior to that of an enemy. At the Saints, part of this superiority was gained through use of a new, relatively short-barreled gun, the carronade, which could be easily handled and was capable of throwing heavy shot at short range with a light charge of powder. For the British style of pistol-range gunfire this cheaper, lighter weapon became very popular. Its severe limitation in range could be disastrous, however, when it was pitted alone against the standard long gun, as both sides were to discover in the famous frigate actions of the War of 1812.

NELSON AND VICTORY

Aside from these improvements in gunnery, the capital naval vessel of 1800 was scarcely changed from that of 1700. The ships themselves were long-lived, Nelson's *Victory* being in its forty-sixth year at the battle of Trafalgar. The organization and manning of the fleets remained basically the same, even after the mutinies of 1797. But as Britain became embroiled in the Napoleonic Wars, British sea power was called upon for more prodigious accomplishments than ever before.

The aristocratic Royal French Navy had been a natural target for Revolutionary reform. The officer class was virtually wiped out and even ships' gunners, the aristocracy of the sailors, were not spared the leveling process; they were dispersed ashore or diluted by new recruits aboard ship. The French fleet put to sea with former merchantmen as captains. But improved morale was canceled by loss of skill, as Lord Howe demonstrated in the battle of the Glorious First of June (1794). Here the French claimed a strategic victory, for their action had saved a large grain convoy from America whose arrival was sorely needed in France; but the French navy never completely recovered from that engagement, even under the later proddings of Napoleon himself.

The British, on the other hand, had never made bolder use of their fleets to further political objectives. By 1796, Trinidad, the Cape of Good Hope, Ceylon, and various spice and sugar islands had been occupied. Napoleon's Italian campaign drove the Fleet out of the Mediterranean, but only temporarily. In good part because of Nelson's tactical skill, Jervis trounced a numerically superior Spanish fleet in the Atlantic off its own coast; and the next year, in the Battle of the Nile, Nelson all but annihilated the French battle fleet that had convoyed Napoleon and his army to Egypt. This last battle had far-reaching strategic results. British exports to Turkey quickly rose to £150,000 annually and imports from Turkey to £200,000. French prestige in the entire Mediterranean suffered. Likewise, Nelson's victory at Copenhagen, claimed by some to have been his greatest, was an important precedent for the use of naval power against the neutrals in the long economic war between Napoleon and England.

Finally, the French dictator, like his continental predecessors and successors, contemplated invasion of the British Isles as the one sure method of defeating England. But his armies were collected on the Channel coast in vain, for the French admirals could not mass their naval power in the face of the British fleets blockading the French coast. The French Toulon fleet, after dashing to the Caribbean and back, was finally caught off the Spanish coast while making a final effort to return to the Mediterranean to support Napoleon's land campaigns there. The defeat at Trafalgar (1805) meant the practical end of Napoleonic sea power. Thorough indoctrination of his officers had enabled Nelson to employ his fleet in flexible divisions, in full confidence that each ship captain would improvise as necessary. His success called for a high order of professional skill, one which was far beyond the means of his opponents. It is worth noting, however, that Trafalgar came a full decade before Napoleon was finally exiled to the South Atlantic.

During those last ten years British sea power was never seriously challenged. Around the world it exerted economic pressure on Napoleon's allies and supported British military efforts. After various misguided efforts, it once again supported amphibious and diversionary operations on the continent, of which the Peninsular campaign had certainly the most tangible results. With great difficulty the fleet was maintained at a strength of slightly over 100,000 men. Desertions were widespread, especially to the American merchant marine; and the British Orders in Council, directing economic

H.M.S. VICTORY. Victory's keel was laid down in 1765. It was Nelson's flagship at Trafalgar in 1805. It is preserved, fully rigged, in the Portsmouth Navy Yard in Britain. (Edward K. Chatterton, *Sailing Ships,* Sidgwick & Jackson, London, 1914, p. 251)

retaliation against Napoleon's Continental Decrees, were particularly resented by the neutrals, especially the United States and Denmark, who profited much from the war. England's economic war was not aimed at starving France into submission, an unrealistic goal, but in arousing discontent among Napoleon's often unwilling allies.

THE WAR OF 1812

In the United States, where more bad blood was stirred up by President Jefferson's embargo on shipping, another form of economic pressure, than by the well-publicized incidents of British impressment, anti-British feeling ran high in the backwoods. The sea ports, having quintupled trade with Britain during the war, knew full well where their profits lay. After all, British sea power had already forced Napoleon to give the United States the greatest real estate bargain of all time, the Louisiana Purchase. It was the Canada-hungry trans-Appalachian states and not those on the coast that swung the United States into war with Britain. Although Americans take pride in their frigate victories, they should remember that in 1814 the Royal Navy block-aded the entire coast, tied up trade, and landed troops to burn Washington. However, in the lake campaigns on the Canadian border, the victories of Perry on Lake Erie and of Macdonough on Lake Champlain retrieved the American army's blunders. But in a strategically important area, the failure of Commodore Chauncey to fight Sir James Yeo to a finish on Lake Ontario left British communications open to the Niagara peninsula and beyond and denied Americans the likelihood of obtaining a decisive victory.

Without a single ship-of-the-line, the United States had not been able to oppose a British fleet in battle; but her resources for ship building and in unemployed sailors had produced a formidable number of privateers which preyed upon Britain's worldwide trade. At the end of the Napoleonic Wars, when the Duke of Wellington was asked what force would be required to defeat the United States, it was American sea power on the lakes that made him place his estimate far too high for the British government to pursue the matter.

The days of the square-rigged ship-of-the-line were numbered. The great age of sail, as we have seen, was the great age of British sail. To protect itself and its commercial interests, an island of rather modest resources had built, manned, and maintained a navy and then had employed it with sufficient courage and wisdom to create the greatest empire the western world had ever known. British naval policy had been the product of a constitutional state in which the merchants possessed great influence. British success had not been owing to any great technical advantage. Apart from improvements in naval artillery, the French usually were ahead in naval architecture and usually had ships, especially frigates, that sailed better than those of Britain. But, even during the age of formalism, the Royal Navy had built up a tradition of sea service and of vigorous action that was superior to that of the French. Therein lay the key to victory on the ocean and to world empire.

11

WAR IN THE NEW WORLD, 1492–1783

THE AMERICAS

The relative ease with which Europeans established themselves in America after the great discoveries should not be allowed to obscure the magnitude of their accomplishment. The Spanish conquistadores subdued, not primitive tribes, but civilized Indian empires. The domain of the Aztecs in Central America contained a population which has been estimated at about 15,000,000; its capital, Tenochtitlan (Mexico City), was a city of 300,000 people; and the Aztec ruler Montezuma also exercised some control over a number of other peoples. The Peruvian empire of the Incas, although not so populous (perhaps 6,000,000 people) was a centralized despotism. Cortés overthrew the Aztecs with a tiny army of 600 men and a few pieces of artillery, partly because he was supported by Indian dissidents; Pizarro seized power in Peru with the incredibly small force of 183 men simply by eliminating the apex of the centralized Inca state.

The most obvious military reason for the collapse of the Indian civilizations was the immense technological superiority enjoyed by the Spanish invaders. Although both Aztecs and Incas had developed ornamental metallurgy to at least as high a standard as that of contemporary Europe, their work was done in precious metals. Their armament, however, was little better than that of Stone Age man. They used javelins, bows and arrows, and wooden clubs with stone blades; they wore armor of brine-soaked quilted cotton and carried wickerwork shields. Against the steel and shot of the Spaniards such equipment was useless. However, the Spaniards won not so much because of the number of casualties they could inflict, but because of the terrifying effect of firearms and mounted men. Moreover, the Aztec and Inca societies had not developed the complex military organization which had accompanied the rise of civilization in Europe. Aztec warfare was largely an extension of religious ceremonial and was not the occupation of any special class or group of individuals. The only Native American advantage was numbers, and this was far outweighed by Spanish armament, organization, and disciplined tactics.

In North America, the conquest of the indigenous peoples was a process which paralleled the western push of European settlement ending, in recent times, with the disappearance of the frontier. There was never any likelihood that the Indians would drive the Europeans back to sea, or even wipe out an established colony. Their numbers were too few; they were broken up into many tribes, united only at rare intervals by an abortive flickering of a "Pan-Indian" nationalism; and they were immeasurably inferior to the Europeans in the tools of war and in the social and economic organization required for sustained and effective fighting.

By 1641 there were 50,000 English settlers on the Atlantic seaboard of what is now the United States; less than fifty years later there were 200,000. The surge to the interior which resulted from population growth was accompanied by the dislodgment or extermination of Native American groups. The basic occupation of the colonists was agriculture; it was the prospect of land which had brought many of them to the New World. To the colonist, the seminomadic Native Americans, who required extensive lands to support tribal life with hunting and crude agriculture, were wasteful obstacles. The strong aggressive drive which land hunger exerted was heightened, especially in New England, by Protestant zeal. America was the New Canaan, a promised land set apart by God for his elect. The Native Americans who encumbered it were pagans "of the cursed race of Ham"; the Reverend Richard Mather, on hearing of the Pequot massacre, rejoiced that "on this day we have sent six hundred heathen souls to hell." The Pequot Wars (1637–44), King Philip's War (1675–76), and the near-extermination of the Susquehannock tribe which accompanied Bacon's Rebellion (1676) cleared the North Atlantic seaboard of most of its aboriginal population; the last substantial coastal group, the Delaware, was deprived of its Pennsylvania lands and forced across the Alleghenies in the 1740s.

Except in regard to the most formidable indigenous power in eastern North America, the Iroquois Confederacy, the relations of New France with its Indian neighbors were different. The Iroquois, or Five Nations, by virtue of their relatively large numbers, more advanced political organization, and commanding geographical position on the Hudson-Mohawk waterway between New France and the English colonies, were able, unlike their weaker Native American brethren, to bargain with the Europeans on nearly equal terms. For the most part, their aims coincided with those of Albany businessmen who wished to siphon off the cream of the French western fur trade; the Iroquois therefore concentrated their efforts against that French economic lifeline. In the middle and latter part of the seventeenth century, the Iroquois, through their fierce depredations, held the tiny French colony on the St. Lawrence in the grip of fear for years at a time. Indeed, not until this unique confederacy exhausted itself through too-constant war could the French relax their vigilance.

The relations of the French with other tribes were much happier. The economic base of the colony was the fur trade, in which Indian assistance and cooperation was indispensable. Moreover, French Catholics regarded the Indians not as pagans eternally damned, but as benighted souls to be rescued from hellfire and guided to salvation. Eventually almost all the tribes of the St. Lawrence, Ohio, and Mississippi basins were bound to New France by economic necessity and missionary effort. It was the unswerving policy of the government in Quebec to reinforce this connection by annual and lavish distribution of gifts, partly to maintain control of the fur trade, but also to enlist the native peoples as allies, in the face of the ever-widening disparity of strength between New France and the English colonies. In view of this vast difference between English and French attitudes and policies, it is not surprising that an overwhelming majority of the tribes supported the French against the English who menaced their homelands. In the same way it was natural for the Indians to ally themselves with the British during the American Revolution and the War of 1812. Encroaching American settlement, and not British officialdom, was destroying indigenous North American culture.

Confronted by American conditions, both French and English soon shed the accoutrements of the battlefields of Europe—armor, pikes, swords, and heavy cavalry. But it should be noted that armored English colonists of the seventeenth century were far more effective against primitively armed Indians than were their eighteenth-century successors against Indians armed with guns and iron tomahawks. The formal tactics of the Old World were replaced by forest tactics which approximated those of the Indians themselves. In the ambush, lightning attack, skirmishing, and sharpshooting from cover of small war, the European eventually surpassed the Indian. But frontier society did not become Indian; far from it. The exigencies of frontier life placed a premium upon qualities produced in a mature, civilized, highly organized society. The discipline and command structure of the Europeans, plus their superior technical knowledge, made possible a level of bush tactics to which the Indians, who rarely rose above the imperative urge of self-preservation, could not aspire. The Indian war party was merely a collection of individual warriors attracted by the fame of an eminent war chief. The chief customarily had no power of command over his braves; after the initial brush, an Indian force lost coherence, since each warrior fought for himself. Indians were always unwilling to attack prepared positions of any kind; with the possible exceptions of the Iroquois's investment of Fort Frontenac and Pontiac's of Detroit, no Indian war party ever attempted a sustained siege.

FRANCE VERSUS ENGLAND

Until well on in the eighteenth century, both French and English colonies in America relied for defense primarily on their respective militias. In the English colonies, the militia structure stemmed from the old English system. The militia was a compulsory levy of the male population which drilled a certain number of days a year and could be summoned in emergency. As with other aspects of colonial life, the control of the home government over military affairs was slight, and each colony therefore developed its own version of the militia. Exemptions from service (which were extensive) varied widely; the size of the unit, whether trainband of infantry or troop of horse, differed from colony to colony; the geographic basis of recruitment for the unit might be the village, the town, or the county; in some colonies, the election of officers was the rule; in others, officers were appointed. Enthusiasm and the sense of urgency necessary for the development of an efficient force were directly related to the likelihood of hostile attack; however, even in Massachusetts, which was exposed to French attack, the amount of compulsory drill had been reduced to four days annually by the late seventeenth century.

Unlike the English colonies in which "salutary neglect" allowed the development of considerable local initiative, the colony of New France, at least after the coming to power of Louis XIV, was closely supervised by the home government. The execution of policy was in the hands of a governor, who was also the commander in chief, and of an intendant; representative institutions did not exist. The colony, according to the tenets of the Colbertian mercantilism, existed solely to provide raw materials and a market for the mother country; as an investment, it had to be defended, but as cheaply as possible. The result was that the militia system was established as a uniform, centrally directed organization in keeping with the authoritarian politics and graded society of this American extension of the Ancien Régime.

The militia was under the direct control of the governor at Quebec, who commissioned its captains in each parish. A populous parish might have more than one company. Because of the much inferior military position of New France, there was a far more decided appreciation of the value and function of the militia than in the English colonies. Exemption from militia service was almost impossible to obtain. The French *habitant* was provided with arms, not issued them for an occasion; he was drilled in musketry at least once every two weeks. In contrast to the Thirteen Colonies, where the militiaman was bound to serve only within the limits of his own colony, the Canadian militiaman was liable for duty wherever his governor ordered.

It was inevitable that French and English should fight one another in North America, but not merely because their mother countries were so often at war. In contrast to that waged in Europe in the same period, warfare in North America was punctuated by atrocities caused by deep antagonisms between the two white groups. Economic rivalries abounded—the clash of interest in the fur trade, in the Atlantic fisheries, and eventually in the rich heartland of the Ohio. To these differences was added religious hatred. The French colonists were products of the Catholic Reformation and remained zealous even when chill winds of skepticism blew from the mother country. The people of New England, who had the closest contact with the French, were fiery Calvinists unalterably opposed to Catholicism.

Neither side hesitated to employ Indian auxiliaries. Neither had much compunction in borrowing the barbaric practices of their savage allies. The Iroquois, incited to take the warpath by their friends at Albany, massacred the inhabitants of Lachine, near Montreal, in 1689; in 1690, in reprisal, a band of French militia and Christian Iroquois slaughtered many people at Schenectady, while other groups of militia and Abenaki Indians raided and killed in New Hampshire and Maine. In 1704, another French and Indian raiding party surprised the sleeping village of Deerfield, Massachusetts, killed 53 persons, and carried off 111 others to captivity in Canada. This barbaric strain, present in American warfare almost from the beginning of European settlement, was to culminate in the excesses which marked the Seven Years' War and the American Revolution.

For the most part the French militia were more effective in forest warfare than their English opponents. Because New France devoted much more of its energies to the fur trade and made more extensive use of Indian allies than did the English colonies, the French militia had a much broader knowledge of terrain, Indian methods of transport, and the technique of living off the country for long periods. Although the men of Virginia and Kentucky, and units like Rogers's New England Rangers, were quite the equal of the French in bush tactics, the English colonial militia was usually ill-trained. In fact, so far were they from fitting the popular picture of sharpshooting backwoodsmen (most of them were peaceful farmers who dwelt far from the frontier) that the British army actually had to detach men from its own regular regiments in 1757 to learn woodcraft and forest fighting. The regiment so formed, under the command of Lt. Col. Thomas Gage, was the first light infantry regiment in the British army.

French superiority in the techniques of small war availed little. Incessant border raids in the period 1690–1713 were only pinpricks, designed to keep the English off balance and disguise the paucity of French resources. The inexorable truth was that the French were only staving off ultimate disaster. The economy of the Thirteen Colonies was much more diversified and healthy than that of New France; the disparity in population was great. (By 1754, there were 1,500,000 English and 60,000 French in North America.) Moreover, the flourishing merchant marine of New England gave the English colonists a formidable weapon, for New France was vulnerable to seaborne attack. The two greatest efforts of New England against New France were combined operations. William Phips failed in 1690; Quebec was too hard a nut for the militia to crack. The capture of the great French fortress of Louisbourg in 1745 was almost entirely a triumph for colonial military and naval arms (supported by four ships of the Royal Navy); its return to France by the Treaty of Aix-la-Chapelle was much resented in the colonies.

However, the full measure of English colonial superiority in numbers and resources was never exerted. Each colony was so jealous of its own authority, and so unable to see that the danger to one colony might well become a danger to all, that the possibility of coordinating effort either defensively or offensively was slight. In King William's War (1689–97), only New York and New England took part, because none of the other colonies felt themselves menaced by the French. In Queen Anne's War (1701–13), in order to preserve Iroquois neutrality, the French did not attack New York, which thereupon allowed New England to bear almost the whole burden of the war against the common enemy. Many prominent Americans saw the necessity of a unified approach to the problem of defense but, because of the strength of local feeling and the suspicion of the political and financial power that any central body in charge of military affairs would have, their proposals came to nothing. Thus Franklin's Albany Plan of Union (1754), which provided that a "Grand Council" of the colonies should "raise and pay soldiers and build forts for the defence of any of the Colonies, and equip vessels of force to guard the coasts and protect the trade on the ocean," was rejected by all the colonial legislatures.

The European regular did not loom large in American war before the middle of the eighteenth century. English regiments were dispatched to the colonies at intervals; for example, in 1676 troops were sent out to help quell Bacon's Rebellion in Virginia. During the conflict from 1689 to 1713, small numbers of British troops were stationed in America, particularly at New York, while a large expeditionary force was used in the disastrous Hovenden Walker attempt on Quebec in 1711. As for New France, regular troops had made their appearance as early as 1665, but thereafter during the Ancien Régime in Canada there was no permanent garrison of French regulars, only the colonial regulars of the *Troupes de la Marine*. Not until the Seven Years' War did European concern for American empire warrant the posting of large drafts of troops.

ORIGINS OF THE AMERICAN REVOLUTION

The history of military techniques in North America has often been portrayed in terms of the different practices of the native-born militiaman, whose tactics were fitted to the environment, and the rigidly orthodox European regular, who was contemptuous of the undisciplined skirmishing of the colonial. In both English and French America the dislike between the "colonials" and the representatives of the mother countries was an inevitable result of profound social differences; in North America distinct societies had been created, different from those of the homelands. This social cleavage, when carried over into military affairs, was accentuated by the contrast between European and American methods.

The Seven Years' War and the American Revolution brought these differences into the open. The Seven Year's War began (unofficially) with a clash between French and American militia in the Ohio Valley and the defeat of George Washington and his Virginians. Until 1757, although the Marquis de Montcalm had at his disposal large numbers of French regulars for some time before the English arrived in great force, a heavy part of the military load, on both sides, was carried by native-born troops. Braddock was overwhelmed near Fort Duquesne by colonial troops and Indians firing from cover, in part because European regular forces did not know how to adapt their tactics to forest warfare, but mainly because the British column had not observed conventional marching procedure and discipline. Nevertheless, Braddock has become a classic example of the inflexibility of the European commander.

Yet the defeat of Braddock is the sole victory of any importance that irregular or provincial troops of either side could claim during the whole war. Increasingly, as the two mother countries became committed more heavily to the colonial struggle, the professional took over from the amateur, the regular from the militiaman. The replacement of Governor Shirley of Massachusetts as Commander in Chief in America by the Earl of Loudoun in 1756 signaled this change. When William Pitt, as secretary of state for the Southern Department, took the direction of the war into his own hands and roused the latent nationalism of the English people, the pace in America quickened, because Pitt believed that the war was to be won overseas. During the first years of the war, the British government had attempted to recruit regulars in America; but, largely because the provincial governments were offering high bounties for service in the militia regiments, the British were not able to fill their establishments. This policy was therefore dropped for one of requisitioning men, money, and supplies from every colonial government; but the traditional colonial unwillingness to cooperate remained. Only Massachusetts, Connecticut, and New York met their quotas, actually contributing 70 percent of all the troops raised in the Thirteen Colonies. By 1758 the British had 20,000 regulars in America. The war was developing into one of orthodox operations and sieges for which the provincial corps were ill-suited.

The war was to be decided by regular troops; and reinforcements from Europe depended upon sea power. New France, deficient in population and industry, incapable even of producing enough food to supply her own people, was desperately in need of assistance. That assistance was not forthcoming. The naval blockade established by Pitt and the supremacy of British naval power conclusively demonstrated by the victories of Lagos and Quiberon Bay (1759) prevented all but a trickle of men and supplies from reaching Montcalm in his extremity. In that critical year he received only a bagful of decorations and promotions and a corporal's guard of three or four hundred men; his request for a diversionary operation against the southern colonies was completely ignored.

Professionalism of European standards marked the final campaigns of the war. In 1758 Louisbourg, a formidable example of military engineering, had fallen to an army of British regulars under the command of

General Jeffery Amherst, in an expertly conducted orthodox siege. With the St. Lawrence gateway opened, the way was then clear for a final assault on the heart of French power. In 1759 the British made a three-pronged assault upon New France. Prideaux attacked Fort Niagara, the key to the West; Amherst cautiously moved up the Hudson-Richelieu gap toward Montreal; General Wolfe and Admiral Saunders ascended the St. Lawrence to Quebec. Montcalm was thus compelled to split his meager defense force. The decisive battle was fought on the Plains of Abraham. Its tactics were indistinguishable from engagements in Europe; but the regiments that fought for France on the Plains of Abraham had been brought up to strength by infusions of half-trained militia who could not fight a stand-up battle in the European manner. The volleys at very close range fired by Wolfe's highly trained troops were among the most effective in history.

The American Revolution marks, indeed, an important transitional step in the history of warfare. The professional military methods of the eighteenth century, which had come to be predominant in the Anglo-French struggle in America by the time of the Seven Year's War, were more important in the Revolution than popular legend admits. At the same time, the war was a portent for the future, pointing toward conflicts quite different from those of the dynastic quarrels of eighteenth-century Europe. This was partly the outcome of military experience on American frontiers since the time of settlement; but, more important, it was the result of a spirit and an attitude on the part of individual Americans which led many to take up arms for a cause. True citizen armies reappeared in western warfare, anticipating the change that came in Europe with the French Revolution.

The Revolution was a war between societies which had grown apart. That of Great Britain was stable and relatively rigid, one in which birth and status counted for much. That of America was much more fluid and democratic. The British military system, although constitutionally controlled, was essentially that of any European state. Commissions in the army were reserved mainly for the aristocracy and gentry, and were obtained through political patronage and purchase. The rank and file came from the lower orders, were voluntarily enlisted, and were professional, long-service soldiers armed and trained according to the prevailing European mode. The military tradition of the Thirteen Colonies, long sheltered by the British fleet and army, was that of the militia, a levy of citizens unschooled in European tactics or standards of discipline.

Yet the Revolution was not won by victories of free citizen–militia over rigidly disciplined British and German mercenaries. It is true that Washington and his generals could not have done without the state militias and the short-term soldiers who accepted a bounty, served for as little as three months, and then went home, often with their army-issued muskets. In the early months of the war, the militia was the only force available; its exploits at Boston and in Canada were remarkable. Generally, however, the militia was used only in emergencies. It was shown again and again that militia were unreliable in the face of an assault by regular forces; General Anthony Wayne declared himself satisfied if he could get three volleys out of them before they ran. Therefore, despite its early successes, the young Republic soon found that in order to grapple on even terms with the enemy, a long-service, professionally trained army was essential. As Washington said, "Regular troops are alone equal to the exigencies of modern war, as well for defense as offense, and whenever a substitute is attempted it must prove illusory and ruinous."

Late in 1776, in an attempt to build a trained army of veterans, Congress authorized the recruitment of men for three years or the duration of the war. But not until 1778, at Valley Forge, did the Continental Army receive any systematic instruction in battlefield drill. A former Prussian officer, General Friedrich von Steuben, introduced exercises in the tactics of the line, a training that became uniform throughout the army and gave the Continental regulars the cohesion needed to stand up against the British. Steuben recognized, however, that the time was too short to apply fully intensive drill of the kind imposed upon Prussian troops; moreover, it did not suit the independence of the Americans. Precision was therefore sacrificed in favor of swiftness of execution in order to bring American marksmanship quickly into play; for the Americans, as Lafayette and others noted, always aimed at an individual target, even when volleying. The crack regular troops of the Continental Army, generally about 10,000 in number, were one of the major elements in the eventual American victory.

It is sometimes suggested that the superiority of the American rifle over the British Brown Bess musket made a significant contribution to American fortunes. The long-barreled weapon of Daniel Morgan's sharpshooting riflemen was certainly superior in accuracy and range to the short rifle used by the Hessian *Jägers*. It took less time to load because the frontiersmen had developed the technique of wrapping the ball in greased cloth before ramming it home. Undoubtedly, too, Morgan's men were the best marksmen of the war; at battles from Boston to Saratoga, they and their like took a heavy toll of brilliantly clad British officers. But riflemen were relatively few. Most of the Americans were armed with muskets. Moreover, the rifle was not equipped with a bayonet, and its rate of fire was much slower than that of the smoothbore musket. At Brooklyn Heights (1776), riflemen firing from cover were bayoneted by British and Hessians who drew their fire and then rushed them before they could reload. Morgan himself conceded that rifles were effective only when supported by muskets and bayonets. His corps was broken up after Saratoga, and its members were used against raiding Loyalists and Indians along the frontier. The day of the rifle was yet to come; but the weapon, and the skirmishing tactics that went with it, made a strong impression upon some British and French officers, and were to have an important influence on war in the future.

With the American Revolution ideological conflict was reintroduced into warfare. In the bitterness which arose out of an irreconcilable struggle over the issue of independence as against the coercion of the colonists back into their old allegiance, the controls that restrained war in the eighteenth century were weakened. Between the regular armies, although no atrocities were committed, the European military code of honor was frequently ignored; the conventions regarding uniforms, the treatment of prisoners, and the rights of noncombatants were largely abandoned, particularly because it was difficult to determine, as when patriots posed as Loyalists at Bennington in 1777, who was a soldier and who a civilian. Excesses were more frequent in the fierce encounters between state militias and Loyalist units. Along the great arc of the frontier, outside the main theaters of the war, American frontiersmen, Loyalists, and Indians waged a savage war of barbarity.

As a result of the ideological issues in the conflict the protagonists wooed public opinion with a flood of pamphlets. In this propaganda battle, the Americans were much more successful than the British. *Common Sense* and the Declaration of Independence had an immeasurable influence on international opinion and in forcing fence-sitters to commit themselves. The heat of the battle of words, and the extreme nature of the issues at stake, drove politicians to utterances quite out of keeping with prevailing moral attitudes. Lord Suffolk, a member of the British government, defended the use of Indians, since "it was perfectly justifiable to use all the means that God and nature put into our hands" to crush those who rebelled against properly constituted authority.

On neither side, however, was there a total preoccupation with the war effort like that later brought about by intense modern nationalism. At the same time, there was on both sides of the Atlantic a division of opinion on ideological grounds which cut across national feelings. In Great Britain, the justice of the war was a party question, with the Whigs supporting some form of conciliation. There was little popular enthusiasm; public opinion tended to split along party lines. The war was fought in the old way by the regular army supplemented by German mercenaries, 18,000 of whom saw American service. The British political system of the eighteenth century functioned efficiently in the prosecution of war only when, as with the elder Pitt in the Seven Years' War and with the younger Pitt in the Napoleonic struggle, it threw up a great war leader or prime minister. This occurred only when national interests aroused widespread support. In the War of American Independence, Lord North was merely the political manager of the House of Commons, whose principal job was to deliver majorities through influence and patronage. With these duties, no ordinary British prime minister had time for war leadership. For England the war was therefore very like the dynastic struggles of the century.

In the American colonies, nationalism was weakened by state loyalties and by the fact that the Revolution was itself directed against central authority in the name of local autonomy. The American population was not of one mind about independence. At the beginning, at least, convinced revolutionaries were outnumbered by the Loyalists and those uncommitted, undecided, or apathetic, but the Tories' failure to organize themselves effectively and the rapid polarization of sentiment in favor of the colonial cause soon left the Loyalists in a decided minority in most areas. Congress, consisting of delegates from the thirteen "sovereign" states, lacked the vital financial powers necessary for the direction of such a struggle. Every state but one boasted its own navy; every state had its militia defense force which was only occasionally made available to the national command; every state conducted a private war with the enemy. The Continental Army reflected the tug-of-war between the urgent need for a unified policy and the antagonism of the states toward centralization and standing armies in general. It was not a truly "national" force, despite the fact that its general officers were appointed by Congress, but a composite army in which the troops of the states retained their identity. The regimental officers of the state "lines" or regiments were named by the state legislature concerned. Owing to the inability of Congress to tax and thus to raise revenue for support of the army, in 1778 it directed the states to issue supplies to their own lines. This increased internal division within the army, because states like Pennsylvania supplied and equipped their regiments much more generously than did others. Despite its lack of consolidation, however, the Continental Army was an important expression of rising American national feeling.

The extent of popular participation belies a simple nationalist interpretation of the war. In 1776 approximately one man in every eight of military age saw service against the British; in the years following, the proportion dropped to one in sixteen, far below modern standards. Of course, it must be remembered that perhaps one-third of the population was Tory. It has been estimated that men engaged in privateering at sea exceeded those in the armed forces in every year except 1776. The only means that Congress had to attract men for long service was to offer bounties. Even here the states often outbid Congress; many men trafficked in bounties; and galloping inflation made civilian life more attractive by reducing the value of the bounty. On two occasions, Congress went to the length of authorizing Washington to draft men, but this was ineffective in the face of state and popular resistance. Despite the war, and in part because of it, the American economy flourished, yet the army was grudgingly supplied by the states and was victimized by profiteers. It was able to maintain a sufficient amount of arms and equipment only because of French help.

THE WAR FOR INDEPENDENCE

The Revolution began with Great Britain holding the advantage over the colonies in almost every respect. Only a few years earlier she had been victorious in a "world war" in which she had no ally but Prussia. Her economic power alone appeared decisive. But during the Revolution eighteenth-century strategical precepts were severely shaken. The British found that it was not enough to menace the supply lines of an American army, or even to defeat it in the field. Once away from the security of their coastal bases, they discovered that not only had they to counter the Fabian strategy of Continental forces but, wherever they passed, the bulk of the civilian population transformed itself into an army of irregulars of a kind unknown in Europe.

Although the Americans might not flock to join the army of Congress, they would fight to defend their own states, and thus the British were exposed to the novel experience of engaging the mass of the population.

The Saratoga campaign (1777) was the clearest example of this new qualification upon strategy. Upon Burgoyne's "magnificent armament" rested the hopes of the British administration. Striking south from Montreal, down the Richelieu, Lake Champlain, and the Hudson, Burgoyne's forces were to take Albany. There he expected to act with the army under Sir William Howe, based in New York City, to cut off New England, the center of rebellion, from the rest of the colonies. His army consisted of 6,500 British and German regulars, more than 1,000 Canadian and Indian auxiliaries, much more artillery than he needed, and all the paraphernalia of a typical European army, including camp followers, servants, and 600 wagons to carry supplies. Moving at a ponderous pace (at one point, twenty miles in twenty days), Burgoyne pushed the outnumbered Continentals from Ticonderoga back across the Hudson. But as his lines of supply stretched to the breaking point, the Americans prevented him from living off the country by methodical devastation along his line of march. Short supply alone did not bring him to a halt. His snail-like progress allowed thousands of militia from New York and New England to concentrate against him. By the time he had reached Saratoga, the original Continental force of 4,000 had been swelled by militia to perhaps 17,000 men. Enveloped by weight of numbers, his own force denuded, and with Howe's army involved in the Philadelphia campaign, Burgoyne surrendered. The mass citizen army of Saratoga has no eighteenth-century precedent; its counterpart is to be found in the armies of the French Revolution.

British armies in North America were severely handicapped by operating at the end of an extremely long line of supply. For this reason, the first concern of British strategy was with the seizure and maintenance of ocean ports, like New York, Philadelphia, and Charleston, which acted as magazines for the armies in the field. When France entered the war in 1778, British communications at once became vulnerable to naval attack, and land operations suffered accordingly. The war became worldwide and Britain faced Spain and Holland, as well as France, in Europe. The American war was but a single theater in a world conflict. The disaster of Cornwallis at Yorktown in 1781 was due to a failure in the use of British sea power. When the French admiral De Grasse won control of the Chesapeake, Cornwallis, having been cut off from New York supplies or hope of evacuation, could do nothing but surrender. Sea power had determined the issue.

12

THE NATION IN ARMS AND
THE NAPOLEONIC WARS

THE FRENCH REVOLUTION

The eighteenth-century professional standing army was as efficient as was possible within the limitations imposed by the political and social structure of the contemporary state. But even while it flourished some men had realized that it did not represent the ultimate development of military strength. The principle of allowing the defense of the state to become the responsibility of hired professionals was convenient insofar as it led to expert efficiency in arms; but it was a source of weakness in that it did not utilize the full resources of the national manpower. Some of the *philosophes,* the leaders of the eighteenth-century intellectual world, had seen this clearly but were puzzled to find a solution. The French philosopher Voltaire had criticized the contemporary military system, scoffing at the fact that the defenders of the state were recruited from the poorest human material. But, typically, he had no other suggestion than that the armies should decide issues by "fighting it out in a field." This was not a solution but was, indeed, exactly what was happening.

Other *philosophes,* critical of the rigid class structure of contemporary society and of its reflection in the "irrational" army of their day, had argued that in an emergency the manpower of the nation should be called out to fight as a militia and should be disbanded when the crisis passed. Montesquieu saw a citizen army as a defense against arbitrary rule; in a republic, he thought, the army must be the people, as in Roman times. Rousseau had carried this a step beyond the normal thinking of the eighteenth century by suggesting that the citizen had a responsibility to fight in defense of his country, an idea which was logical for a democratic form of government but not for the autocratic monarchies that prevailed in Europe. Switzerland, where Rousseau lived, had such a militia. Elsewhere it might have endangered the regime.

Some military thinkers had put forward similar ideas, among them Count Jacques de Guibert (1743–90). Guibert, who died early in the French Revolution, was the author of a treatise on tactics published in 1772 which preached a war of movement and proposed the recruiting of a citizen army. He had criticized the limitations of the contemporary supply system and had suggested that the army should be freed from the restrictions which the magazine system imposed and should march with a minimum of baggage through the enemy's territory, living off the country. He believed that a nation which was sound in government could create such an army and would be able with it to dominate its neighbors. He thus forecast the great revolution in military organization which began soon after his death. But he apparently did not realize that it could not take place until a political revolution had occurred.

The effect of a political revolution on military organization in America has already been discussed. In France the same thing happened in rather different circumstances. The monarchy had an antiquated fiscal system which, failing to utilize the full resources of the French nation in the wars of the eighteenth century, had bankrupted the state. The Estates-General, which met in 1789 in an attempt to solve the financial tangle, recommended constitutional changes which would have sharply curtailed the power of the Crown. Population growth had led to economic distress, especially in the cities. Crises led to mob violence, and the royal armies, their rank and file unpaid and disaffected, could not be used safely. Of the whole army, the real prop of the despotic state, only the King's Swiss Guards could be relied on in emergency. The constitutional regime set up by the men of 1789 was terminated by the consequences of the French declaration of war on Austria in 1792. The war thus launched was not only a preventive one against counterrevolutionary intervention by the great powers but also an ideological crusade for the liberation of other peoples. Within a few months, the French monarchy was overthrown and a convention was summoned to design republican institutions. Under the convention, foreign invasion, internal counterrevolution, mob pressures, and economic problems led to the establishment of the "Terror" by the radical Jacobins, a dictatorial regime intended to crush counterrevolution. This regime established the citizen army, large-scale war production, mass indoctrination, and economic regimentation characteristic of modern states in wartime. The old structure of "privilege" in the state and of a hereditary absolutism was gone, and the republican principles of popular sovereignty and of "liberty, equality, and fraternity" were henceforward to be powerful forces in French politics. A state in which the people, and not a monarch, seemed to hold supreme power had won the hearts of the masses.

THE REVOLUTION AND THE ARMY

When the Revolution occurred in France and unhinged the state, a new type of army emerged. The most obvious sign of the change was a vast increase in the size of the military forces resulting from the conscription of the citizenry in 1793. But the new army was not to be merely a citizens' militia. It rapidly took upon itself the efficiency of professionalism. Aristocratic monopoly of commissions, consolidated in 1791 by a royal decree, was swept away, thus opening the way for a new kind of officer. Most of these were former NCOs of the old royal army. They quickly became competent professionals. But the army of Revolutionary France did not become merely a larger version of the eighteenth-century standing army with more competent professional officers. As a result of the circumstances of its creation it remained different in kind as well as in size and professionalism. The "nation-in-arms," the "armed horde," as it has been well called, was born of the Revolution, and it reflected that fact in its morale. It was moved not only by discipline but also by ideological and patriotic fervor. The vote and national military service were corollaries. The people had, in theory, seized control of the state from the dynasts and now had to defend it; and the new form of state had harnessed public opinion to the machinery of national defense, with the result that its military power was vastly increased. Domestic unrest was thereby defused, which was important especially when monopolies, corruption, and racketeering became endemic as they did soon after the establishment of a new form of government, the Directory. Generals of the French Revolution and, later, Napoleon showed that these forces could be used for aggression. The emperor was able to requisition supplies as well as men. Price controls, severe punishment for hoarders, and the nationalized manufacture of arms were new features of what had become a totalitarian military state. They thus fulfilled Guibert's prophecies.

It must be noticed that in addition to this sociological explanation of the new might of the revolutionary state, there is also a technical explanation which some writers have suggested was the real one. During the long period of wars in the eighteenth century, military efficiency had been increased by improved weapons, by new forms of military organization, by greatly improved communications, by better mapmaking, and by the growth of productive capacity. One technical development in particular is most frequently mentioned as a cause of the revolution in warfare at the end of the eighteenth century, namely, the further development of artillery. The French had attempted to regain the military prestige lost by their final defeat by England by greatly improving their ordnance. Inspector General Jean-Baptiste Varvette de Gribeauval (1715–89), who, like Guibert, died at the outset of the Revolution, had introduced a series of changes after 1763 which have been described as the foundation of the military achievements of Revolutionary and Napoleonic France. Guns were made with interchangeable parts for mass production; carriages were built to a standard model; the mobility of the guns was improved by harnessing the horses in pairs instead of in file; hardwood axles replaced heavy iron ones; and accuracy was promoted by the introduction of a "tangent sight," a graduated brass measure which enabled the gunner to "lay" the gun on a target. The greater mobility of the guns made it possible for the artillery to accompany divisions. Divisional formation, which was to become a feature of the warfare of the Revolutionary age, facilitated strategic maneuver.

The building of roads and canals, which had gone on apace in France and Western Europe in the eighteenth century, had made possible the deployment and movement of divisions of an army in complicated maneuver. Furthermore, the development of the science of mathematics and of engineering (the latter had been chiefly promoted in the military colleges in the eighteenth century), and the great improvement in cartography (which had been made possible by greater skill in mensuration), undoubtedly contributed to the increased power of the new armies.

But these technical developments do not explain the great increase in the size of armies or their new morale. If they had been of primary importance, they would have provided the Ancien Régime with a new military strength which might have prevented the Revolution. The technical achievements were secondary to the morale factor. It is reasonable to argue, then, that although the technical developments of the eighteenth century strengthened the military power of the revolutionary state, they did not create it.

The history of the first years of the Revolutionary Wars shows exactly how the political and social upheaval affected military organization and methods. When the Revolutionary government of France first declared war on the Austrian and Prussian monarchs who thought to restore Louis XVI to power, it had to rely on armies which consisted mainly of regulars from the armies of the Ancien Régime with an intermixture of volunteers. They were far from successful. Then, in 1792, came the triumph of Valmy, the beginning of the victories of the French Revolution. Valmy is sometimes said to have been won by a cannonade of massed artillery. It is true that the opposing infantry never came into contact; but the victory was not caused by a cannonade. The French won at Valmy because their enemies were weakened by indecisive command and by dysentery. However, the victory set France afire with nationalism, and, as a result, the Revolutionaries wildly invaded the Low Countries, claiming to bring freedom to oppressed peoples everywhere. But once again their armies proved unequal to the task. Defeats, and the desertion of their only experienced military leader, the former royal general Dumouriez, brought France and the Revolution to the brink of disaster.

It was this crisis which produced the Terror in France and with it a complete overhauling of the military system. The Committee of Public Safety, set up to root out treason, became in effect a war cabinet with dictatorial powers. Thus was remedied one serious defect in the system of government which had replaced the despotism of Louis XVI. Hitherto, the Revolutionary state had lacked an efficient executive. Now it possessed one of the most powerful ever known. Within a few weeks there came the decree of national mobilization and then one of requisition. The effect of the enhanced power of the Revolutionary state was thus to organize the resources of the nation more fully in its service. Lazare Carnot, one of the greatest war ministers of all time, proceeded to reverse the defeats of the early years.

REVOLUTIONARY TACTICS

France soon discovered a formula whereby numbers and zeal could bring victory. Highly trained light infantry had been used as skirmishers in the eighteenth century to cover the advance of the line. By the time of the Revolution, agricultural developments had increased their value: in some parts of northern Europe open fields were giving way to enclosures, and walls or hedgerows provided greater cover for the skirmish line. Thus at Hondschoote in the Low Countries in 1793, walls and dikes prevented the enemy from forming line and maneuvering against the French. Captain J. L. A. Colin, an authority on eighteenth-century infantry tactics, shows that, although most French armies retained the capability of maneuver in the field, the Army of the North, which fought the British, lost maneuverability, because its cadres were heavily infiltrated with recruits. The French army was short of trained light infantry in 1793 and 1794 and therefore used poorly trained troops for skirmishing.

These circumstances led to a mistaken interpretation of the tactical innovations of the Revolution. Sir Charles Oman, a distinguished British military historian, and General J. F. C. Fuller, a British advocate of mobile warfare, have said that, as a consequence of using inferior troops, Revolutionary generals resorted to massive columns. However, after Dumouriez ordered an attack in columns-in-line at Jemappes in 1792, that formation was apparently not used for a year or two. It is also true that the raw levies of the Revolution could not have advanced far in line and would have been thoroughly disorganized by the debris of front lines falling back to the rear. Partly for this very reason, but even more because small columns-in-line would not mask the new mobile field artillery inherited from Gribeauval's reforms in Louis XVI's army, the Revolutionary generals favored advancing in columns-in-line for attack. Artillery was the decisive factor in Revolutionary victories. It was not massive columns but swarms of highly trained skirmishers and small assault columns-in-line, with a powerful artillery bombardment, that constituted the Revolution's tactical innovation. Revolutionary zeal was also important, but perhaps not so important as a new ruthlessness. This affected foraging—not a new practice, but now carried on with greater severity. It also affected discipline and the willingness of commanders to drive their men and take risks: for revolutionary generals who failed were liable to be summarily dismissed or even shot. Furthermore, the system of conscription introduced in 1793 gave them a glut of lives to squander.

Just as the Revolution produced new tactics and tactical formations, so it also affected the organization of armies. "Divisional" organization had, in fact, been developed in the French Army shortly before the Revolution. By

1794, the French had called up over half a million men. Not since the great
barbarian invasions had there been so many men in arms. Armies of that
size could not be directed and fought in the old way, as a single whole,
commanded by a single general. It was now imperative that they be divided
into divisions, which were portions of an army composed of all arms, able to
operate either in cooperation with other parts of the army or, if need be, by
themselves.

Divisional formation revolutionized strategy. Armies were now composed of detachable parts which could fight the enemy alone until the rest
of the army came up in support. Divisions could also be detached to carry
out encircling movements; and on the defensive, could be used to prevent
them. Divisions could advance along parallel roads and concentrate immediately before making contact with the enemy. Thus generalship was made
more complicated, and staff work became more important. Highly detailed
maps, showing natural features, were now required. Improved cartography
made them available.

At the same time, the Revolutionary fervor of the conscripts caused them
to operate in a manner that would have been more difficult for the regulars of
the eighteenth century. They marched faster, with the Revolutionary quick-
step of 120 paces to the minute; they bivouacked in the open; and since they
lived on the country, they carried less baggage. They thus freed themselves
from the restricting magazines which had often held eighteenth-century war
in a straitjacket.

Furthermore, there was a new spirit in warfare. When the mass citizen
army became a reality, it became imperative to instill in it a nationalistic
and Revolutionary zeal. Under the Convention established in 1792, deliberate efforts were made to indoctrinate soldiers in Revolutionary patriotism
with handbooks and manuals containing military "catechisms," inspira-
tional stories of heroic Republican soldiers and sailors, and Revolutionary
songs instead of the hymns and prayers of former days. The troops of the
Revolution thus marched for a cause, primarily to defend their country,
secondarily to free the oppressed. They used psychological weapons as well
as physical ones. They sought to win the minds of their enemies and to turn
the people against their rulers. Ideological warfare, unknown in Europe
since the end of the period of religious strife, returned once more.

But it is important to realize that the Revolution had produced a
greater professional skill as well as greater morale. The wild revolutionary
soldiers of the early years rapidly became veterans without losing all of
their early zest. Once more, as in England in the days of Cromwell, professional skill and belief in a cause were joined to produce an unbeatable
combination. The Revolution had been preserved by military action; and
military efficiency was therefore highly prized by the Revolutionary state.
Indeed, the armies of the Revolution were vastly more effective than those
of the past, not only because all ranks were united by a common enthusiasm, but because the old limitations had been broken down by a powerful,
impersonal state.

THE REVOLUTION AND MILITARY TRAINING

The political revolution brought an important revolution in military training. Ever since the coordinated operations of pikemen had overcome the individual skill and courage of knights, the importance of training had been recognized; but, through the reforms of Maurice of Nassau, the training process from the time of the Swiss phalanx of pikes, down to the "line" of muskets and bayonets, had been directed at solid formations of troops. The employment of skirmishing tactics in the era of the Revolution meant that the soldier now had to be trained to operate as an individual as well as part of a group. The whole training process was thus affected; and it brought a parallel revolution in military discipline. This can be most clearly demonstrated in the British Army; but similar changes occurred in the armies of Revolutionary France, of the United States, and in due course in all armies everywhere.

Jägers and Pandours, highly mobile troops of a semiregular nature from the woods and mountains of Southeastern Europe, had fought in loose formation in German armies in the eighteenth century. The experiences of the British in North American fighting in the eighteenth century had led to the formation of a "Rifle Corps" and of companies of "Rangers" in which a completely new form of discipline was asserted. In place of the lash, which had been used brutally to enforce discipline in the old armies, these new regiments fostered the self-confidence and pride of the men and developed a high morale. The British experiment, one not popular with conservative military leaders of the old type, had lapsed after the end of the American Revolution, but during the Revolutionary Wars in Europe it was revived. In 1800 there was formed the "Experimental Rifle Corps." A little later came Sir John Moore's Light Brigade, which he trained by new methods at Shorncliffe. In the words of military historian Sir Basil Liddell Hart, Moore gave "a new spiritual meaning to discipline." His training methods and the new discipline were amply justified in the Peninsular War when the Light Division proved itself.

The appearance of new methods of training and a new discipline was due to a combination of technological and sociological factors. The adoption of the rifle for light infantry (which had been delayed for centuries because rifling impeded the speed of loading) was partly responsible. Much more important was the impact of the political revolution which had occurred first in America and then in France. The citizen-soldiers who composed the "rabble in arms" could be trained to fight together only under a system of discipline which was less degrading to the individual than that of the eighteenth century. Competence in the mastery of more complicated weapons and tactics could not be developed by the brutal methods of the drill sergeant; and the masses, especially those who were beginning to advance from the lowest levels of untutored ignorance, required more enlightened methods of training when they were enlisted in armed forces.

The impact of the Revolution, and at the same time of technological progress, had an even more immediate and obvious effect on the training of line officers. Military engineers and gunners needed some knowledge of science, particularly mathematics, and this had led in the seventeenth and eighteenth centuries to various schemes to give formal education to potential officers. The American and French Revolutions emphasized this trend toward a professionally trained corps of officers. Neither the United States nor Revolutionary France could rely on an aristocracy for a supply of officers, and each was therefore compelled to set up military academies in which entry and graduation alike depended on merit and hard work. L'Ecole Polytechnique was founded in 1795; the year 1802 saw the birth of military colleges in France, the United States, and Great Britain which eventually became established at St. Cyr, West Point, and Sandhurst, respectively. Thus similar forces were operating also in countries which had not experienced a revolution at first hand.

The social and technical revolution which came at the end of the eighteenth century had profoundly affected the nature of armies and of warfare. Along with the fuller employment of the resources and manpower of the state came a new kind of army based on a new spirit and discipline and trained more fully than ever before in the art and science of warfare. So, the increasing participation of the nation in national armies and warfare did not bring an inevitable decline in military proficiency. On the contrary, sounder methods of training the men in the ranks, and improved selection and training of officers, made possible a great increase in the effectiveness of military power in the nineteenth century.

THE RISE OF NAPOLEON

The new mass armies of the Revolution were inherited by Napoleon Bonaparte, who used them to climb to power in France and to perpetuate and spread the political upheaval which had overturned the Ancien Régime. The "little Corsican," a former royal artillery officer, had learned the way to power as a result of an episode of 5 October 1795. Placed in charge of the defense of the Convention against an attack by civilian insurrectionists, Napoleon used field guns ruthlessly in the streets of Paris to repulse the assault. As a result, he came to realize that an ambitious man with few scruples using military force could dominate the state. It was a lesson he was never to forget and one which was to have a profound effect on human history.

The Directory, which had now taken over the government of France, glad to divert the army by foreign ventures, sent Napoleon to command the Army of Italy, which was ragged, half-starved, mutinous, and therefore dangerous to the government. He was ordered to drive the Austrians back across the Alps, a formidable assignment that might well have spelled the end of his career. But the Revolutionary spirit was strong in the mass

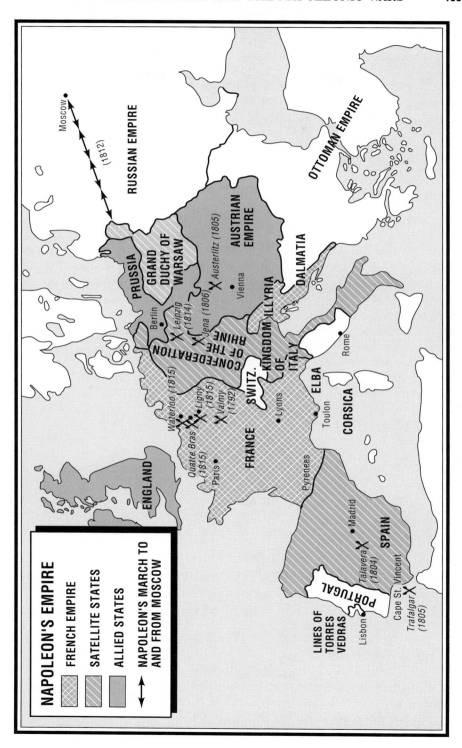

NAPOLEON'S EMPIRE

FRENCH EMPIRE

SATELLITE STATES

ALLIED STATES

NAPOLEON'S MARCH TO
AND FROM MOSCOW

RUSSIAN EMPIRE

OTTOMAN EMPIRE

(1812)

Moscow

PRUSSIA

GRAND
DUCHY OF
WARSAW

AUSTRIAN
EMPIRE

DALMATIA

Austerlitz (1805)

Berlin

Leipzig
(1814)

Jena (1806)

Vienna

ILLYRIA

CONFEDERATION
OF THE
RHINE

KINGDOM
OF
ITALY

Rome

SWITZ.

ELBA

Waterloo (1815)

Ligny
(1815)

Valmy
(1792)

Lyons

CORSICA

Quatre-Bras
(1815)

FRANCE

Toulon

Paris

Pyrenees

ENGLAND

Madrid

SPAIN

Talavera
(1804)

Cape St. Vincent

PORTUGAL

LINES OF
TORRES
VEDRAS

Lisbon

Trafalgar
(1805)

armies which the Revolution had called into being, and Napoleon at once showed both of those personal qualities that made him a great leader of men, and also his genius for strategy. He almost always tried to attack where he had local superiority. In 1796 the ragged army followed him into Italy, forced the Piedmontese to make peace, and expelled the Austrians. At the engagement at the bridge at Lodi, he deliberately blooded his troops by a head-on assault when he knew that there were fords which would have provided an easier route. By that time his personal magnetism had won the hearts of his men, and they were prepared to follow him anywhere. In due course, the armies which the Revolution had created made their new leader emperor of France and thus turned the clock back to autocracy. For nearly two decades they fought to conquer Europe in his name.

WHY NAPOLEON TRIUMPHED

Napoleon's campaigns and battles have been studied by a multitude who have sought the secret of his genius. No one can deny him a high place among the great captains of world history as a master of the art of strategy; and his personal military reputation is, indeed, only enhanced by the fact that he inherited all his weapons from his predecessors. The arms which Napoleon's troops used were not new and therefore do not in themselves explain his achievements. The musket and bayonet of the infantryman had changed little since Louis XIV's time; Napoleon's artillery came to him as a result of the reforms of Gribeauval in the last years of the Ancien Régime, and he made little improvement in it; and although he introduced heavier defensive armor for his cuirassiers and furnished more of his light cavalry with lances to overreach the bayonets of opposing infantry, his mounted troops were armed little better than those of earlier centuries. Napoleon mastered Europe with weapons available to other men before he rose to power.

Indeed, in some ways he was an arch-reactionary toward new weapons and technological progress in the matériel of war. Although science and research were applied to war under the new regime, Napoleon in 1799 disbanded the balloonists who resulted from it. They might have shown him the direction of Blücher's retreat after Ligny in 1815 and thus have warned him that he would meet the Prussians as well as Wellington's army at Waterloo. Shrapnel had been invented by a British major of that name before 1803. It would have made short work of Wellington's squares at Waterloo, but it was a carefully guarded secret that Napoleon never learned.

There is therefore a stronger case for saying that Napoleon triumphed through a more efficient use of well-known weapons, and by his personal strategic flair, than by the application of science and technology to war. His usual practice on campaign was to sleep until 1:00 A.M., by which time his cavalry had brought in full reports of the enemy's movements. It is significant that he made extensive use of cavalry for reconnaissance. (Frederick the

Great, half a century earlier, had done so only sparingly, partly because he was not able to trust his troops when too far from his immediate control.) The information obtained was carefully outlined on the largest available map of the area and the emperor's plans were then added. Frequently Napoleon himself was to be seen crawling about on the map to indicate dispositions. The tactical handling of the battle was then left to subordinates.

Napoleon was a past master in the art of confusing the enemy by striking where he was not expected. Pierre de Bourcet, a chief of staff of the royal armies in both the War of the Austrian Succession and the Seven Years' War, and from 1764 the director of the school for staff officers at Grenoble, had taught that an enemy could be misled by moves of different units which appeared to be disconnected but which were actually part of a connected plan. The aim was to compel the enemy to divide his forces, then to fall upon one part before it could be reinforced. This was the basis of Napoleon's strategy. He used it with signal success against the Austrians and Italians in his first campaign; and he rarely departed from some variation of this basic technique. One variation he learned by accident at Marengo in 1800. The battle was won for him by the timely appearance of a division which had failed to concentrate in time for the beginning of the battle, but which arrived fresh when both the French and the Austrian armies were weary. Thereafter, Napoleon often held reserves back until the enemy forces were worn out; and at Austerlitz he deliberately repeated the strategy which had triumphed accidentally at Marengo.

As a rule, Napoleon engaged the enemy in front with a holding attack while a corps (that is, a group of divisions) swung in a wide flanking movement to fall upon the enemy's rear and threaten his communications. When Napoleon heard the guns of the detached force and so knew that it was attacking, he would deliver a hammer blow at a weak spot in the center of the enemy's line. This blow, the decisive one, was normally made possible by a tremendous concentration of artillery fire. It was frequently delivered by a massed cavalry formation. He kept the bulk of the cavalry for use on the charge and sent only divisional cavalry with the flank attack. It was his practice to keep the heavy cavalry, the cuirassiers, in reserve until the later stages of a battle for the final shock charge. For infantry assault he always employed heavy columns, the tactical formation more properly called "column of divisions," which derived directly from the methods of the generals of the French Republican armies. Column of divisions was a formation in which the three battalions of a regiment advanced one behind the other with a front of one hundred muskets and a depth of nine ranks in each battalion. It was normally preceded by a screen of *tirailleurs,* or skirmishers, whose fire protected the column until the moment of impact.

These strategical and tactical methods were not entirely his own. He borrowed them from earlier theorists, and in some cases they had already been tried out in action. Mobility, divisional organization, heavy concentrations of artillery, and (according to some interpreters) a tactical formation called *ordre mixte* were the keys to Napoleonic victory. Each of these had been advocated earlier. Guibert, in his *Essai général de tactique* (1770) and *Défense du système de guerre* (1779), had stressed the need for greater mobility and had advocated divisional formations. It has been shown that divisions had been used by Marshal Saxe in the War of the Austrian Succession. Guibert had also proposed the *ordre mixte,* in which the battalions were drawn up alternately in line and in column, thus combining the shock of the column with the fire of the line. Napoleon borrowed it from Guibert but appears to have used it only for those parts of his battle line which were detailed to hold the enemy troops opposed to them. For the assault, he used his heavy columns.

It is frequently said that Napoleon won his battles by the weight of his bombardment. He was, of course, an artillery officer and had first attracted attention at Toulon in 1793 by his command of the guns. His belief was that while the infantry was the main arm of an army, it could not stand up to superior artillery. But even here Napoleon was a borrower rather than an innovator. The Chevalier du Teil, Napoleon's superior in command of the artillery at Toulon, had urged in his *L'usage de l'artillerie nouvelle dans la guerre de campagne* (The Use of the New Artillery in Action) that artillery be concentrated at the point of attack and not dispersed along the whole line. This practice was followed by Napoleon, and he used the big guns to blast a hole in the enemy's line into which the infantry could penetrate. As time went on, and as the quality of French conscripts deteriorated, he increased the proportion of artillery in his armies and relied more and more on bombardment. In 1801, he replaced the civilian drivers of the gun teams by soldiers. Napoleon raised artillery from the status of an auxiliary to that of equality with infantry and cavalry. Nevertheless, his use of artillery was only a contributing factor. It is not the sole key to his domination of the battlefields of Europe for nearly two decades.

The increase in the size of armies and the practice of operating with several detached corps or divisions greatly increased the difficulties of command. Napoleon had a chief of staff, but that officer acted more in the role of what would today be called an adjutant general. He also had chiefs of artillery, of engineers, and of the quartermaster department. Each of these officers had a large subordinate staff. But there was no organized coordinated general staff as it is understood today. Napoleon did much of the staff work, both operational and supply, himself. He was, in effect, his own chief of staff in the modern sense. It was his genius to be able to direct the operations of his armies in circumstances in which lesser men would have failed.

However, the explanation of Napoleon's meteoric career springs not so much from his own military genius, which is unquestionable, as from the Revolutionary and nationalistic spirit inculcated by the events of 1789. His long run of victories was due to the mass armies which the Revolution had produced and in which it absorbed the surplus population of disorderly city mobs. As Revolutionary fervor faded, those armies were maintained by the use of the frightening power which the new "democratic" state possessed.

The Revolution had conferred upon the government a power of coercion far beyond that enjoyed by the autocratic monarchs of the eighteenth century. That power was used to compel Frenchmen, and even the men of conquered countries, to fill the ranks in the armies of Napoleon as fast as they were depleted by casualties and disease. The First Empire conscripted about three million men for all arms; and few of them returned to civil life while the emperor reigned. He said that he had an income of 200,000 young men a year and, on another occasion, that he "did not take much heed of the lives of a million men." He was apparently never aware of the danger of growing public opposition to his demands, even when he had to employ an army of troops to round up those who evaded military service. Before the Revolution, field armies of 40,000 or 50,000 were already large, but the French imperial armies, freed from the magazine system, grew in size until by 1809 they were over 100,000 strong, very much larger than those put in the field by Louis XIV and the Grand Alliance. In 1812, on his futile invasion of Russia, Napoleon took over half a million men from all Western Europe except Spain, the biggest armed horde on the continent up to that time.

A secondary factor in Napoleon's military success was increased industrial production. Although, as we have seen, he made no great use of new advances in military technology, it must not be forgotten that the huge armies upon which he built his empire could be provisioned and supplied only because production and communication had greatly improved toward the end of the eighteenth century. While the full blast of the Industrial Revolution had not yet reached the Continent from the British Isles, there had already been an expansion in productive capacity. Gribeauval had learned the principle of interchangeable parts from Austria, and in 1785, Nicholas Le Blanc anticipated Eli Whitney's use of it in the manufacture of arms on a mass-production scale, although still largely by hand rather than by machine. Furthermore, distribution was vastly improved by new roads and canals. The emperor left to France the best road system in Europe, one which could be used for both military and commercial purposes. He paid great attention to problems of production and supply. When the war with England and the Continental Blockade cut him off from British industrial resources, he strove hard to further the industrial development of his empire. But he did not understand economics and how to stimulate business, and the net effect of his efforts was insufficient to meet Britain's rapid commercial expansion.

Napoleon's reliance on masses of men and materials remedied a serious omission in eighteenth-century thinking about war. So much emphasis had been placed upon tactical and strategic maneuver by the generals of the Ancien Régime that it had become accepted that relative strengths counted for little. Indeed, Napoleon often won when he did not possess absolute superiority in numbers; but it was usually by the achievement of a temporary local superiority over a part of the enemy forces. On the whole, it is true to say that his victories were due to the intelligent use of mass. His successes caused opinion to turn away from the eighteenth-century tendency to ignore mass and to swing to the other extreme. As the Napoleonic legend grew, the concept of the primacy of mass grew with it.

A fourth consideration in Napoleon's achievements was the ideology which he deliberately spread. Like the armies of the Revolution, those of the Empire always marched "to free the oppressed." Napoleon deliberately sought to destroy the fiber of his enemies by seducing their soldiers. In Italy, he preached liberty and the expulsion of the foreigner even though he was bringing chains and a new foreign dictatorship. After the Battle of Jena, in 1806, the people of Berlin received the emperor with smiles and cheers. He seemed to them to represent liberty, not oppression. The spreading of beliefs and opinions, true and untrue, was not a new weapon in warfare, but after the political and social revolution of 1789 it had come to possess a new importance. Here again Napoleon benefited from the Revolution which he always claimed to be preserving.

As a result of the new emphasis on ideas, the very nature of warfare was changed. The purpose of war came to be the complete overthrow of the enemy instead of, as in the dynastic and commercial struggles of the eighteenth century, the gaining of a limited advantage. Whereas formerly men had fought to ensure a succession or to seize an outlying province, now the goal of war was to carve up the state by major amputations, to revolutionize it, or to drive out its legitimate government; and the ultimate aim of the victor was complete annexation. The partitions of Poland (1772, 1793, 1795) had already shown what could be done to destroy a state completely. The warfare of the eighteenth century, limited both in manner of operation and in objective, thus gave way to war in which military action and objectives alike were "total."

NAPOLEON'S DOWNFALL

The same factors which had made possible Napoleon's conquest of most of Europe led inevitably to his downfall. In successive wars he had defeated the Austrians and the Prussians and had conquered Spain and Italy. In 1807 at Tilsit, he had forced the czar of Russia to come to terms of peace which in effect recognized Napoleon's possession of most of continental Europe except the czar's own territories. Only the British remained hostile. They had destroyed the combined French and Spanish fleets at Trafalgar (1805) and Napoleon, foiled in his hopes of invading the British Isles, had sought to subdue them by decrees cutting them off from trade with all his territories and satellites. However, the attempt to impose this economic policy upon Portugal and Russia aroused forces which eventually helped to destroy him.

In 1794 the British government had suddenly realized that the country, if it had to face the possibility of invasion, would be practically defenseless. In the following years the commander in chief, the Duke of York, undertook a

program of reform which brought more changes in the army than had occurred in the previous two centuries. He increased its pay, improved its drill
and equipment, introduced corps of riflemen, and reorganized military administration. The regimental system that was to become the chief source of
pride, tradition, and morale in the British Army, was retained; but a new
method of feeding recruits into the regular army through the militia was introduced. As compulsory militia service was based on the ballot, the country
had selective service for home defense until the end of the war. The British
attacked through Portugal and Spain with an army which had been thoroughly overhauled since its defeats at the hands of the French republicans.
Led by Arthur Wellesley, later Duke of Wellington, who had learned his craft
in wars in India, this revitalized army defeated Napoleon's marshals in Spain
and in 1813 invaded France.

In 1812, Napoleon had been compelled to retreat from Moscow. Russian
Cossacks, aided by "General Winter," practically destroyed the remnants of
his army. Inspired by this, the Prussians and Austrians rose against their
conqueror, and their arms were strengthened by nationalistic passion which
French arrogance had stirred up. At the Battle of the Nations at Leipzig in
Saxony, the aroused nations employing armies that were professional in nature inflicted a crushing defeat on the emperor. When Paris fell to the invader, Napoleon abdicated. His return from Elba in 1815, and the Hundred
Days which culminated in Waterloo, were an anticlimax. The nations were
aroused and in arms and even had Wellington suffered defeat instead of winning victory at Waterloo, Napoleon could never have regained his old hegemony in Europe.

Napoleon was struck down partly because he had become intoxicated by
his own success and by power to command armies which had actually out-
grown his control, but even more because the ideological fervor which had
raised him up had spread and cast him down again. The nations he had conquered adopted his methods. Mass armies became used universally and
were no longer a monopoly of the French. Thus when Prussia, after Jena,
was limited to an army of eighteenth-century proportions, the chief of staff,
August von Gneisenau (1750–1819) invented a system of reserves by which
the trainees remained liable for mobilization and so available to a sudden
army expansion. This device was possible only because the Prussian people,
particularly the young men, were willing to accept it. Their patriotic hatred of
France had been inspired in a way not possible before the Revolution. The
recruiting of foreign soldiers had to be abandoned because of the poverty of
Prussia. In 1813 and 1814, the *Landwehr* (a universal militia) and the *Land-
sturm* (a kind of *levée en masse*) were instituted. Prussia thus achieved an
army of the new mass type without undergoing a political revolution herself,
but rather as a result of the pressure of the Revolution in France. When
Napoleon was defeated in 1814, his armies had been wasted by casualties and
by shortage of conscripts to a mere 214,000 effectives, while those of the allied
nations numbered more than 325,000. The "big battalions" had changed sides.

Along with mass armies, the continental nations copied French tactics.

Thus, whereas the French had formerly achieved victory by hurling columns against a weak spot in the enemy line which opposing generals, following eighteenth-century practice, had spread to cover all points, the continental European armies all began to use columns covered by skirmishers and backed by concentrated artillery. A penetration no longer achieved the same deadly result.

The British army, however, had adhered to the linear formation in the belief that the new French system was fundamentally wrong and would be successful only against unsteady troops. When Wellington, who had been employed in India after his first meeting with the French Revolutionary armies in Flanders in 1794, went to take command in Portugal in 1808, he was determined to test his theories of how the "line" could defeat the French tactics that had overthrown all the armies of Europe. He was convinced that most European armies were beaten before they started and that well-trained troops, confident in their leaders, could not be defeated by the crude methods developed by the exigencies of the Revolution.

His conflicts with Napoleon's marshals in Spain, and afterward with the emperor himself at Waterloo, were accompanied by, and have been followed by, a long and sometimes bitter dispute about the relative merits of the line and the column. The debate has tended to oversimplify the issue and to exaggerate the difference between the tactical formations used by the two opposing sides. In particular, it has led to an erroneous belief that the French attacked with a whole army, or with whole divisions, in long narrow columns. To some extent this belief is based on the fact that occasionally a French column on the march did bump into a deployed British line before it had time to deploy into its tactical formation of columns of divisions and *ordre mixte*. The weakness of the column attack, as Wellington and the British realized, was that a battalion deployed against it in line could overlap it with musket fire on each flank. The men in the rear of the column could be shot at but could not shoot.

Wellington was never able to muster as many field guns as the French. His plans for the use of linear formation were based on the idea that the line should not be exposed until the enemy came within musket shot, that it should be protected by a skirmish line of riflemen, and that it should be flanked either by natural cover or by cavalry or artillery. His ideal position, as at Waterloo, was behind the crest of a low hill where his men could stand or lie until the enemy, toiling up the slope, came within range. Sheltered by the ground, Wellington's troops were out of artillery fire until the time came for them to advance over the crest and fire deadly volleys. Thus, the British retained the linear formation of the eighteenth century; but there was one significant difference. They reduced it to two ranks instead of three, to spread the effect of its fire power. Despite all the victories of the French based on column formation of nine or more ranks, the thinning of the line was the proper trend for tactical development. Heavy phalanxes had been obsolescent ever since the pike had given way to the musket as the decisive infantry weapon. Battles were now won by fire power and not by push of pike or bayonet. Napoleon himself admitted that the sword and pike of ancient times meant deep formations, but that modern firearms had made shallower formations imperative. Wellington's resolute adherence to linear formation was a sound recognition of that essential truth.

Napoleon had correctly emphasized mass, numbers, artillery fire power, and the psychological weapons given him by the new popular state; but the methods which he used on the battlefield were not really suitable to the new age. He had endeavored to drill the revolutionary mob into massive phalanxes. Such a development could not proceed far in face of the weapons then available.

THE WAR OF 1812

The War of 1812 between Great Britain and the United States might logically have been expected, as a result of the place and time of the conflict, to reproduce many of the features of the new warfare of the Old World. War in North America had never been marked by restraint; the American Revolution had anticipated many of the military innovations of the French Revolution and the Napoleonic Wars; and the ideological bitterness of 1776 had not significantly abated. In Europe, for almost two decades, Great Britain had been engaged in a life-and-death struggle with France in which she had developed new methods of fighting to meet those of Napoleon.

Yet the War of 1812 was a comparatively mild conflict in which the number of participants was relatively small and the temper and pace much less than that of the war in Europe. The explanation for this apparent anomaly is simple. The British were involved in a major European war; their garrison in Canada was never large, even when veterans were diverted to North American service in 1813 and 1814. Canadian militia units were never more than auxiliaries of the regular forces. For the most part, the British contented themselves with holding the Canadian border, blockading the American coast, operating against American shipping, and indulging occasionally in pin-prick raids such as the descent upon Washington in 1814.

The initiative rested with the United States, whose population, wealth, and resources represented a military potential far in excess of what the British could muster in Canada. But the United States went into the war divided. British interference with neutral American vessels on the high seas, the seizure of American sailors who were said to have deserted from the Royal Navy, and an alleged British manipulation of the tribes of the Northwest had been seized by the land-hungry, nationalist "War Hawks" of the West and South as cogent reasons for war. The time seemed opportune for rounding out the Union by the inclusion of those British possessions that had remained within the empire at the time of the Revolution. "The militia of Kentucky are alone competent to place Montreal and Upper Canada at your feet," declared Henry Clay. But to the Federalist party, whose strength lay in New England, the war was a gross error of policy, since it interfered with maritime trade, the lifeblood of that area. New England opposition to the war rapidly led to a movement for secession from the Union. When the war was thus a party question threatening the very existence of the nation, an all-out American drive to win Canada was impossible.

American strategy reflected the sectional origins of the war. The key to Canada was Montreal; if it fell, communications with the entire western portion of the colony would be cut off, and Upper Canada could be taken bloodlessly. Throughout the war, however, American planning was directed more toward the west, that is, to the Detroit and the Niagara frontiers. In part, this was due to the frontier desire to eradicate once and for all the menace of the tribes now united by Tecumseh (1763–1812). In addition, there was a yearning for the fine farmlands of the Ontario peninsula, and also a mistaken belief that the American settlers, who had been moving there in large numbers in recent years, could be counted on for help. Thus, despite an invasion of Upper Canada at Niagara, the American effort was wasted on peripheral campaigns. Montreal went virtually unchallenged. American thrusts to capture the city were defeated in minor, irresolute engagements at Châteaugai (28 October 1813) and Crysler's Farm (11 November 1813).

On both sides of the border there has persisted a legend that militia took the most prominent place in the fighting. The fact is that the contribution of the American and Canadian militia was not great, and the conduct of the Americans, as at Detroit and Queenston in 1812, was notoriously poor. The major exception, the brilliant victory of Andrew Jackson's militiamen over Pakenham's veterans at New Orleans in 1815, came about because Jackson realized his men must not be exposed on the open field. The British foolishly attacked riflemen and artillery sheltered behind strong breastworks and suffered over two thousand casualties for their pains, while the Americans lost only seven killed and six wounded. The true significance of the War of 1812 was that the American army had begun to shake off the faulty militia tradition of the American Revolution; it was regulars trained by Winfield Scott who fought British regulars on their own terms in the fierce battle of Lundy's Lane (1814) in the Niagara peninsula, where both sides claimed victory.

13

THE NINETEENTH CENTURY'S
ILLUSION OF LIMITED WAR

POST-NAPOLEONIC RESTRAINTS ON WAR

A t the beginning of the nineteenth century all the signs seemed to point to an intensification of warfare and of its impact upon society. The industrial and agrarian revolutions had already made possible a greatly increased national war effort; bounding population figures had provided the manpower for larger armies; improvement of communications, even before the introduction of steam railways, had increased military mobility; and the new order of society thrown up by the political revolution had led to embittered international rivalry.

Nevertheless, after the upheaval of the Napoleonic Wars, the century settled down to a long period almost completely undisturbed by major strife. For over fifty years most nations abandoned the mass armies which had grown up during the struggle against the French. For nearly a century British sea power made it unlikely, and perhaps impossible, that a minor international conflict should become an international struggle for world hegemony like those in the days of Emperor Charles V, Louis XIV of France, and the Emperor Napoleon. The nineteenth century was a time of decision in the history of human society or at least in the history of those European nations which had come, by that time, to control much of the world and whose policies therefore affected all humanity. During most of the century there was widespread hope that a way could be found to limit the impact of war upon society. Naturally these hopes and efforts became more urgent as the century progressed and small wars, the threat of greater wars, and the increasing power of destruction became evident. But these hopes and efforts were not new and were not at first caused by frightening portents. They had their roots deep in the past; they represented an alternative to the path to general strife and total war which was eventually chosen. Not until 1914 did it become apparent that the prospects for the limitation of combat were illusory.

Two interesting but very different attempts to restrict war made early in the period must be noted: the European Congress system set up at Vienna in 1815 and the North American Rush-Bagot Agreement of 1817. After the defeat of Napoleon, the victorious allies, to prevent a recurrence of the French threat, arranged for meetings of the great powers whenever the Vienna settlement was menaced. They were designed to make the world safe for monarchy; they made no allowance for possible social change. The Congresses soon came under the domination of the reactionaries. But at Congresses at Aix-la-Chapelle (1818), at Troppau (1820), at Laibach (1821), and at Verona (1822), it was found impossible to obtain agreement upon concerted action against revolutionary movements which had overthrown regimes set up in 1815 in Italian and Iberian states. Austria and France, therefore, intervened independently. When the question of reasserting Spain's authority in her South American colonies was raised in 1823, and Russia was pressing claims to the West Coast of North America, a "Doctrine" announced by President Monroe of the United States, powerfully backed by British Foreign Minister Canning and the Royal Navy, prevented intervention and at the same time destroyed the whole system of Congresses. Spasmodically during the rest of the century, when an international war seemed imminent, the idea of a "congress" of the "Concert of Europe" was revived. In 1878 a congress was to be called at Berlin to settle the question of the disposal of the Balkan territories, which were slipping from the sickly grasp of the Sultan of Turkey; and a second congress at Berlin in 1884–85 dealt with the fate of Central Africa. But these revivals were only a shadow of the grand scheme evolved in 1815 to rule Europe by congresses and so to preserve peace and the status quo. The Concert of Europe had no great permanent value as a check upon war in the nineteenth century.

The second move to restrict war, the Rush-Bagot Agreement, although pregnant with possibilities for the future, had even less immediate effect on the renewal of major warfare. After the war of 1812–15, Britain and the United States, in order to avoid the expense of a naval building race in peacetime on the Great Lakes, such as had developed during the late war, agreed in 1817 to reduce their naval establishment on each lake to a single gunboat. Recollection of this agreement has endured as a magnificent example of the manner in which international problems can be settled by mutual agreement; but it was an unusual case involving only two powers and in a remote area. It was limited to only a small part of the possible causes of tension between the two powers concerned. And even within the area covered by the agreement, the land fortifications which were erected in the ensuing years and are still extant are mute witnesses to the weakness of international agreements as a means of preventing martial preparations.

The decline in war and in preparation for war which followed the defeat of Napoleon cannot be attributed to any organizations or agreements specifically designed for that purpose.

THE REACTION AND REVOLUTIONS

The lull in warlike activity which marked the first half of the nineteenth century was a result of many factors. Primarily, it was a reaction against the prolonged military efforts of two decades of wars which had been the greatest conflict in western history up to that time. People everywhere were weary of war; and the nations were physically and financially exhausted. Europe needed a long period of peace for reconstruction.

Secondly, in the minds of the legitimate monarchs now restored to power and of the upper classes on whom they depended, war was inextricably mixed up with Jacobinism and was therefore dangerous. The suppression of liberal and nationalist movements, the first requisite of their policy, might require warlike measures; but a major military action that might once more arouse popular zeal was unthinkable. At all cost, a new world conflict must be avoided. International police actions, many of which were necessary up to 1849, were moves to forestall the possible rise of another militarist like Napoleon, as well as to check constitutionalism and liberalism.

Nevertheless, the political upheaval which had been spread widely through Europe by French example and French arms could not easily be suppressed. Agitation for constitutional reform and for the expulsion of alien rulers, driven underground by the reactionary Quadruple Alliance of the three eastern monarchs and France, flourished in secret societies like the *Carbonari* in Italy and flared up again in 1830 and 1848, when revolution spread across Europe with the speed of a prairie fire. As time went on, political unrest was increased by social dislocation stemming from the disruptive force of the Industrial Revolution. The condition of the new industrial proletariat eventually drove some workers to think of the seizure of political power, either constitutionally or by violence, as the only palliative. Their masters, the industrial middle classes, were also excluded from power until, aided by worker agitation, they gained political influence by violence in France in 1830 and by constitutional means in Britain in 1832. In each case the workers, feeling themselves cheated, became more restive. Elsewhere in Europe, power remained in autocratic and aristocratic hands, at least until midcentury.

Industrial unrest was not a cause of war at this time. The dangers from social disturbances occupied the western nations internally. At the same time these countries were engrossed with the expansion of their industrial production. In the long run increased output would be applicable to war purposes; but not until later in the century did the competition for materials and markets become a potent cause of war. Meanwhile, the bourgeois regime of Louis Philippe set up by revolution in France in 1830, the free

Belgium which threw off Dutch control in the same year, and the states of North Germany tied together economically by the Zollverein agreements negotiated between 1828 and 1834 made great strides in their industrialization and thus absorbed energies which might otherwise have been devoted to international strife. For many years after 1815, no nation threatened the interests of its neighbors to a degree likely to provoke a general upheaval.

The Napoleonic Wars had demonstrated in various ways the importance of economic factors in modern warfare. However, political philosophers and economists who often led middle-class thought were neglectful of the economic components of military strength. Adam Smith, the high priest of free trade and private enterprise who had published *Wealth of Nations* in 1776, when the revolutionary phase began, had not been guilty of this cardinal sin. Indeed, he had regarded war as the "noblest of all arts"; he had believed in the necessity of a standing army, provided there were adequate safeguards for the preservation of liberty; and he was even prepared to accept the Navigation Acts and other elements of protectionist legislation for the sake of military security. But his successors in various intellectual fields—for instance, David Ricardo, Joseph Hume, Jeremy Bentham, Thomas Malthus, and Herbert Spencer—thought of war only as a malignant disease which impeded the operation of the social laws they strove to formulate. Bright and Richard Cobden, the Manchester Radical politicians who worked to put the principles of free-trade economics into practice, were pacifists. For them war was the greatest of all destroyers; they were not prepared to recognize that military preparations were necessary and entailed adjustments in the national economy in peacetime. In fact, free-trade policy was eminently suited to Britain's position in midcentury, when she had a long lead in industrial production and held command of the seas. She needed cheap food and raw materials and had little use for a large army. The industrial expansion and light taxation which free-trade policies furthered were directly beneficial to her potential military strength.

One of the most powerful deterrents of war in this period was undoubtedly the prevalence of liberal ideas. Liberalism is hard to define, especially in these days, when it has come to be identified in some quarters with extreme left-wing philosophies which are its opposite. In the nineteenth century liberalism implied freedom and was identified with constitutionalism and with opposition to the absolute form of government which had prevailed in the preceding century. This political creed was paralleled in the economic sphere by liberal doctrines which demanded freedom for the businessman from governmental restriction and control and from onerous fiscal policies.

Liberalism, both political and economic, acted as a powerful check upon the war-mindedness of the peoples of Europe in the first half of the nineteenth century. Economic liberalism meant free trade; and liberals, believing that trade barriers were a nonviolent form of international conflict, believed that their abolition would make actual warfare less likely. In general, liberalism stimulated humanitarianism which regarded war as barbarous; and while not all liberals were thoroughgoing pacifists, some were active propagandists of an anti-war doctrine. One of the best examples of pacifist liberalism in action was the negotiation by Richard Cobden of the British free-trade treaty with France in 1860. Even later, when it had become apparent that a new conflict of major proportions was in the making, it was widely believed that, with the establishment of an effective system of manhood suffrage, no nation would freely vote itself into a war that would bring death, maiming, or hardship to a large number of the voters. This line of thought was long persisting. Thus, later still, with universal suffrage, when women electors became a majority, it was expected that they would vote solidly against war. Meanwhile, universal education would in due course have turned the minds of the masses to more civilized pursuits.

TRADITIONAL MILITARY SYSTEMS RECONFIRMED

The fear which the entrenched ruling classes had for rabid nationalism and the preoccupation of the middle classes with the pursuit of wealth were reflected in the military system of the nineteenth century, which in many ways was reminiscent of that found before 1789. Most states discarded the national conscript armies which had beaten Napoleon and returned to regular standing armies, limited in size, and closely attached to the regime.

After Waterloo, the British army was slashed from its strength of 685,000 to 100,000 by 1821; 50,000 of these were "hidden away" in the colonies from the economizing eyes of members of Parliament. The desire for economy was almost the only interest that Parliament took in military affairs, despite the fact that as the years wore on, the organization of the army, and the conditions of life in it, became shockingly bad. A parliamentary commission of 1837 recommended that the thirteen independent agencies of army administration be centralized under the secretary at war (a bureaucrat, not to be confused with the secretary of state for war, who was a politician and a member of Parliament) as an efficiency measure; but this proposal was dropped because it incurred the displeasure of the Duke of Wellington, who felt that the secretary at war would become a "new leviathan." Thus the British began the Crimean War in 1854 with the army organization of 1815.

The army was left to the guidance of those classes who had traditionally led it, the aristocracy and the gentry; it was protected from criticism by the immense prestige of the Duke of Wellington, who lived until 1852. There was little penetration of the officer corps by middle-class elements. Rates of pay for officers were low; only members of the aristocracy and gentry, to whom military distinction appealed, were prepared to buy commissions and expend their own money to keep up a suitable position. The rank and file of the army were even more distinct from society at large. Enlistment was voluntary, but for life; the only way to get a discharge was through remarkably good or remarkably bad behavior. The social structure of the army was thus identical with that of the dynastic armies of the previous century; if anything, discipline was stricter than before, and punishments up to 300 lashes were not unknown. The only training was of the parade-ground variety, with little practice in musketry; the guards regiments, for example, fired only thirty rounds every three years. The British army in India, "afflicted with oriental rot," outdid eighteenth-century armies in cumbrousness. At the siege of Bhurtpore (1808), General Gerard Lake's army was well supplied with meat: 12,000 bullocks on the hoof. Sir Henry Fane invaded Afghanistan (1839) with a retinue of 40,000 camp followers. The camels of his Bengal contingent alone numbered nearly 30,000. His officers' requirements were prodigious: one brigadier had sixty camels to carry what he thought necessary for comfortable campaigning.

The British Army was completely out of step with the unfolding industrial might of Great Britain. Defense had a low priority, and Prime Minister Sir Robert Peel spoke for a majority of Englishmen when he favored "husbanding our resources in time of peace" and trusting to "the latent and dormant energies of the nation."

In France, under the restored Bourbons, there was an absolute reaction from the Napoleonic army. The Charter of 1814, while preserving many of the political gains of the Revolution, sharply restricted the franchise and at the same time renounced conscription. Although the peacetime strength of the army was only 150,000, less than under the Ancien Régime, voluntary enlistment did not produce the required number of recruits. In 1818 a limited form of conscription was instituted. Out of the annual class eligible for call-up, 40,000 were selected to serve for six years. Shortly after the middle-class revolution of 1830 the size of the army was doubled and the period of service lengthened by a year. The French Army embarked on North African conquest; but no other important alteration was made. The only other military force in France was the civilian National Guard, a poorly trained middle-class part-time reserve. Not even in 1848, when the Orleans monarchy was ousted and Louis Napoleon emerged as president, did any significant support appear for a military establishment of Napoleonic grandeur. When a motion for the restoration of universal military service was proposed to the Constitutional Convention (1850), it was rejected by a vote of 663 to 140.

Of the great powers, only Russia and Prussia retained the principle of the mass army. In Russia, where there was no semblance of a middle class or of constitutional rule, military policy was an emanation of the czar's will. It was Czar Alexander's belief that the Russian army should always be equal in size to the armies of Prussia and Austria combined. Thus selective conscription was continued; the army was maintained at a strength of about 750,000; and this immense establishment consumed a third of the total revenue of the state. Russian power, however, although feared throughout Europe, was much less formidable than it appeared because the Industrial Revolution scarcely made itself felt in that country until the twentieth century.

In Prussia, the reasons for the retention of the principle of the nation-in-arms were more complex and were related to her position in the German Confederation and in the European power balance. Only by virtue of military strength had she become a force to be reckoned with; in population and resources she was still far behind other major nations. Prussia's need for security confronted her conservative leaders with a dilemma. They feared the revolutionary potential of the mass army so much that many of them were prepared to sacrifice a strong foreign policy in order to secure internal safety. The leaders of the left, however, were strongly nationalist as well and desired the retention of the reforms of Karl Baron von Stein (1757–1836), General Gerhard von Scharnhorst, who established the Prussian General Staff in 1806, and Gneisenau, his collaborator and protector. The result was compromise. The principle of the nation-in-arms was adhered to by conscription, but the military budget permitted the recruitment of only a third of the class. The *Landwehr,* the militia, into which the conscripts passed after completion of three years' service, was, because of its democratic tradition, kept separate from the regular army. The officer corps of both forces was virtually closed to all those whose fathers had been enlisted men, workmen, or retail tradesmen. Thus the dominance of the aristocracy was ensured. Prussian military power was relatively great. The regular army was maintained at a strength of 125,000; in addition, a third of the male population received sufficient military training to enable its rapid commitment in time of war. Prussian expenditure on military matters was much higher than that of other European powers, amounting at times to more than half of the annual revenue.

INTERPRETERS OF NAPOLEON'S SUCCESS

The restoration of the conservative military system of the eighteenth century in place of the mass nationalist armies of the Revolutionary era could not entirely exclude the influences of the recent past. Napoleon's genius had irrevocably transformed warfare and could not be effaced. Since he had neither set out his ideas in any complete form nor analyzed his methods, his many unrelated utterances about war tended to suggest that his victories were the result of the manipulation of the brute force of the Revolution by a prodigy, rather than of scientific military operations. In the eighteenth century men had studied tactics and administration but, as Saxe had pointed out, had not uncovered the basic principles of strategy. Not until Napoleon destroyed the eighteenth-century system of making war did military men strive to discover the principles of war by analyzing the campaigns of the master and of other leaders. The two most important studies of this kind were Clausewitz's *Vom Kriege* (On War), 1831, and Jomini's *Précis de l'art de guerre* (Digest of the Art of War), 1836. The former, which will be discussed later, had a delayed influence leading to a new view of the nature and impact of war. The latter, on the other hand, won immediate acclaim and directed the attention of soldiers to what were really pre-Napoleonic concepts.

Antoine Henri Jomini, a French-Swiss who began his career as a bank clerk in Paris, had entered the French Revolutionary army as an unpaid supply officer and had attracted the attention of Marshal Ney by his agility of mind. Ney encouraged him to publish a volume in which he compared the generalship of Frederick the Great with that of Napoleon. Jomini served as Ney's Chief of Staff in Spain but later defected to the Russian army. After the war he studied the campaigns of Napoleon and became recognized as the leading interpreter of the emperor's military methods.

Jomini analyzed and defined the various kinds of warfare and described different types of military operations. He disentangled strategy from tactics and logistics, showed the importance of strategic planning, and established its fundamental principles. He was the father of modern methods of teaching military science, setting down the general divisions of operational study and distinguishing and describing specific types of operations in a fashion that became universally adopted.

Jomini believed that war was not a state of confusion out of which a leader like Napoleon arose as a result of personal genius, but a field of human activity in which rules obtained much as elsewhere. He argued that Napoleon had simply used, successfully, principles that had always been applicable to war. He sought to set down those principles for the guidance of future commanders. He believed that a commander should seize the strategic initiative; should maneuver so as to impede the enemy's lines of communication and supplies without endangering his own; should concentrate the bulk of his forces against the decisive point, taking care to attack only a portion of the enemy's forces; should achieve victory by use of mobility and surprise; and should follow rapidly in pursuit of the defeated foe. In the course of his discussion he laid emphasis on such military concepts as lines of operation and zones of operation, concentric and eccentric maneuver, and interior as opposed to exterior lines of attack. He was the first military theorist to note the distinction between interior and exterior lines; and he regarded the former as being much the more likely to bring victory.

By his emphasis on lines of operation Jomini, in effect, returned to the eighteenth-century method of approaching the study of war as a geometric exercise. He was fond of diagrams. This emphasis on the diagrammatic approach to strategy tended to depreciate the importance of psychological factors in war, even the factor of surprise, which Jomini had, in fact, included in his discussion. It led him also to underestimate the worth of such new military formations as skirmish order. He thought the value of skirmishers was merely to make a noise; it was the heavy columns which followed them up that carried a position.

Jomini had never fully realized that with the Revolution and Napoleon a new age had dawned in warfare. In emphasizing the continuance of traditional features, he missed the things which were new. There can be no doubt that this interpreter of Napoleonic warfare actually set military thought back into the eighteenth century, an approach which the professional soldiers of the early nineteenth century found comfortable and safe.

MIDCENTURY REVOLUTIONS AND LIMITED WARS

The revolutions of 1848–49 showed conclusively that the existing military establishments were capable of snuffing out the crude military efforts of amateur ideological zealots. But soon after the midcentury a succession of wars demonstrated that the old order was fast disappearing. The factors which had prevented a major European conflict were wearing thin. The nations had long recovered from the exhaustion caused by the Napoleonic struggle. Buoyed up by increasing industrial prosperity, they had begun to take hesitating steps to ameliorate the worst aspects of their social problems. After the firm suppression of the revolts of 1848, the possibility of a successful rising against autocracy was remote; and while some workers took up subversive plotting in secret, more eventually turned to industrial trade unions as a solution for their grievances. However, a gradual increase in general prosperity took the edge off the bitterness of class conflict. At the same time international problems diverted much dangerous energy into patriotic channels. Inevitably, problems which defied settlement by other means now led to armed strife between nations.

The first conflicts between major powers which broke the long nineteenth-century peace, although revealing some features which were in the future to bring about a revival of great nationalist struggles like those of the Revolutionary era, were in the main limited wars of the pre-Revolutionary type. The first of these wars, that fought in the Crimea against Russia by England, France, Turkey, and Piedmont, was caused by the disintegration of the Ottoman Empire, which left a power vacuum in the Middle East. It is sometimes described as an "unnecessary war" because the pretext for Anglo-French intervention, the Russian invasion of the Balkans, had ended before the fighting began. The war continued because public opinion in England and France had been inflamed and would not be appeased without bloodshed. Whether this war was necessary to delay eventual Russian expansion into the Mediterranean to challenge British and French sea power is still a matter of contention. But the war was limited. The contestants were too far apart for major engagements; action in the Baltic was limited to naval operations; and elsewhere Russian sea power was too weak to prevent the allies from taking the initiative. The Russians, fearing Austria, kept most of their troops in Poland. The allies made a major raid to eliminate the new Black Sea naval base, Sevastopol, and found themselves in a winter campaign far from their bases, assaulting a city which had not first been blockaded. But, in the absence of railroads, the Crimea was equally far removed from the heart of the Russian Empire.

With such obstacles, this war could only be limited in objective. There was no prospect that it would end in anything but a negotiated peace.

The Crimea exposed astonishing weaknesses in the professional armies of the great powers, rusted through long disuse. The British had had a kind of staff college for over fifty years; but 206 staff jobs out of 221 were given to men who had not taken the course. Command appointments were made on the basis of wealth and birth rather than merit, experience, or even seniority; personal feuds lessened any hope of effective cooperation between units; allied cooperation was poor; the army had not been trained to operate in any formation larger than a battalion; it had some difficulty in finding its way in the Crimea because it had no adequate maps; and tactical blundering led to one of the most famous military failures of all time, the charge of the Light Brigade. The supply organization faltered and almost collapsed. Horses were landed without adequate fodder because of shortage of transports. The British, despite their great lead in industrial engineering, had not expected to have to lay a light railway to carry supplies from the port of Balaklava to the front five miles away. They possessed no ambulances and no adequate medical supplies or organization: one hospital had only twenty chamber pots for 1,000 men with diarrhea. Though the French had wooden barracks, better hospitals, and litters for their wounded, they were in little better condition, but suppression of the press by Napoleon III hid the extent of their privations. It was only because the Russians (whom Lord Raglan persisted in miscalling "the French"), with no railroads south of Moscow, could not take advantage of the invaders' weaknesses that Sevastopol was ultimately captured.

There were some signs, even during this war, of change. Florence Nightingale's by-passing of senior officers, who could terrorize men but could not handle an accomplished lady of gentle birth, brought about the provision of good nursing and cut the death rate in hospitals from 42 percent to 2 percent. A new feature in war were the detailed reports which William Howard Russell sent by telegraph to his paper, the *London Times.* This powerful medium provided information which army commanders described as treasonable because it revealed the incompetence of military authority. The stories brought about the fall of the Aberdeen government and began a new era in which military commanders in remote areas were increasingly subjected to supervision by political authority.

Within a few years a second limited war again demonstrated that the restraints upon war had not yet been entirely removed. In the Italian War of Liberation in 1859 the objective of Piedmont was the liberation of Italy, an objective which was clearly total in nature. The Italians had no hope of overcoming their Austrian oppressors without the aid that they obtained from Napoleon III of France. He plunged into the war to satisfy his thirst for glory and his vague liberal aspirations. The French emperor found, however, that he had alienated many conservative Catholics by allying himself with the Pope's rival for authority in Italy, the King of Piedmont. Hence, after the battle of Solferino, he withdrew from the war, leaving the Piedmontese to make a peace which won them only part of their desires.

Yet this war in Italy, like that in the Crimea, although limited in result, showed some signs of the changing forces in warfare. Napoleon III and his generals believed that they had beaten the Austrians by the weight of their columns of bayonets. Actually the attack was pushed home only because the enemy had been demoralized by the devastating bombardment of new French rifled artillery and because the columns were protected by clouds of riflemen firing from cover. Colonial campaigns in Algeria had given the French experience in skirmishing tactics which proved more effective than the traditional mass attack of the Napoleonic pattern.

By far the most important result of these two limited wars was that they brought home to many Europeans a realization that international conflict was likely to be far more devastating than the great struggle against Napoleon. It became clear that liberal ideology, when combined with nationalism, was a force which could work for war instead of for peace. In all the midcentury wars liberalism, or the frustration of it, was an important *casus belli*. The Italian war was a war of liberation. The Crimean war appeared to some people as a war between western liberalism and eastern autocracy. The American Civil War was hastened by the movement to free the slaves. The Prussian wars were skillfully planned by Bismarck to recruit liberal nationalist sentiment for his policy of Prussian expansion.

As a result there was an adjustment in the liberal attitude to the problem of war. In place of a vague belief that great wars would be impossible if liberalism triumphed, there grew up a positive and vigorous attempt to curb the impact of war upon society. The sufferings of the combatants in the Crimea, caused more by political and military bungling than by the destructive powers of weapons, had aroused the humanitarianism of Victorian England. Henri Dunant, a Swiss observer at the Battle of Solferino, seeing the greater range and accuracy of the new rifled cannon used by the French, and the dreadful effect of smoothbores firing grape and canister at close range, wrote a famous book, *Un Souvenir de Solferino* (1862), which was an important factor in bringing into existence the Red Cross by international agreement. At Geneva in 1864, twenty-six nations agreed to respect regulations governing the care of the wounded, the rights of prisoners of war, and the protection of medical supplies and of hospitals covered by the new Red Cross flag. About the same time, in America, rifled weapons and mass armies were producing new horrors which taxed beyond capacity the limited military medical facilities of the day. Conditions in field-dressing stations and hospitals of both Federal and Confederate armies were as bad as in the Crimea. British observers said they were actually worse. But lessons were learned in the bitter school of experience, and military medical services were expanded and improved. An American Nightingale, Clara Barton, fought for better conditions for the wounded and went on to campaign for American acceptance of the Red Cross Convention, a feat accomplished in 1882.

CODES FOR MILITARY CONFLICT

Thus, the mid-nineteenth-century wars, largely as a result of liberal agitation, led to the first of a long series of international conventions or "laws" relating to the conduct of war and designed to protect combatants and noncombatants by imposing limitations on the use of military force. These rules were introduced at the very time when wars were becoming more frequent, more bitter, and more destructive. They may be regarded, therefore, as a reaction against that trend and an effort to curb it.

The idea that the conduct of the participants in war was subject to the control of law was not new. The laws of war are a part of "international law," the long history of which has been referred to earlier. They were different in kind from "municipal law" (the law of individual countries) because the latter is enforced by sanctions, or penalties imposed by a superior authority, and no such superior authority existed in the international sphere. International law is, nevertheless, a real and significant force. It is derived from the writings of great international jurists, from the decisions of international courts, and from agreements between two or more states.

Modern "laws and usages of war," which have developed from these sources, are subject to several influences or principles. In the first place, it is arguable that a belligerent is justified in using any amount and any kind of force to overcome the enemy; but humanity prohibits the use of more violence than is necessary to achieve this end. In addition, a tradition that all soldiers are brothers-in-arms has served to introduce the idea that there are "fair" and "foul" methods of fighting; and a spirit of chivalry has meant that incapacitated combatants and noncombatants, especially the aged and women and children, should be given reasonable treatment. Within the limit set down by these principles, the rules of war which gained acceptance during the latter part of the nineteenth century were specifically designed to moderate the incidence of war upon society. Actions which were harmful to individuals were prohibited, subject, however, to the overriding qualification, "insofar as military necessity permits."

Generally speaking, the nineteenth-century movement to impose restrictions on warfare served to confine hostilities to the armed forces, to prevent wars of attrition, and to localize wars. Inevitably these restrictions tended to make wars short but frequent, and to favor aggressors. They were on the whole acceptable to military men as well as to the liberals. The net result of the coincidence of the desires of the humanitarians and the militarists was that the movement to impose some restraints on warfare enjoyed a certain degree of success. Quarter was given in battle. As in the Middle Ages and the eighteenth century, rules were set down for the capitulation of beleaguered fortresses in which bodies of troops were hemmed in by greatly superior forces. Even in the American Civil War, which was far from being a limited conflict, the parole system was regularly employed. (A suburb of Annapolis is named Parole because it was the site of a large Civil War parole camp.) But Lincoln ended parole and exchange in 1863 when he realized that the war was one of attrition. In the Franco-Prussian War prisoners were released on the condition that they should not serve again. The definition of the rules of war had led, however, to the summary punishment of those who contravened them. Released prisoners who were captured again in battle were shot out of hand. *Francs-tireurs* who fought without uniform were arbitrarily executed.

CODES FOR SEA WARFARE

Because the rights of neutrals were more likely to be infringed as a result of naval operations, the attempt to limit warfare by international agreement had a more important, and in some ways more successful, history when applied to warfare at sea. At the outbreak of the Crimean War, in order to prevent a recurrence of the blockade which had throttled Europe half a century earlier, all belligerents were called upon to state their policies. All announced that they would not issue letters of marque to privateers. Britain stated that she would not seize enemy goods on neutral vessels, while France promised not to seize neutral goods on enemy vessels. The concessions by the western allies were severe handicaps upon their superior naval power. It is probable that they were made only because Russia was unlikely to present a real challenge at sea.

These limitations upon the methods of conducting naval warfare were originally intended only for the duration of hostilities. However, at the Peace Conference in 1856, the powers present issued the Declaration of Paris, which laid down permanent rules for the conduct of naval operations. Privateering was abolished for all time; a neutral flag was held to cover and protect all enemy goods except contraband of war; and neutral goods, except contraband of war, when carried in enemy ships were not liable to confiscation. It was also ruled that naval blockades must be effective to be recognized as legal—that is to say, they must be maintained by sufficient force to prevent access to a blockaded coast. This was a repudiation of the British doctrine and practice of "declaring" blockades which were maintained only intermittently.

The temporary renunciations made by the powers for the duration of the Crimean War, which were made only because that war was not primarily a naval conflict, were converted into general rules to limit all future naval operations. These rules were produced just at the time when steam power, ironclad ships, and rifled guns were revolutionizing war at sea. Before long, the submarine was to appear, to assume the part formerly played by the privateer in naval strategy. It is probably safe to say that the outlawing of privateering remained effective simply because privateering was no longer considered by governments to be a useful method of making war. Other limitations upon naval operations—for instance, blockade—remained subject to dispute because of difficulty of definition and because technical developments modified the conditions of naval warfare.

The first real test of the efficacy of rules to govern naval warfare came with the American Civil War, but in circumstances which distorted the validity of the test case. Traditionally, the United States had been nurtured on the doctrine of the "freedom of the seas," even in time of war. During the Napoleonic Wars the British blockade of Europe had been strenuously opposed by the United States, whose sailors rejected British claims to search American vessels for enemy goods and for deserters. When the Civil War broke out, British vessels were engaged in trade with Southern ports blockaded by the United States. American statesmen were now put in the position of having to justify the search and seizure of neutral British vessels, thus reversing the roles the two countries had had during the Napoleonic Wars. Both the United States and the British, however, were aware that their actions must be governed not only by the circumstances of the moment but also by the fact that precedents were being created. The British, therefore, did not give unqualified expression to the doctrine of the freedom of the seas.

Two important naval incidents during the Civil War led to diplomatic exchanges that had an important bearing on the history of attempts to limit warfare. Britain had allowed a Confederate cruiser, the *Alabama,* which later took a tremendous toll of Federal shipping, to be built in a British port. On the other hand, a Federal warship had halted the *Trent,* a British packet, on the high seas to remove two Confederate agents, Mason and Slidell, who were, however, soon released. This incident nearly led to international war. After the Civil War, the *Alabama* case was submitted to international arbitration. A decision was given in favor of the United States, and the award of $15,500,000 was paid by Great Britain. This was a precedent for the enforcement of international law by the process of arbitration. It followed the settlement in the Treaty of Washington (1871) of outstanding disputes between the United States and Great Britain and Canada, aided by the conviction that war between the North Atlantic powers was unthinkable. Neither country offered any real threat to the other's vital national interests or way of life.

Meanwhile, efforts were being made to extend and strengthen the international codes restricting war both by land and by sea. On 11 December 1868, the Declaration of St. Petersburg, signed by seventeen states, forbade the use of explosive charges in projectiles under fourteen ounces. The cannon shell was not to be matched by the explosive anti-personnel bullet. In 1874, following upon the horror inspired by the siege of Paris by the Germans and the ensuing brutal suppression of the Paris Commune by the new French government, an international conference at Brussels agreed that the bombardment of towns should be prohibited. All civilized peoples approved this restriction. However, during the following decade, a group of French naval officers known as the *Jeune Ecole* (Young School), which advocated commerce raiding as a war strategy, argued that sentimental restrictions should not stand in the way of any method by which a war might be brought to a speedy conclusion. Therefore, sea powers should be allowed to bombard seacoast towns or hold them to ransom. Continental naval powers began to include mock bombardments in their maneuvers. As this practice appeared to favor superior naval power, Britain followed suit. Hence, when the bombardment of open, undefended towns came up again for discussion at the Hague Conference in 1899, bombardment by land guns was prohibited, but the question of bombardment from the sea was referred to a future conference.

Other rules of naval war which needed clarification were the arming of merchantmen and the nature of blockade. After the abolition of privateering, it was generally accepted that the state could acquire and arm merchant vessels and operate them as cruisers. But this still left unsettled the difficult problem of when such a vessel could be converted. Thus, during the Russo-Japanese War, two Russian merchant ships passed through the Dardanelles (which were closed to warships by international agreement) and through the Suez Canal, and then declared themselves to be warships and seized neutrals carrying contraband. Clearly, if vessels could change their status at will, no rule could be effective. A second problem was to arise during the First World War. By international law, merchantmen which were unarmed were protected against attack but forced to submit to search in proper form. When German submarines sank merchantmen, sometimes without any formal warning, the British began to put guns and gun crews on merchant ships for their protection.

The problem of blockade was similarly affected by changes in weapons and in ships. In the days of sail, neutrals and weaker sea powers alike had claimed that, to be legal and effective, a blockade must be "close." Right up to 1914, the French actually argued that the blockading ships must be anchored in such a manner that vessels entering the port could not safely pass. However, with the development of long-range coastal guns, of submarines, and of torpedoes, a "close" blockade of this type became manifestly impossible, and, indeed, some naval authorities argued that blockading was gone forever. What actually happened was that a new technique of long-distance blockade was evolved, operated sometimes from the home ports of the blockading fleet. Such a blockade, created by declaration, was of doubtful legality under the rules that had been accepted. But as a result of the changing nature of war this legal difficulty was overcome by a wide extension of the list of goods classified as "contraband of war."

ATTEMPTS TO LIMIT CONFLICT

At the end of the century, a Polish banker, Ivan S. Bloch, wrote a book forecasting the nature of warfare with new weapons which technology had made available, especially machine guns and heavy artillery (which will be dealt with in a later chapter). The czar of Russia claimed to be so impressed by the horrors that Bloch portrayed that he proposed that the nations should limit armaments, mitigate the horrors of war, and provide a system of arbitration for the settlement of international disputes without resort to war. His real reason may have been a desire to slow down the arms race to give Russia time and capital to industrialize. He summoned a conference of the nations, which met at the Hague in 1899. No support could be found for the proposal of arms limitation; but some progress was made in other respects. However, the decisions on which agreement was most easily reached were those which, in fact, provided loopholes for evasion of any rules that might be laid down. Thus, it was agreed that the laws of war applied only when both sides were parties by ratification to the particular international agreements concerned; and it was agreed that reprisals were permissible in retaliation against enemy breaches in the laws of war. A second conference was held at the Hague in 1907, when dumdum bullets (with soft expanding noses) and poison gases were outlawed. An attempt was also made to forbid the use of projectiles or explosives from airplanes. The only really important result of the Hague Conferences was the establishment of an international panel of justices for the settlement of disputes; and even in that sphere the achievement was limited. Recourse to this International Court was optional and not obligatory. Codes to define the rights of neutrals by land and sea were discussed in 1907 but were not accepted by Great Britain, the power most likely to resort to blockade in the event of war. Accordingly, another conference, held in 1908–9, produced the Declaration of London to govern blockade, contraband, and the rights of search. By 1914, no power had ratified its terms.

Throughout all these endeavors to restrict, restrain, or abolish war, it was evident that no nation was willing to tie its hands in any way that would limit its freedom of action. All paid lip service to the general principle of preventing or mitigating the horrors of war; all issued manuals to their troops outlining the international law of war as they saw it. But every step taken toward limitation of war invariably appeared to one or more of the nations as a potential danger to national security. Any attempt to prohibit a useful weapon was difficult to enforce.

Much of the initiative had come from Russia, which was not the most advanced or liberal of the great states. It is not unlikely that one reason that the czar took the lead in measures to limit war was his belief that Russia was less able than other nations to fight a modern war successfully. However, the more "progressive" powers, for instance, the United Kingdom and the United States, which might have been expected to support schemes to check the barbarism of war, were at least as cautious as other powers. Thus, British caution had rendered ineffective the agreements made at Brussels in 1874. The United States did not accept the Geneva Convention of 1864 until 1882. Captain Mahan of the United States, at the Hague in 1899, voted against a motion to prohibit poison gases. At the same conference the British opposed the abolition of dumdum bullets on the grounds that such bullets had stopping power much greater than that of ordinary bullets and that this was necessary against a rush of uncivilized tribesmen. The plain fact was that the western democratic nations were wary of subscribing to measures which might limit the potential military value of their superior industrial organization. Measures to restrict and limit the impact of warfare in the century between the Great Wars of Napoleon and of Kaiser Wilhelm II thus failed, primarily because national sovereignty prevented any real possibility that adequate restrictions would be adopted or, if adopted, would be effective. Many other nations expressed surprise at a statement in a German military manual of 1904 that the "law of necessity" would justify breaches of the international laws of war; yet, in dire emergency, any belligerent power would probably take any action that it deemed necessary to win a war or prevent defeat.

A similar fate met the preaching of minority groups who held pacifist views and called for the abolition of war. After the middle of the nineteenth century, the industrial workers of Europe were organized into labor and socialist parties whose aim was to obtain a greater share of the wealth they produced. One of the basic tenets of socialists, whether gradualists like the Fabians or revolutionaries like the communists, was that the real enemy of the worker was the bourgeois capitalist employer against whom all the workers of every nation had a common cause. Socialist parties were therefore opposed on principle to imperialist wars between nations. The most outspoken group to take this stand was the German Socialist party, which seemed to be sufficiently powerful to be a real obstacle to any warlike plans the kaiser might concoct. The strongest socialist literary indictment of war was produced by an English writer, Norman Angell, author of *The Great Illusion.* Angell attacked imperialism and argued that colonies were a financial burden and that the conquest of colonies did not pay.

All these movements against war proved singularly ineffective in 1914. When war came, the German Socialists abandoned their former principles and fell into line with the nationalists. In England, sincere conscientious objection to fighting was allowed as a reason for exemption from combatant service; but the number of conscientious objectors was not large enough to hamper the war effort. Everywhere, despite all the prophecies of the horrors of a modern war, the crisis of 1914 was greeted by scenes of popular enthusiasm. Public feeling against war was in fact shallow. Democracy showed that it could be as belligerent as autocracy. The carnage of 1914–18 had to take place before there was really effective popular support for movements to restrain warfare. But the long period of relative peace in Europe had made possible economic expansion, industrialization, and the consolidation of Europe's imperial conquests across the globe. Some of these developments were to be the primary causes of the international conflicts of the next century.

14

THE PAX BRITANNICA

SEA POWER AFTER THE DEFEAT OF NAPOLEON

At the peace conferences which terminated the Napoleonic wars, British statesmen, adhering to a maritime policy, had abstained from acquiring territory on the continent and so avoided continental commitments. By supporting the creation of the Kingdom of the United Netherlands in order to ensure that Antwerp and the Scheldt were not possessed by a major power, they had found a satisfactory solution for an old problem which had frequently embroiled Britain in continental wars. The revolt of Belgium in 1830, and its establishment as a separate, independent kingdom ruled by the future Queen Victoria's uncle and guaranteed in its neutrality by the powers was, for Britain, an even happier solution. In 1837, the accession of Victoria brought an end to the personal union with Hanover (in which the succession followed the Salic Law of male descent) and removed another source of entanglement in European wars, one which, however, had already ceased to have any real importance.

At Vienna in 1815 the British attitude toward acquisition of territory outside Europe was governed by the prevailing doctrine that colonies were, as Disraeli said later, "millstones around our necks" costing far more to administer than their trade was worth. All political parties and groups in Britain in the first three-quarters of the century were influenced by this disenchanted view of empire. However, some reformers saw overseas settlement as a way to ease population pressures, and India, owned by a commercial corporation, the East India Company, was in a different category. The Whigs and Manchester Radicals were confident that British industrial supremacy made tariffs, and the closed imperial commercial system that accompanied them, obsolete; the Tories, although clinging until the 1840s to a policy of agricultural protection, were convinced Little Englanders who saw dependencies as political and military, as well as economic, liabilities. It seemed to most that the ultimate destiny of the white colonial empire of settlement was independence; the bitter lesson of the American Revolution seemed to be confirmed by the abortive rebellions in the Canadas in 1837–38; it was this attitude which made possible the Confederation of Canada in 1867 and its evolution as an autonomous nation within the Empire.

Therefore, after the Napoleonic Wars, colonies seized by British fleets during the war were returned to their former owners at the conference table; but there were certain important and significant exceptions. The British realized that the war had been won by "those far distant, storm-beaten ships upon which the Grand Army never looked [but which had] stood between it and the dominion of the world" (Mahan). They therefore retained those places which, as strategic bases, controlled the world's sea lanes: Heligoland in the North Sea, Gibraltar and Malta in the Mediterranean, Mauritius in the Indian Ocean, the Cape of Good Hope (for which the Dutch received compensation), the South American settlements that later became British Guiana, and Ceylon.

With Sir Stamford Raffle's acquisition of Singapore in 1819 Britain was to be found astride every major sea lane except the Dardanelles; and during the remainder of the century she continued to pick up supplementary naval bases throughout the world, especially those on the routes to India and the Far East. Her most important additional acquisitions were Aden (1839), Hong Kong (1841), and Cyprus (1878). Control over the Suez Canal (opened in 1869) was obtained by the purchase of the Khedive of Egypt's shares in 1875 and by the establishment of a protectorate over Egypt in 1882. In the nineteenth century Britain, and Britain alone, possessed the far-flung bases essential to world maritime predominance.

Other factors contributed to the increased importance of sea power as opposed to land power and hence to the British hegemony. Three centuries of European expansion had augmented the relative weight of overseas possessions in the balance of power; with the American Declaration of Independence there had appeared the first of the non-European powers to have influence in the modern world, and others followed later; sea power, exerted to bridge the oceans of the world and to exploit distant resources, became proportionally more significant. The Industrial Revolution, by increasing productivity, by intensifying the quest for markets and materials, and by improving the means of communication, including those across the oceans, made control of the seas imperative to a power with imperial possessions. British industrial leadership implied maritime leadership as well. Finally, the revolutionary upheavals and Napoleonic Wars had weakened the continental land powers. Trafalgar had thus left the Royal Navy supreme in the world at a time when sea power was of increasing moment and when Britain was the only power able to wield it.

For nearly a century after Trafalgar she had no dangerous rival upon the seas. France was unable to repeat what she had done after the Treaty of Utrecht when she had rapidly built a new fleet of wooden ships and had soon been able to rechallenge British sea power. Her task had now become more difficult because the design of warships, which had changed little since the sixteenth century, began in the nineteenth to undergo a process of rapid transformation; and her industrial capacity was not yet adequate to threaten the British in the new types of war fleets. Russia, whose land power was unreasonably exaggerated and feared from 1815 to 1914, was locked up in the Black Sea until 1871. The United States, whose sailors and merchant marine had been a valuable part of British naval strength when the Thirteen Colonies were British, led the world for a time in the design of merchant sailing ships, those beautiful clippers which represented the climax of sailing-ship development; but, as a result, America lagged behind in steam.

The United States assigned small squadrons to protect her flourishing trade against piracy in the Far East, the Caribbean, and the South Atlantic; in 1845 she established the Naval Academy at Annapolis, Maryland, to give the professional and essentially engineering education that technological change was making desirable for naval officers; in 1846–48 she employed her naval forces extensively in the Mexican War, which brought California and Texas into the Union and, in a number of amphibious operations in the Gulf of Mexico, showed the advantages of steam auxiliary power; and in the next decade she used naval power to back her diplomacy with China and Korea and in the opening of Japan. All this American activity between the War of 1812 and the Civil War was made possible, in large part, because of the Royal Navy's control of the seas, and it was on such a relatively small scale that it constituted no challenge to British naval supremacy. American naval power greatly increased during the Civil War. Then, for a full generation, it fell into the doldrums. National energies were devoted to reconstruction, domestic economic expansion, railroad building, industrial growth, heavy immigration, and opening up the interior of the continent to agriculture.

During the century most British governments were influenced by the numerous disciples of Adam Smith who taught that free trade was the best way to commercial greatness and argued that wealth did not depend upon the possession of empire. The Navigation Acts, upon which British mercantile and maritime strength had formerly depended, were repealed. These artificial props were no longer needed. Such was the margin of Britain's lead in industry, trade, and shipping that her merchant marine and commerce continued to expand. Despite the passivity of her governments, the flag followed trade. The British Empire grew from 20,000,000 people in 1,500,000 square miles in 1800 to 390,000,000 people in 11,000,000 square miles in 1900. Its foreign trade in the same period grew from £80,000,000 to £1,467,000,000. This phenomenal development was in part the product of quietly pervasive strength upon the seas.

British leaders were conscious of their heritage of sea power and of Britain's position as keeper of the seas. After the United States had given the lead by standing firm against the exactions of the Barbary pirates and by defeating an Algerian fleet off Cartagena (20 June, 1815), an Anglo-Dutch squadron commanded by Lord Exmouth bombarded Algiers and put an end once and for all to this scourge of peaceful trade in the Mediterranean. After the slave trade was abolished for British subjects (1807), Britain worked for international cooperation to suppress it altogether; and the task of policing the seas fell largely upon the Royal Navy. The Lords of the Admiralty regarded the oceans of the world so much as a British concern that they undertook the task of surveying and charting them, excluding only the north coast of Africa, which was left to the French, and the east coast of North America, which was done by Americans (many of whom were West Point graduates).

The influence of sea power was exerted without major campaigns or battles, and its most important effect was the negative one of silently preventing the conflicts of continental land powers from developing into worldwide wars. But it was also exercised deliberately to further British policies by the direct use, or the threat, of force. The South American colonies of Spain and Portugal won their independence largely through the work of the British Admiral Cochrane commanding South American squadrons (1808–25). The Monroe Doctrine (1823), proclaimed by the United States to forestall the plans of the Holy Alliance to aid Spain in reconquering South America and against Russian penetration down the west coast of North America as far as San Francisco Bay, was effective to a large extent because it was openly backed by the Royal Navy. The Greeks gained independence from Turkey when Britain, cooperating with France and Russia, annihilated a Turkish fleet at Navarino Bay (1827). Sixty-five out of eighty-one Turkish ships were destroyed, and Muslim power was checked for a century. British sea power ensured the independence of Belgium (1830), bombarded Acre (1840) to prevent Mohammed Ali from conquering or dismembering Turkey, and blockaded the Piraeus (1850) to compel the Greek government to compensate a Gibraltar Jew, David Pacifico, who had claimed protection as a British subject. Sea power made possible the invasion of the Crimea (1854–56) and, as seen in an earlier chapter, was more effective than the land communications within Russia upon which the defending armies depended. Sea power watched benevolently over the unification of Italy from 1860 and forestalled a Russian advance upon Constantinople (1878), compelling the Russians to negotiate the disposal of the disintegrating Ottoman Empire. Throughout the century, in less spectacular places, when British traders were threatened on their lawful business by unruly elements or by hostile, anti-foreign regimes, the dispatch of a gunboat was the usual, and the effective, method of restoring peaceful commerce.

Most of these examples show that British sea power was exercised to foster British trade, but that it also frequently acted in the name of liberty. Small powers, and national groups seeking freedom and independence, often found comfort under the guns of the Royal Navy. The Disraeli-Gladstone debates about the Eastern Question, in which Britain had to choose between supporting the territorial integrity of the "unspeakable Turk" or aiding independence movements of Balkan Christians who, being Slavs, might fall into the Russian orbit, dramatically illustrated the strength of feeling in Britain that her navy should be used in the cause of liberty. This association of sea power and freedom was not illogical, for navies differ from armies in that they can less easily exert influence within the territories of subjected states beyond the reach of their guns. Sea power can pervade the oceans but can only occasionally operate in the interior of continents where rivers are navigable. Only by the use of supplementary military forces can sea power be translated into an oppressive imperial power; and Britain's concentration upon the navy limited the size of her military forces. In these circumstances it is not surprising that the second British Empire, built up after the collapse of the first in 1776, was, like its predecessor, remarkable for the growth of autonomy for white colonial settlers and also for its relatively liberal treatment of other races. Furthermore, British global predominance during the nineteenth century was an important factor in creating the illusion that war could be kept under control.

CHALLENGES DURING THE CENTURY

The Pax Britannica could be maintained as long as England, the greatest industrial nation of the century, could retain her naval supremacy through the flood of revolutionary technical developments in naval warfare that separated Trafalgar from Jutland. The Royal Navy was conservative by nature, and not without reason. Trafalgar had confirmed British confidence in her maritime prowess; and the Nelson tradition was a source of naval strength. Having proved itself in the superior handling of the square-rigged ship-of-the-line, the Navy was reluctant to encourage any change which might make its greatest ships obsolete. (Lord Melville, the First Lord of the Admiralty, wrote in 1828 to the Colonial Office that "steam was calculated to strike a fatal blow at the naval supremacy of the Empire.") Nevertheless, the British had to counter every innovation by a possible enemy. Because of the growth and efficiency of Great Britain's heavy industry based on excellent coal and iron resources, and because for decades the cheapest iron production, the best engines, and the most skilled engineers were all British, effective response to challenge was long possible.

Among the nations which strove to overtake Britain's lead upon the seas, especially the French, the Americans, and the Germans, naval inventiveness sometimes found greater stimulus. The outcome of this naval rivalry was a tremendous revolution in the nature and design of warships—from the three-decker sailing ship like the *Victory* (1759) to the first iron-hulled, armored steam warship, the *Warrior* (1861), and thence to the first battleship of the World War I era, the *Dreadnought* (1905). Unprecedented changes affected the architecture of the ship, its hull and power plant, and its armor and armament. New naval weapons were developed, such as the mine and the torpedo. These developments sprang from advances in technology; and they led from new ships and weapons to changes in tactics and in the composition of fleets, and to important alterations in the strategic role of sea power.

Within the half-century after Trafalgar, improvements in ordnance and in steam propulsion eventually made the square-rigged ship-of-the-line obsolete. The first steam warship, Robert Fulton's *Demologos* (1814), built for the United States Navy to break the British blockade of New York, was something of a freak. It was a catamaran with engine in one hull, boiler in the other, and paddle wheel between the hulls. Armed with thirty heavy 32-pounder guns protected outboard by extremely thick wood sheathing, with two 100-pounder submarine guns which fired under water, and capable of a speed of six knots, it might have achieved the purpose for which it was built had the war not ended before it could be put to the test. Not a seagoing vessel, and with a small fuel capacity, the *Demologos* drew attention to, rather than solved, the problems of adapting steam power to the fighting ship.

Early engines were large and heavy, low in power and high in fuel consumption. Paddle wheels were obviously vulnerable; and their installation replaced a good part of the ship's broadside battery; early boilers were dangerous and a liability in combat; and fuel was so quickly exhausted that the steam engine was long used only as an auxiliary to sails. For these reasons steam was at first adopted only experimentally for tugs and special craft. When an effective screw propeller was invented (1836), it at last became feasible for the three-deckers to use steam as auxiliary power. Tests in which ships were matched in a tug-o'-war demonstrated that the screw was more efficient than the paddle wheel. The first propeller-driven warship was built by the United States, the frigate *Princeton,* in 1844. With no paddle wheels to obstruct the line of fire, she had full broadsides, and with her engines protected by being placed below the waterline and coupled directly to the screw, she was a more serviceable auxiliary steam warship than any of her predecessors. In 1840 the Royal Navy had only 29 steamers; but by 1849 this number had increased to 121 out of a total of 460 ships.

Meanwhile, hoping to challenge British sea power once more, the French had developed a new weapon, the Paixhans shell gun (1824). Its horizontally fired explosive shell threatened to make unarmored warships obsolete. By 1837 the Paixhans shell had been adopted by the French navy; and within two years the English and Americans also armed their ships with shell guns. The 8-inch shell gun took its place alongside the old 32-pounder as the standard heavy naval weapon. By 1840, British ships were also armed with a 68-pounder; and many other experimental weapons were appearing, such as the 12-inch wrought-iron guns of the *Princeton.* The explosion of one of these, named the Peacemaker, killed the American secretary of state and the secretary of the navy and thus tragically demonstrated that further progress in ordnance depended on improvements in metallurgy. These were not actually forthcoming until the Bessemer process of steel manufacture, introduced in 1856, had been sufficiently improved to control the content of carbon.

The first operational test of the changing navies was provided by the Crimean War. At the battle of Sinope (1853) a Russian fleet proved the efficacy of the shell gun by blowing a Turkish fleet of wooden ships out of the water; but when the British and French fleets arrived in the Black Sea, the Russians offered no further naval resistance. Russian admirals were understandably unwilling to match their ordnance against the heavier guns and greater volume of shellfire of the Allied fleets and their softwood ships against oak and iron. Of equal importance was the fact that the Crimean War demonstrated conclusively the operational superiority of the steam auxiliary. In both the Black and the Baltic seas mixed Allied forces found that the sailing ships could be left behind while the steamers went forward about their business. More dependable and more maneuverable, the steam warships were outstanding in the bombardment of Odessa, circling about so as to present a more difficult target to the Russian gunners. They were also used to tow sailing ships into position for bombarding the naval base of Kinburn, where armored French gunboats first revealed the ability of the ironclad to withstand bombardment.

Great Britain now found that her world-circling naval bases were of much greater importance even than she had expected because her steamships depended upon them as coaling stations; and her lead in the iron industry soon was seen to be the true fount of naval power. When France launched the *Gloire* (1859), the first seagoing armored ship, England countered with the iron-hulled *Warrior* (60 guns), protected by 4.5-inch side armor. England's shortage of ship timber, the structural limitations of wood, the increasing weight of ordnance, and the increased displacement necessary to float an armored ship, its ordnance, and its coal bunkers, all combined to make the iron-hulled ship inevitable. The *Warrior* was of such great size, 420 feet overall, that she could not have been built satisfactorily of wood. But wooden hulls persisted for some time, particularly for use in tropical waters, where a copper-sheathed wooden hull would not foul so rapidly as iron, and for smaller ships, especially those of the United States, where there was an abundant supply of wood.

The iron hull, however, not only solved major constructional problems but gave Britain an additional advantage. Iron ships were built by private contractors; and thus the skill and facilities of her great merchant shipbuilding industry were made available to the Royal Navy. The Royal Dockyards, where England's wooden navy had been built, became merely maintenance bases for the fleet. Orders for warships from foreign nations, necessary to make the shipyards profitable, brought further profits to British shipbuilders and made British designs and methods prevail throughout the lesser navies of the world. Later in the century Britain's shipbuilding industry was to retain its world leadership, even when her iron and steel production had fallen to third place. A further change, once again to Britain's relative advantage, was the rate at which warships became obsolescent after 1859. The *Warrior* was launched in the exceptionally brief period of two years after keel laying, but within that short time improvements in ordnance had made a more heavily armored ship desirable. Each decade required a new fleet, an expensive program, which Great Britain could afford more easily than her less-wealthy competitors.

H.M.S. WARRIOR, the first British three-masted, steam-propelled, iron-hulled warship; launched December 1860 to counter the French *La Gloire,* a wooden ship with iron plating laid down in 1859. (Alan More, *Sailing Ships of War, 1800–1860,* Minton Bacch, 1926, plate 89)

NAVAL OPERATIONS IN THE AMERICAN CIVIL WAR

The American Civil War provided the next operational test of machine-age sea power. In that conflict the agrarian Confederate states were severely handicapped, for they had been dependent on the Northern states and Europe for almost all their manufactured goods; and the South had neither merchant marine nor navy. Northern naval strategy was accordingly based on an "Anaconda policy" of economic strangulation of the Confederacy; but could such a blockade be maintained effectively against steam-propelled blockade runners? It became immediately apparent that the long Southern coastline and its many inlets could not be closed unless the Federal fleet possessed conveniently located coaling stations and operational bases. Thus arose the strategic necessity for the capture of Port Royal (between Charleston and Savannah) and of New Orleans, the recapture of Norfolk and Pensacola, and, finally, the attack on Fort Fisher guarding Wilmington, N.C., from which supplies moved by rail directly to Richmond. The completion of this ring of bases, isolating the South from all important foreign aid, contributed indirectly, but significantly, to her defeat.

The Confederacy's efforts to break the blockade were concentrated in a few heavily armored ships capable of demolishing the wooden ships of the Union navy. A first effort was made by the *Virginia,* an iron-casemated ram improvised on the razed hull of the former frigate *Merrimac.* Little more than a floating battery, she demonstrated again that shell guns could destroy wooden ships and so discouraged a naval attack on Richmond; but her encounter with the Union ironclad *Monitor* left the impression that ironclads could not destroy one another. If the Confederate gunners had been prepared

MONITOR AND MERRIMAC—"Naval Engagement in Hampton Roads. . . . ." Engraving (New York; Virtue, Yorkston, & Co., n.d.); copy in Special Collections, Perkins Library, Duke University.

to use solid shot instead of shell, it is possible that they could have destroyed the *Monitor;* and if the Union gunners had known that their 11-inch Dahlgren guns were capable of taking twice the weight of powder charge used, they could certainly have destroyed the ram. Neither ship was a true seagoing vessel, and neither contributed much to the progress of naval architecture. The revolving armored turret of the *Monitor* was the single feature which was to be incorporated in the modern warship. From this and later engagements with Confederate ironclad rams, the obvious lesson was learned that only armored ships could defeat armored ships. This was not a new discovery in England and France, each of which had warships much superior to those used in the Civil War.

The Confederacy also experimented in shallow rivers and harbors with mines which sank more Union ships than did gunfire, and with primitive submarines and spar-torpedoes. Vastly more significant were the successes of her steam-auxiliary commerce raiders. One of them, the *Alabama,* came uncomfortably close to sweeping clipper ships from the seas; and another, the *Shenandoah,* devastated the Union whaling fleet. But fewer in number than the German surface raiders of World Wars I and II, Confederate cruisers had no measurable influence on the outcome of the war. While the American merchant marine was much diminished and never again regained

the place it had held during the first half of the century, its continued decline was more a result of economic factors than a consequence of Confederate commerce raiding. During the war, the United States could not replace
the tonnage lost or transferred to foreign flags to a sufficient extent to
compete with the British.

By contrast, the naval support of the Union armies in their Mississippi
Valley campaign materially shortened the conflict in the up-river fighting.
Ironclad gunboats maintained water-borne logistic support, provided heavy
artillery fire for the troops, and contributed to Farragut's capture of New
Orleans by drawing Southern military power away from this greatest Confederate port. Then the fresh-water and salt-water navies joined in continued support of the land campaign. The result was that the Confederacy was
split apart and the way was prepared for Sherman's great wheeling movement, which contributed greatly to the final victory.

In the year following the American Civil War, the Battle of Lissa (1866)
again illustrated the relative invulnerability of ironclads to gunfire. An Austrian fleet under Tegethoff steamed in wedge formation right through an
Italian fleet of superior armament and put it to flight, having sunk one ship
by ramming. This return to galley tactics led to a mistaken emphasis on the
ram which temporarily influenced all navies. Actually, Lissa had no clear
lessons to teach. A wooden ship, the *Kaiser,* withstood as much punishment
as the ironclads. Only one ship was sunk by ramming, and that was stationary at the time.

As armor was thickened, guns were made more powerful. The
government-owned Woolwich Arsenal's monopoly of gun-manufacture was
ended in 1859, and a global armaments industry emerged in the 1860s, in
which manufacturers in several countries competed. England's development
of the Armstrong built-up gun, of Whitworth's armor-piercing projectile,
and of the Franco-American "interrupted screw" breech mechanism kept
the great gun as the primary weapon of naval warfare. In 1873 the Royal
Navy commissioned the *Devastation,* the first mastless, armored, turreted,
screw warship. She had a waterline belt of foot-thick armor and carried four
35-ton muzzle-loading rifles protected by 14-inch armor plating. In 1881 the
Inflexible carried 24-inch side armor, the thickest ever used, and 16-inch
guns. From that point on armor was improved in quality and reduced in
thickness. A temporary peak in gun size was reached three years later when
some English ships actually carried 16.25-inch guns weighing 111 tons each.
A reaction then set in against size in ordnance. During the 1880s, when
slow-burning powder was introduced, it became possible to fire projectiles
from long-barreled naval rifles with far greater range and velocity. The
naval engagements at the end of the century in the Sino-Japanese War
(1894) and the Spanish-American War (1898) emphasized gunnery and the
need for methods of improving accuracy and range. It was recognized that
a practical balance between armor and armament in any particular ship
must be determined by the overall fighting functions for which its type was
designed.

Behind these technical developments lay constantly expanding industrial capacity and wealth from commercial enterprise. The opening of the Suez Canal in 1869 was a prelude to the new imperialism of the 1880s and 1890s. Great Britain, isolated from European conflicts, remained neutral even during the Franco-Prussian War of 1870. In Europe, favoring maintenance of the status quo, she regarded her fleet as essentially defensive. Elsewhere, she sought new outlets for her products and led the imperial penetration of Africa and Asia under the inverted slogan "Trade Follows the Flag." As late as 1889 Russia and France were still her only important rivals in sea power. In that year her "Two Power Standard" was officially adopted and announced, by which the Royal Navy was to be kept at a strength to match the combined strength of the next two maritime powers.

THE RISE OF PUBLIC INTEREST IN SEA POWER

A new surge of imperialism in the form of worldwide colonial and trade rivalry among the European powers induced Continental states to undertake expanding naval programs, and these of necessity produced larger appropriations for the Royal Navy. It was then, when she seemed to be at the peak of her power, that Britain lost first place as the leading industrial nation. Her wealth was greater than ever; but her share of world trade was relatively less than before. German industrial capacity was increasing rapidly and in direct competition. The United States in 1890 took over England's position as the world's greatest iron producer. In the Orient, Japan was developing a textile industry that would lead to a further challenge to England's commercial supremacy.

All the world's ambitious maritime nations were stimulated in the 1890s by the publication of an American's analysis of British success in an important work, *The Influence of Sea Power upon History, 1660–1783.* In this and subsequent books Rear Admiral Alfred Thayer Mahan set forth a philosophy of naval warfare which, by setting up naval power as the key to imperial success, was powerful "big navy" propaganda. His work was only one feature of an intellectual renaissance occurring in the United States Navy. After the post–Civil War doldrums, the navy had been reinvigorated by a group of officers led by Admiral Stephen Luce, who had reformed the education of enlisted personnel and in 1884 had founded the Naval War College for the advanced instruction of officers. Mahan had been appointed lecturer on tactics and naval history and became its second president. His studies of the influence of sea power had been delivered there as lectures.

Mahan discovered nothing that was not already known. The British had been putting his theories into practice for a long time. But hitherto, naval history had been written merely as a series of thrilling episodes; no one had set down the principles of naval warfare in popular, yet scholarly, form. Mahan described the operation of sea power, analyzed the elements upon which it was based, and examined critically various kinds of naval strategy. He taught that sea power worked silently to build great empires. It was in England that Mahan first attracted the imagination of the general public. Mahan put into words what Englishmen had long subconsciously realized, and gave a plausible explanation of British supremacy. Nevertheless, by stirring up interest in sea power in other countries, including his own, Mahan furthered the growth of rivals to the Royal Navy. His work was translated into many languages and had great influence upon naval development in Germany, Japan, and France.

Everywhere popular interest in naval strength was on the increase. The new imperialist rivalry for trade and colonies was paralleled by competition in naval-building programs. In the 1880s the group of French navalists known as Jeune Ecole argued that a weaker naval power could offset an enemy's superiority by resort to commerce raiding. Up to then the general public had little knowledge of naval affairs, even in Britain, which depended so much on sea power. But this French challenge inflamed British opinion. Captain John Fisher, an ebullient naval officer, leaked information to W. T. Stead, a journalist who published an article in the *Pall-Mall Gazette* which helped to secure increased naval appropriations. Fisher got his way again in 1886 when the Admiralty was permitted to approach engineering firms directly to request the development of a quick-firing gun that the government's Woolwich Arsenal could not supply. The door was thus opened for private arms manufacturers, and to rivalry for contracts by what a later generation was to call the "military industrial complex." In his *The Pursuit of Power* (Chicago: University of Chicago Press, 1982), Professor William H. McNeill has labeled it "command technology." In the ensuing decade, through magazine articles, through the new penny press, and through propagandist organizations like the British and German Navy leagues, popular interest in naval affairs was stimulated. Thus was prepared the ground for the dramatic Anglo-German naval building race which preceded the First World War, and also for a global armaments industry, with baneful effects, but it also stimulated industrial development generally.

At this time American naval men were usually admirers of things British. The British government, for its part, regarded the United Sates as a friendly rival, omitted to take account of the United States Navy in its "Two Power Standard," and cultivated good relations with America, for instance, by accepting arbitration of issues arising out of the Civil War (in which Canadian interests were surrendered). The resulting Treaty of Washington (1881) solidified cordial Anglo-American relations, with subsequent long-term advantage for Canada. America's own brand of the prevailing imperialism made itself manifest in the annexation of the Hawaiian Islands, Guam, the Philippines, and Puerto Rico, and in the so-called Roosevelt Corollary to the Monroe Doctrine, under which the United States policed the Caribbean states and, with the help of the navy and the marines, acted as debt collector for American, and incidentally for foreign, investors where necessary. Theodore Roosevelt encouraged the building of an ocean-going navy which he wished to be second only to Britain's. Later the satisfactory settlement of the Panama Canal question with Great Britain ensured the continuance of amicable relations between the two countries. The net result was that Britain withdrew some naval units from the West Indies and thus was reinforced in European waters without endangering her Central American interests.

In Europe, alarmed by the expensive naval race and the prospects of war, the Russian czar had proposed an international conference to consider naval limitations, restrictions on war, and arbitration. The resulting Hague Conference (1899), as has been seen, accomplished a little in every category except that of naval limitation. Neither Mahan nor Admiral Sir John Fisher of the Royal Navy was surprised at this failure, for to each it was clear that the growing rivalry on the seas could be resolved only by world acceptance of England's dominant position or by England's acceptance of a relatively weaker position. In the atmosphere then prevailing neither alternative was realistic. The spirit of the times was better illustrated in the German Naval Bill of the very next year, by which Admiral Alfred von Tirpitz obtained a "risk fleet," a naval force so powerful that Britain would be able to accept the challenge of Germany only at the risk of weakening her naval strength to the point where she could be defeated by the third naval power, presumably France.

SHIP AND WEAPON DEVELOPMENT

The British reaction to these events was vigorous, if reluctant. During the 1890s, by comparison with energetic American, German, and Japanese naval leaders, the British Admiralty had seemed at times complacent. A myth about remarkably accurate shooting by American ships in their war with Spain (the fact was that, under almost ideal conditions at Santiago, less than 4 percent of the shots were hits) aroused interest in gunnery. International censure of Great Britain during the Boer War, coupled with an invasion scare, likewise caused the Royal Navy to look to its laurels. Beginning in 1901 came an era of reform. Instead of expending ammunition by throwing it overboard, as one critic alleged was done, the fleet turned seriously to gunnery practice. In Captain Percy Scott it found an expert who, like Captain Douglas in the eighteenth century, proved that much more could be done with modern naval guns than most officers had realized. His idea of "continuous aim" was followed by the development of the director system for salvo gunfire. By that system the accuracy of naval guns at ranges of 10,000 yards was to be improved tenfold by the time of World War I. The German, Japanese, and American navies were making comparable improvements on their own account with equal, or even greater, success.

In the field of weapons, the improvement of the self-propelled torpedo—designed as early as 1860 by Robert Whitehead—and its use as the
offensive weapon of the submarine, extended naval warfare into a new dimension. Although the idea of the submarine was ancient, the practical development of a submersible ship was not achieved until almost the end of
the century. By 1898 French submarines took part in maneuvers. Two years
later the American Holland submarine was purchased by both the United
States and Britain. In tests both the submarine and the torpedo boat appeared capable of revolutionizing tactics and of giving a lesser sea power a
weapon which could force a major fleet to take up a purely defensive strategy. The fast torpedo-boat destroyer was developed in reply; but only in
actual warfare could the relative merits of these vessels be properly tested.

After the Boer War, Britain's diplomatic relations improved with every
nation except Germany, which had become Britain's chief rival. In 1902 a
treaty with Japan counterbalanced Russian naval strength in the Pacific.
Less than three years later an entente with France gave England a free
hand in Egypt and effected a general reconciliation, possible in large part
because the French had dropped from the naval race by default. With the
German menace in mind, Britain made all the moves.

The Russo-Japanese War demonstrated what twentieth-century naval
power could do. For Mahan the war was a showpiece of naval strategy,
illustrating the dangers of a divided fleet and of confusion in objectives. The
Russian fleets at Vladivostok and Port Arthur were eliminated in detail.
When Admiral Rozhestvensky's Baltic fleet, burdened with coal and convoy-
ing supply ships, arrived in the Pacific, it was overwhelmed by Admiral Togo
at Tsushima (1905), a naval victory comparable with Trafalgar. For Fisher,
the vigorous newly appointed First Sea Lord at the British Admiralty, a man
of great prescience, the tactical aspects were of immediate practical interest.
Both sides had scored hits at up to 10,000 yards with the heaviest guns doing
the most effective firing. Neither the torpedo nor the submarine had played
any conspicuous part, although elsewhere mines had been employed in shallow waters with considerable success. In every engagement in the war superior speed permitted the Japanese to control the course of the action.

Of speed Fisher said, "It is the 'weather gage' of the olden days. You
then fight just when it suits you best." He proposed a fleet of 21-knot
battleships, 25-knot cruisers, 36-knot destroyers, and 14-knot submarines.
In his view, with such ships, and holding the five keys that lock up the
world, Singapore, the Cape, Alexandria, Gibraltar, and Dover, Britain
would be secure. As a result of the experience of Tsushima the main instrument of his new navy was to be the "all-big-gun" ship, the first truly
twentieth-century battleship, and the prototype of all battleships built
since. With a displacement of about 18,000 tons, Fisher's *Dreadnought* carried ten 12-inch rifles in five turrets, three on the centerline and one on
each beam, giving her a fire power equal to that of any two other battleships afloat at the time she was launched. When the improved fire-control
director system was installed, her guns could score hits at ranges up to

20,000 yards. Her eleven-inch armor belt was two inches thicker than that of other battleships, and the first battleship turbines drove her at 21 knots, two knots faster than any foreign battleship building or afloat. Extensive watertight compartmentation and facilities for burning oil or coal (oil would give one-third greater cruising range) marked her as the most advanced warship built up to that time. Fisher's innovations revolutionized naval warfare and readied Britain for war.

The tactical developments of World War I, culminating a century of experimentation, were clearly envisaged by Fisher; and his views were shared by the most competent leaders in all the world's major navies. The backbone of the fleet was the battleship which would, for maximum gunfire concentration, be fought in line of battle but at great distance from the enemy. It would be supplemented for scouting and special duties by the battle-cruiser, a ship with the high speed and relatively light armor of a cruiser, but with the guns of a battleship. Cruisers would be the eyes of the fleet or could be detached like the frigates of the days of sail. Destroyers would screen the fleet from attack by submarines and torpedo boats and would use their own torpedoes as offensive weapons. The submarine, Fisher believed, would revolutionize fleet actions. However, while the submarine extended offensive operation even into the anchorage of the battle fleet, the mine and the torpedo gave weapons to the defensive.

Great changes had come also in the field of strategy. Close blockade was no longer possible, but "distant surveillance by the main fleet in well defended bases" would be supplemented by mine fields, submarine pickets, and aerial observation. The vulnerable flank was seaborne commerce; for no modern western nation was self-sufficient. Allied exclusion from the Baltic and Black seas was to contribute to the defeat of Russia just as surely as Germany was herself to be brought to defeat by the Allied blockade. Hence, as long as the elimination or control of all neutral channels for supplies was possible, close blockade was not necessary. In a war of attrition, a lesser power, through its submarines, could strive to deny the use of sea communications without having first obtained positive control. Technology had added a new negativism.

THE END OF BRITISH
NAVAL HEGEMONY

But it was not technology that brought the Pax Britannica to an end. Tirpitz's Navy Law of 1898, giving Germany a fleet to challenge Britain in the North Sea, was "applied strategy." Many Germans had come to believe that two great commercial rivals could not coexist in peace.

After the challenge of German sea power was definitely accepted in 1905 by the laying down of the *Dreadnought,* and by the concentration of the Home Fleet in the North Sea as a result of the entente with France, the Pax Britannica was for the first time seriously endangered. Germany

H.M.S. DREADNOUGHT, the first modern, all-big-gun battleship, laid down in 1905 and launched in 1906, revolutionized naval warfare in the first half of the twentieth century. (R. A. Fletcher, *Warships and Their Story,* Cassell, New York, 1911, p. 314)

responded by laying down several *Dreadnought*-type battleships; Britain
replied that she would build two ships for every one. British policy in building the *Dreadnought,* which made every other type of battleship obsolete,
was sharply criticized on the grounds that it scrapped the advantage
Britain already enjoyed in numbers; later this criticism was countered by
the argument that battleships of her great size delayed the German menace
because they entailed the deepening and widening of the Kiel Canal,
through which ships could pass from the Baltic to the North Sea and, so, to
the Atlantic. But the most important feature of the bitter Anglo-German
building race was the antagonism which it revealed. The British simply
could not understand why Germany, with little coastline and with relatively
few overseas colonies, desired to enter into a maritime armaments race. The
obvious inference was that she had ulterior motives.

The end of the Pax Britannica cannot, however, be attributed solely to
the naval-building race, which was only a manifestation of worldwide economic competition and imperial rivalry; nor was it due to the failure of England's maritime policy. Germany had taken the place held by France in the
eighteenth century; but her industrial might and larger population made her
a greater menace, and it increased the magnitude of the resultant war. However, the power which permanently upset Britain's predominance, and with it
the century-long balance, was the United States. In wartime neither England
nor Germany could accept the American doctrine of the "freedom of the
seas." The seas had never really been free, a fact learned and forgotten by
the United States in her cooperation with the Royal Navy in the quasi-war
with France (1799–1800), in her wars with the Barbary Pirates (1803–5), and
in her war against England (1812–15). During the Pax Britannica the freedom of the seas had served the interests of Great Britain and had been enforced by her sea power. When war came in 1914, if the United States wished
to keep the sea lanes open for her own trade, she had either to seize control of
the seas herself or enter an alliance which had the ability to control them.
The American policy of isolation, and of avoiding entangling alliances, had
become incompatible with her own economic interests. After the First World
War President Woodrow Wilson's policy, expressed in 1916, of building an
American navy "incomparably the greatest navy in the world," created a
naval situation entirely different from that of the nineteenth century, one in
which Britain accepted American equality on the seas and in which the
United States was content not to use her superior industrial power to achieve
the maritime preponderance which was within her reach. But sea power then
operated in a much more unstable world than that of the nineteenth century
and one in which its function was complicated by the development of air
power, a new world created by the vast social and technological changes
which had occurred during the nineteenth century.

15

APPROACH TO
TOTAL WARFARE

CLAUSEWITZ

The intellectual link between the warfare of the Revolutionary-Napoleonic era and the total warfare of the twentieth century takes us back to the beginning of the nineteenth. That century was covered in different contexts in each of the two preceding chapters. The link is to be found in the writings of Carl von Clausewitz, a Prussian general who fought against the French in both the Prussian and Russian armies and who served under Scharnhorst, Blücher, and Gneisenau. When Clausewitz was appointed the director of the Military School in Berlin in 1818, he found that his position gave him ample time for the study of military history. He spent so much time in private reading and contemplation that he was suspected of secret drinking. His magnum opus, *On War,* was still unfinished when he died of cholera in 1831, but it was given to the world by his widow.

Whereas Jomini had attempted to distill from Bonaparte's campaigns the pure essence of Napoleonic strategical and tactical doctrine and in so doing had actually reverted, to some extent, to eighteenth-century concepts of warfare, Clausewitz, although he did not develop a system of Napoleonic strategy, did grasp the new spirit bred into warfare by the political revolutions. He wove his conclusions into a philosophy of war as well as a guide for action. His book, though a study of the nature of war, was still, to some extent, a text for operations. Since its conclusions are largely unaffected by changes in the nature of weapons, his work is applicable in all ages. By revealing the foundations upon which strategy must be built, he has come to exercise a powerful influence upon strategic doctrine in the modern world.

According to Clausewitz, "War is nothing else than a continuation of political transactions intermingled with different means." The core of his teaching is that these different means entail the full utilization of the moral and material resources of a nation to bring about, by violence, the complete destruction of the enemy's means and will to resist. He declared that warfare must always tend to become "absolute," by which he meant the use of the utmost violence to achieve complete annihilation; but he was well aware that absolute warfare was only an abstract concept and that, in practice, war must fall short of it.

Clausewitz held that the destruction of the enemy's armed forces is the first aim of generalship and that the best method of bringing it about is by direct attack. He placed great emphasis on the battle as the ultimate goal of strategy. He wrote, "Let us not hear of generals who conquer without the shedding of blood. If bloody battle is a horrible spectacle, then it should only be a reason for treating war with more respect. . . ." He emphasized the importance of mass and concentration in Napoleon's methods but missed the significance of elastic deployment, which was an equally important part of the Emperor's technique. His disciples carried this distortion of Napoleonic strategy still further to the point of undue reliance upon numbers and weight.

Clausewitz demonstrated the weaknesses of those eighteenth-century theorists and practitioners of war who had sought to achieve victory without fighting; and he disagreed with thinkers like his contemporary, Jomini, who sought to discover laws of the science of war. The greatest contribution which Clausewitz made to military thought was to show that there can be no single tactical pattern or strategic system by which victory can be ensured. The moral factor, and the element of chance, cannot be reduced to a law of movement. Nevertheless, in the future, Clausewitz was accepted as the prophet of a strategic doctrine that was as rigid as any which he had challenged.

On War was actually a delayed-action time-bomb. When it was first published, Jomini held the field as the recognized interpreter of Napoleon's genius. Clausewitz was known only in Prussia until the elder Moltke applied his ideas against France. The French army "discovered" *On War* in the 1880s. By the beginning of the twentieth century Clausewitz's phrases had found their way into the thinking and writing of the general staffs of all the great armies of the world, even in Britain, where, although an English edition of *On War* had appeared in 1873, officers still studied strategy on Jomini's lines rather than war with Clausewitz. Responsibility for the bloody slaughter in the "mausoleums of mud" on the Western Front in the First World War has been attributed to Clausewitz by many authorities, including the British military writers Liddell Hart (who coined the phrase) and General J. F. C. Fuller.

As Liddell Hart pointed out, much of the blame must really fall to those military leaders who read Clausewitz's startling sentences out of context and without the qualifications that invariably accompanied them. The passage in *On War* which most fully expresses the idea that war is a continuation of politics by other means is actually directed toward showing that the beginning of a war does not mean the end of political relations between the belligerent nations, since war in itself is a form of political transaction. Thus that vital passage, the key to all Clausewitz's doctrines, should properly be interpreted as an argument for the supremacy of political over narrowly military policies and for the exercise of moderation once victory has been ensured. Instead, it has too often been used to justify the very opposite policies; according to Clausewitz, when this occurs, "we have before us a senseless thing without an object."

ECONOMIC AND MILITARY STRENGTHS

Just as Clausewitz carried the doctrines of totality in war from the Revolutionary era into the twentieth century, another contribution to the achievement of totality was being made by thinkers in nonmilitary spheres. In countries which did not enjoy a guaranteed security and prosperity, men were searching for a formula for national power and wealth. They found it in the reiteration of a simple truth which the mercantilists had emphasized a century earlier, namely, that military strength is, in part, a product of economic factors. Thus, in 1791 in the United States, Alexander Hamilton turned his back on Adam Smith and advocated the protection of infant industry for the avowed purpose of increasing national security. Against the sentiments of most of his fellow countrymen he also favored a standing army and navy. In Germany, members of the "Romantic-Nationalist" school, which included Adam Müller, Heinrich von Treitschke, and Johann Gottlieb Fichte, likewise rejected doctrinaire laissez-faire economics and espoused autarchy, or self-sufficiency, as a means of providing strength for war. The German economist Friedrich List preached the economic union of the Germanies as a step toward political unity, which in fact was borne out by the creation of the Zollverein (customs union) 1834. Fully realizing that they could probably not be carried out without war, he advocated programs of unification and expansion for Germany. For some members of this German school, the warlike spirit was itself a vital element of national strength and efficiency. They inaugurated the worship of the state and nation as a philosophy of life.

These nationalist arguments, calling attention to the close connection between economic and social forces and military strength, received powerful, but unsolicited, support at midcentury from the founder of communism. In 1848, in *The Communist Manifesto,* published to rally the workers to support widespread insurrections against autocracy, Karl Marx issued a call to arms. The fundamentals of his argument as elaborated in his massive work *Das Kapital* (1867–94) were that political power must inevitably pass to the proletariat which, through its labor, was the creator of wealth; but, unlike earlier socialists, he believed that violence would be necessary to effect the transfer of power.

Marxian theory appealed strongly to the lower classes whose lot had been made desperate by the harsh conditions created by the Industrial Revolution.

But the Industrial and Agrarian Revolutions had a much more direct and immediate effect upon warfare than that resulting from Marxist propaganda. They provided the material, and some of the techniques, by which the impact of warfare was greatly increased and total warfare became possible.

In the eighteenth century, English farmers had pioneered improved methods of animal husbandry and tillage. A few years before the Revolution of 1789, Arthur Young, an English agricultural journalist, while touring rural France, had noted that French agriculture was backward by comparison with English. The Revolutionary leaders propagated his ideas of improvement; but it was not until the early nineteenth century that improved agricultural methods began to be adopted on a wide scale on the Continent. Increased yield per acre in Europe was accompanied by the opening up of the continents of America, Australia, and Asia for the production of food. McCormick's reaper, invented in 1831, and Appert's perfection of the art of canning were responsible for making available a greatly increased supply of grain and meat for industrial workers in time of peace and for armies in time of war. The use of citrus juices and potatoes to prevent scurvy, introduced into nautical diets from 1795, was equally important in land warfare to prevent a decimation of armies on campaign or in sieges, often a result of this debilitating disease. By the mid-nineteenth century, it was possible to provision armies far larger than those of the Napoleonic era and also to keep them healthier.

INDUSTRIALIZATION

At the same time the Industrial Revolution made possible the production of many forms of goods on a vast scale, including, of course, uniforms, weapons, and other military supplies. The application of power machinery, of capitalist organization, and of the factory system to manufacturing processes had first appeared in the sixteenth century, but it made no significant impression on warfare and society until the late eighteenth and early nineteenth centuries. A remarkable series of inventions in the English textile industry between 1733 and 1785 enormously increased the output of cloth; Eli Whitney's cotton gin (1792) furnished vast supplies of raw cotton to the mills; and from 1800 steam power and iron machinery in great factories poured forth an ever-increasing flood of cheap cotton and woolen goods into the markets of the world. Elias Howe's invention, the sewing machine (1845), clothed and shod the great armies of the second half of the nineteenth century.

For military purposes, the revolution in the iron industry was even more important. The development of the technique of using coke instead of charcoal for smelting iron ore had been perfected about 1750, just in time to prevent a serious shortage of metal in England in the Seven Years' War. At that time Britain produced only one-third as much iron as France; but by 1840, because of the proximity of British iron mines to coalfields, the ratio had been reversed. Meanwhile methods for producing artillery by mass production had been worked out in Woolwich Arsenal in England, in the Carron Ironworks in Scotland, at Le Creusot in France, and elsewhere, and

CONGREVE ROCKETS. Rockets have a long history of use in war. There were rocket companies in British armies in the Napoleonic Wars and in the War of 1812. These latter are immortalized in the words of the national anthem of the United States, " . . . the rockets red glare, . . . ". (Massey Library Collection, Royal Military College of Canada)

had made available the great quantities of ordnance used in the wars at the end of the eighteenth century. Specific military and naval demands stimulated important new technical processes in the manufacture of iron. In 1784 Henry Cort, a purchasing agent for the Royal Navy, patented the reverberatory furnace and "puddling" in order to purify and toughen iron, and the rolling mill to work it more easily. In 1854, on the outbreak of the Crimean War, W. G. Armstrong offered to the British War Office his method of making guns built up from cores of steel; and this made possible more accurate boring, rifling, and breech loading. The Bessemer steel process, in which air was forced through molten metal to purify it, was introduced in 1856 in answer to a request by Napoleon III for a steel capable of withstanding the new explosive shells. The Bessemer, open-hearth, and Gilchrist processes of eliminating impurities in steel manufacture greatly lowered costs of production and led to the general use of steel, in place of iron, for heavy artillery by the last quarter of the nineteenth century.

Similar improvements had been made in the manufacture of small arms. Whitney had introduced the principles of interchangeable parts and division of labor in his Connecticut factory for the manufacture of firearms. These techniques were applicable to industry generally and were the basis of all future mass-production methods. The percussion cap, invented by a Scots clergyman in 1807, made practicable a breechloader. In 1830 the "sugar loaf bullet" was introduced in the United States and, while it still required the use of the greased cap, it simplified loading the weapon. In Britain Captain Norton (1842), and in France Captain Minié (1833), perfected the cylindro-conoidal bullet which, in rifled barrels, increased the effective range of small arms to 1,200 yards. By the middle of the nineteenth century a revolution in armaments had been inaugurated. Thus in 1840, the Brown Bess musket, which had been the standard small-arms weapon of the British infantryman since 1690, began to be replaced by the rifle. Gatling's machine gun (1862), used on a small scale in the Civil War, was the forerunner of the automatic small arms which were to have enormous importance in future warfare. But the massed fire of American riflemen had already begun to revolutionize tactics.

The Industrial Revolution affected strategy as well as tactics, principally by the improvement of transportation. The steam engine, in addition to increasing production by driving textile mills, operating steamhammers and boring tools, lifting coal from deep shafts, and pumping water from mines, was belatedly applied to locomotion. Watt's first efficient stationary engine had appeared in 1768; but Fulton's first practical paddle wheeler steamed up the Hudson only in 1807; and Stephenson's first passenger locomotive did not begin operation between Stockton and Darlington until 1825, and then only at walking speed. The realization of the military importance of steam locomotion came much more quickly than the locomotive itself. By 1833 a German, Friedrich Harkort, had calculated that strategically placed defensive railways could save much time in the transportation of troops and would bring them into battle unfatigued by the march. Nine years later a scheme had been worked out for covering Germany with a network of railways which a French deputy described as "aggressive." List's plans for German union were based partly on the building of strategic railways which would not only unite the country but would reverse its historic role as the battlefield of Europe by permitting German military power to move outward. In 1846 the Prussian VI Corps, with all its equipment, was moved 250 miles in two days, a journey which hitherto had taken from ten days to two weeks. Within a few years of the running of the first passenger train, clear-sighted men had compared the effect of the invention of the steam locomotive upon society and war to those of printing and gunpowder.

One important result of rail travel, and of another associated development, cheap postal services, was the binding of nations into more tightly knit communities. The railway made possible the opening up of the United States, Canada, and Russian Siberia; and the railway and steamship transported their products to world markets. At the same time the railway marked the beginning of a greater mobility of national populations; and the railway and post office maintained family ties which might otherwise have been severed. Thus, Prince Albert's Great Exhibition in London in 1851 attracted thousands from the north of England by cheap excursion trains. This invasion of the metropolis by northern "barbarians" was feared at first as a possible source of disorder, but proved to be a merry, peaceful occasion. Everywhere railway building led to closer national unity and greater national strength.

Only the continuation of peace prevented steam locomotion from having an earlier effect upon military operations. Even so, Russian troops were moved by rail to suppress revolution in Austria in 1849; steam transports took British troops to the Crimean War in 1854; and the French made use of railways in the Italian campaigns in 1859. In the last case, although the French thereby gained a temporary advantage, they lost it because they failed to organize the transportation of adequate supplies for the troops and for the horses moved quickly by train. Hence, it was not until proper techniques were worked out for their use in the American Civil War and in the Prussian wars of the following decade that railways played a really significant part in military operations.

POPULATION GROWTH

Along with the great increase in the supply of material for war and the greater mobility furnished by the railway and steamship went an unprecedented increase in the populations of the countries of Europe and North America, brought about by improvements in methods of production and by better hygiene and enhanced medical knowledge. During the nineteenth century the population of the United States increased fifteenfold; but a large part of this was the result of immigration from Europe. In the same period, however, the increase in England, Scotland, and Wales was more than threefold; and the population of Europe as a whole grew from 185,000,000 to

400,000,000. With less than a quarter of the area of Asia, in 1800 Europe had a third of Asia's population; by 1900 that had been narrowed to one-half.

One result of this great increase of European peoples was to provide the manpower for the more frequent wars of the second half of the century. Secondly, the fact that some countries increased their population by natural and other causes more rapidly than their rivals led to a disturbance of traditional balances of power. The rise of Germany and the decline of France during the nineteenth century are partly explained by the inferior French birthrate.

The increase in populations was accompanied by a redistribution of people into new or different areas, for instance, into the Middle West in the United States, and into the north of England and away from the south. It also led to the growth of great urban centers. It was in the new areas of the West and the new industrial cities that movements for the extension of the franchise were strongest. Manhood suffrage had been achieved in most of the United States by 1827 (for whites), in France by 1848 (although only as an empty form until 1871), in Prussia by 1850 (subject to constitutional impediments), in the German Empire by 1871, and in Britain by 1884. Even when not accompanied by real representative government, the democratic franchise carried with it an obligation upon the citizen to share in the defense of the community. Therefore, in the latter part of the century, conscription became the rule on the continent of Europe. But socialists preferred a militia, and in 1911, Jean Jaurès's *L'Armée Nouvelle* (The New Army) made that the basis of the leftist concept of national defense.

The democratic franchise implied the necessity of general public education. Compulsory universal elementary education had appeared in Prussia as early as 1794. It was introduced in the United States between 1852 and 1880, in France in 1882, and in England between 1870 and 1880. Widespread literacy was an important preliminary step in the education of young men for modern war.

The concept of popular sovereignty had been reintroduced into Europe by the revolutions of 1848, only to be checked in France by Louis Napoleon, not by the use of artillery in the tradition of his uncle, but by manipulation of the ballot in popular plebiscites. Similarly, in Prussia, the constitutional movement of 1848 was betrayed and dissolved by being switched into patriotic channels. In both countries, war was used, in part at least, to divert the masses from domestic constitutional problems: by the Crimean and Italian campaigns in the case of France and by the wars with Denmark, Austria, and France in the case of Prussia. One point of close connection between war and the growth of the "popular" but undemocratic state was thereby made apparent. In 1859 in England, fear of the rise of Napoleon III and of a second French attempt to establish hegemony in Europe led to the organization of "volunteer corps" by middle-class citizens. War, and preparation for war, was not a monopoly of autocracy.

THE AMERICAN CIVIL WAR

A few years later, the American Civil War clearly demonstrated many of the features which were to mark the warfare of the age of democracy and pseudodemocracy. It was the first great war to be fought in the era of the Industrial Revolution, and so it showed the effects of the technological advances in industry and agriculture which were to revolutionize warfare. It showed also the growing importance of economic factors in modern warfare, in that in a struggle between two different types of economy, agrarian and industrial, the ultimate result was a victory for the power that was stronger industrially and financially. Although in the territories of both combatants there were large dissident minorities, the Copperheads in the North and the Unionists in the South, the Civil War was an ideological conflict, a struggle of rival "nationalisms" and cultures. Abolitionist sentiment acted as a powerful stimulant in the North, and propaganda was a useful weapon. This was not a war simply of hostile armies, but of hostile peoples, in which both sides were equally convinced of the absolute justice of their cause. The North was fighting for the preservation of the Union, the South for independence. Emotions kindled by the slavery issue intensified the bitterness. For both sides the objective was total. There could be no treaty, no compromise, no argument, no terms except unconditional surrender or separation.

The Civil War, the first great war to be fought by modern democratic states, raised many of the problems which trouble a democracy at war. The North, believing in democratic liberty and freedom of the press, fought the war in a blaze of publicity which endangered military security. The problem of command was even more difficult. Whereas in autocracy civil and military power is centralized, democracy cannot adopt that arrangement easily without endangering the sovereignty of the people. Democracy at war needs to produce a political leader as head of the state whose grasp of grand strategy enables him to guide war policy. The Constitution of the United States, and after it that of the Confederacy, by making the president the commander in chief of the armed forces, ensured that the higher direction of the war was a civilian responsibility; but it was difficult to draw a clear line of demarcation between major policy decisions in grand strategy to be made by the president and the planning and carrying out of military operations which must be the task of professional soldiers.

In the Confederacy a solution was found by choosing a president with much military experience. Believing himself to be a great military leader, Jefferson Davis would have preferred a military command. But he lacked the ability to get on with men over whom he had no disciplinary control.

Davis was much less successful as a war leader than Lincoln, who had fewer obvious qualifications for the task, but who had a flair for human relations and was able to learn from experience. At first, Lincoln concerned himself in the direction of the war at all levels. But he was continually seeking military leaders whose ability he could trust, and also a suitable system of command. He found the ideal solution in March 1864, by appointing Henry Halleck as his chief of staff in Washington and Ulysses Grant as general-in-chief of the army. Halleck thus served as interpreter between the president and the leader of the army. Although the United States did not actually evolve anything quite like a modern general staff during the Civil War, it did create a system of command which proved efficient in practice.

It was inevitable that Congress should seek contact with the army in order to obtain information on military affairs about which, through its legislative powers, it was compelled to make decisions. Therefore it created a Joint Committee on the Conduct of the War. Opinion has differed about the value of congressional interference. Most of the Committee's activity took the form of legitimate inquiry and of suggestions and advice; but democratic control of the military effort also brought some intrigue.

The armies on both sides in the Civil War, although largely composed of volunteers, were a reversion to the democratic principle of the nation-in-arms. The Confederacy numbered about five and a half million whites. Only about five hundred of the regular army of the United States, most of them officers, left to join the South. The Confederate government therefore had to build an army from the bottom up. It called for quotas of volunteers from the states, men who were enlisted at first for one year and, later, for three years or the duration of the war. Conscription acts in 1862 and 1864 required the registration and enlistment of every fit man between seventeen and fifty. In all, some 900,000 whites were enrolled, with a maximum in 1863 of something less than half a million.

The North, which was more than four times as populous as the South, began the war with the regular army of the United States of about 15,000 men and supplemented it by volunteers levied by quotas upon the states. In 1863 it also resorted to conscription to fill out quotas not completed by volunteers. Throughout the war it recruited about 2,375,000 men, with a maximum strength in April 1865 of a little over a million. Thus nearly 20 percent of the total population in the South, and over 10 percent in the North, was recruited for military service, either voluntarily or by conscription. Individual armies often ranged from 60,000 to 130,000 men.

At the outset of the Civil War, some troops still had smoothbore flintlocks, but the basic infantry weapon was a muzzle-loading rifle, a Springfield or an Enfield; and toward the end of the war a breechloader was introduced. A magazine rifle and a primitive machinegun also made their appearance. Field artillery included a smoothbore 12-pounder "Napoleon" gun and the 3-inch rifled Parrott gun. Casualties were high. At Gettysburg, where 81,000 Confederates fought 100,000 Federals, the killed numbered 20,451, while the wounded and missing numbered an additional 23,059. And Gettysburg was only one, although one of the biggest, of some 150 major encounters in less than five years. These casualties were evidence of the resolute behavior of the soldiers of democracy given what was, by European standards, a negligible amount of military training.

Heavy casualties, and the employment of amateur soldiers, hastily recruited and trained, led to important innovations in tactics. Prior to the war American army manuals described various formations, but commanders preferred attacks with regiments of about a thousand men on a narrow front of two companies in column. In the war these mass tactics gave way to attacks in two-rank line with a frontage of perhaps a thousand men. Experiments were made with skirmishers thrown out ahead of the attack, and with leapfrog rushes in which half of the attackers provided covering fire before rushing forward in their turn. One difficulty experienced in these tactics was caused by the necessity for the soldier to stand while reloading his muzzle-loader. In addition, it was sometimes difficult to induce men to advance from the prone position. But as the war went on, and as the amateurs became seasoned veterans, the *élan* of the troops on both sides grew. They pressed home attacks with a resolution that astonished foreign observers who, nevertheless, commented adversely upon the Americans' lack of precise formation. Attackers and defenders alike relied heavily upon fire power. Indeed, so little use was made of the bayonet that according to a later authority, that weapon was often thrown away as a useless encumbrance. In summary, the American Civil War revealed the growing importance of fire power and the folly of the traditional close-order charge in the conditions of modern warfare. It showed that the infantryman was destined to seek shelter in trenches from the devastating power of rifled weapons. The spade and the ax had become major weapons of war. Breastworks and rifle pits were already widely used in the field.

Cavalry tactics were similarly affected by the greater killing power of weapons. Foreign professional soldiers noted that the American horse failed to carry a charge home with the sword and believed that this was due to inadequate training. What they did not see was that rifled weapons were making the old kind of cavalry warfare impossible. American cavalry acted as mounted infantry, using their horses to carry them rapidly to the scene of an action and then dismounting and taking cover to use their breech-loading carbines, weapons which, although not equal to the infantry rifle, were much superior to the musket. Thus cavalry was used to prevent a stagnation in tactical mobility which was fast developing. Independent cavalry raids deep into enemy territory, strategic operations of a kind which Frederick the Great would never have dared to use and which Napoleon never contemplated, exploited opportunities created by the armies' reliance on the railroads.

RAILROADS

The most important way by which mobility was retained in the Civil War was by the strategic use of the railroads. The theater of operations was enormous, as large as the whole of Europe, and thinly populated. Railways had been laid down in both regions, although on a much greater scale in the North than in the South; but in neither area had they been built for strategic purposes. The use of railroads for war on any important scale was unprecedented. The commanders of the Civil War had to work out principles and techniques for their military use. Some of the problems connected with the use of railroads for military purposes were solved by the North in this war; and the more effective use of the rails was an important factor in the ultimate victory. It was achieved by leaving most of the lines in private hands unless the companies proved unresponsive to military needs.

It was found that the railway made operations possible at great distances from supply bases, that troops and materials could be carried quickly to threatened points, and that a great increase in the fighting power of troops in the line accrued from the correct use of the steam engine. Special corps were created for the restoration, operation, and destruction of rail lines. Hospital trains and armored trains appeared, and cuttings, embankments, and railroad lines generally became tactical objectives, often more important ones than the natural features of the ground.

At first, Herman Haupt, a West Point graduate who had civil experience in railroading, was appointed in the North to supervise the construction of railways and the transportation of troops and supplies and was fairly successful. But difficulty in coordinating the activities of rival private companies led to the appointment of Daniel C. McCallum, a railroad man, as director of military railroads with full power to requisition. He employed civilian railroad personnel. Some of his feats of transportation in this war were remarkable. In September 1863, 23,000 men with artillery and road vehicles were moved 1,200 miles in seven days to save Rosecrans after his defeat at Chickamauga. This operation would have taken three months on the march. Thus the North worked out a technique of operating the railroads under the full authority of the state but by making use of the experience of civilian railwaymen.

The failure of the South to organize its railroads as effectively for war purposes was a factor in its defeat. To some extent this was unavoidable. The South lacked the industrial capacity to maintain and extend its lines and rolling stock. Early in the war Jefferson Davis ordered the completion of a vital link to maintain the armies on the northern front; but no leader saw the necessity for a thorough railway policy and administration. The South was deeply committed to the principle of laissez faire. Left to themselves, the railroads, which had mainly been built to carry goods from the interior to coastal ports, languished while waiting for the breaking of the blockade. Their lines and rolling stock might have been more advantageously employed to supplement the work of the lines running to the war fronts in Virginia and Tennessee. In February 1865 the Confederacy took over control of the railroads and thus ended suicidal private competition; and by that time it was too late. The railways were worn out. Lee's army in Virginia starved because, although there was food in Alabama and Georgia, it could not be moved to the troops.

While the Americans had been pioneers of the use of the railway in war, their experiences had not had a revolutionary effect on the structure of military command. But it was in Prussia that the realization of the importance of railways in war contributed to the perfection of the General Staff system to plan and direct the great mass armies of modern times.

GENERAL STAFFS

France under the Ancien Régime had had a general staff for the collection of military information and the preparation of war plans; but it had been abolished in 1790. Napoleon had felt no need for aid from such an institution. However, the growth of armies, the use of divisional formation and of improved field artillery, and the evolution of skirmishing tactics in place of the traditional line of battle meant that command must be exercised from a distance rather than by visual oversight. Plans must be prepared on a map; orders must be given in writing, instead of orally; and the telegraph, perfected by S. F. B. Morse in 1844, made possible remote control. In Prussia, the Quartermaster General's Department of the seventeenth and eighteenth centuries, which had been made responsible for survey and cartography, was reorganized in the early nineteenth century by Baron Christian von Massenbach and Scharnhorst. It consisted of the Great General Staff and the General Staff with the Troops, both under a single chief of staff. The former collected and collated information for planning, and the latter acted as adviser to the commanders of armies in the field. Staff colleges were set up to train staff officers. The reason for the growth of general staffs was not merely the necessity of creating a system of command capable of handling mass armies but was also a result of the experience of the Napoleonic period that inadequate preparation for war might bring national extinction. Prussia, because of her relative weakness, had the greatest need of planning. As her king and his family held field command by law, a general staff provided necessary expert assistance.

Field Marshal Helmuth von Moltke (1800–1891), appointed Prussian chief of staff in 1857, saw that railways had made possible much more precise and reliable calculation of the movement of troops and supplies, with the result that accurate and detailed planning for the mobilization and commitment of the armed forces to war could now be undertaken. He therefore set up a railway subsection of the General Staff in 1864, in time for use in the Danish War of that year.

In 1866 the subsection became the separate Railway Section of the Great General Staff, a recognition of its importance. The lightning mobilizations of the Prussian army which struck down Austria in 1866 and France in 1870 were almost completely done by rail under the supervision of this section of the General Staff. There were mistakes, even in the war against France. In their enthusiasm, railway officials, supply officers, and contractors sometimes rushed goods forward in excess of requirements, with the result that lines and stations became choked. But Austrian and French plans for the use of the railways were, by comparison, ill contrived and defective; and there can be no doubt that Prussian exploitation of the railway was one of the most important weapons of victory.

Seeing the Prussian victories, other nations quickly adopted the general-staff system, although not necessarily in exactly the same form as in Germany. Some form of the capital-staff system was set up in Russia in 1863, in France in 1874, in Austria-Hungary in 1875, in Japan in 1879, and in Italy in 1882. Everywhere these general staffs changed the functioning of military command. They enabled a chief of staff to supervise and coordinate the operations of armies in several different theaters; and within each theater of war, the staff system provided a method by which armies of great size could be controlled. Whereas in the eighteenth century a force of about 40,000 had been about all that could be conveniently handled by a commander, in the Franco-Prussian War the Germans mobilized and committed 480,000 men at the outset. Moltke saw that the general staff system, coupled with improved methods of transportation, made possible huge concentric movements and thus gave the advantage to external lines of operation, in place of the inner lines favored by Frederick the Great and even by Napoleon.

At the same time, the growth of capital staffs inevitably affected the relationship between civil authority and military command. General staffs, corporate entities planning for future eventualities, came to exercise a powerful influence over national policy to the disadvantage of hereditary autocratic rulers, and even more of democratic institutions. In Germany under Kaiser Wilhelm II, the general staff came to rival the civilian executive and legislature as a formulator of policy.The urgent nature of its task, the security of the state, made it the ultimate arbiter in emergency.

PRUSSIAN VICTORIES

The Prussian victories were the result of other factors besides the railways and the improved system of command. Even without the advantages derived from the scientific use of railroads, the Prussian system of mobilization was superior. While France and Austria had to organize their army corps when war was imminent and thus lost valuable time, the Prussians had maintained the corps organization first set up during the great wars of liberation. What is more important, the Prussians had never completely abandoned the concept of the nation-in-arms which Scharnhorst had established by his army reforms after the disastrous defeat at Jena. In 1862, Bismarck obtained appropriations to extend conscription by overriding the constitutional powers of the lower house. Hence, within two weeks of the beginning of hostilities in 1870, the Germans raised 1,180,000 trained men as against France's 330,000. While military experts in France and elsewhere were convinced that a smaller army of long-service veterans which had seen recent service in North Africa, in the Crimea, and in Italy would prevail against a horde of short-service conscripts, quantity proved on this occasion to be superior. It must be said, moreover, that the Prussian officer corps was superior to the French in morale, ability, and training, and that this probably compensated for deficiencies of the rank and file. The strength of Prussia, which so surprised European opinion in 1870, was thus based upon a retention of principles discarded by other nations after 1815, and also upon the militarist traditions of the Prussian monarchy, in which popular clamor for constitutional government had been transformed into patriotic zeal for German unity and power.

In addition, the Prussians had the advantage of superior weapons. In the war with Austria, the Prussian infantry had a breechloader, the needle gun, which not only had a more rapid rate of fire than the enemy's muzzle-loader but also could be fired from the prone position. The Austrian infantry, having to stand up to reload, presented good targets. Realizing the importance of this advantage in weapons, the French hurriedly introduced the *chassepot* which, with a range of more than 1,200 yards, was twice as effective as the needle gun. But as their troops had not had enough training with the new rifle, the advantage of a superior infantry weapon was lost. The German breech-loading artillery was superior to the French muzzle-loading rifled cannon, was more numerous, and was better handled. The French hoped to counter this superiority by the *mitrailleuse,* a machine gun with thirty-seven rifled steel barrels which fired simultaneously. It looked like a fieldpiece and was misemployed by being used as one instead of as an infantry weapon; but at close range with good cover it was effective. The Germans also possessed rifled siege artillery, which fired shells up to four miles, and howitzers.

Although Moltke had learned something about the value of rail power from the American Civil War, he learned little else. He described this struggle as a contest in which huge armed rabbles chased each other around a vast wilderness. European soldiers generally were convinced that the American conflict had little to teach. As a result, the Prussians and French alike learned the effect of new rifled weapons in the bitter school of war and were compelled to modify their tactics under fire. The French infantry, relying on their superior rifles, made use of rifle pits and exacted a heavy toll from German infantry attacks in column. Under stress of battle, the Prussians deployed into skirmish order. They also adopted, probably not by direct borrowing, the American system of attack by which one half advanced while the other provided covering fire. Whereas the Americans, using muzzle-loaders, had been able to employ this form of attack only when the ground provided some shelter for the covering parties who had to stand up to reload, the Germans with their breechloaders could fire from the ground. The Franco-Prussian war showed that all infantry must fight as skirmishers, that cavalry could no longer charge unshaken infantry, that the bayonet attack was an anachronism, that positions must be carried by fire power, and that the full power of concentrated artillery must be used in every battle. The last

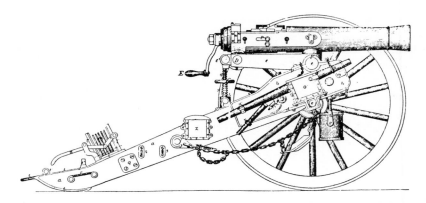

THE FRENCH MITRAILLEUSE OF 1870. Early machine guns were crank operated. Some had many barrels. They were first misused by being deployed as artillery pieces. (W. M. Folger, *The French Mitrailleuse,* GPO, Washington, 1873)

fact meant that, ultimately, industrial potential would be all-important in a major war.

The Prussian military victories in the 1860s, and the political results which followed, led many European nations to inspect their armories and reexamine their military organization. In Britain, ever since the Crimean War, it had been realized that the army needed a more thorough overhaul than that which had been achieved during that short, sharp conflict. But military inertia blocked this until Prussian military might cast a sudden shadow over the continent. The British Army was then reorganized and centralized under a political head responsible to Parliament; the iniquitous system of the purchase of commissions was abolished; military training and military education were investigated and improved; and regiments of the line were associated with the militia units of particular districts to foster recruiting and to combine those twin fountains of morale, regimental tradition and local pride. Garrisons were withdrawn from overseas colonies, like Canada, which could now stand alone, and the Cardwell system (named after Edward Cardwell, secretary of state for war, who was responsible for many of these reforms), by which the battalions of each regiment alternated in home and overseas duty, provided forces to guard the expanding British Empire in an age when international tension was manifestly increasing.

Bismarck had given the lead to his generation by using war deliberately as a tool of policy. The most important outcome of the premeditated wars with Denmark, Austria, and France was the unification of Germany under Prussian leadership. This provided wider scope for industrialization, and Germany shortly became a major industrial power. At the same time, two other new industrial powers appeared on the scene, the United States, where industrialization had been greatly furthered by the triumph of the North, and Japan, which had undertaken a conscious program of Westernization

Coal —Increase in Production, 1893–1913*

	Percent of Increase
United States	210
United Kingdom	75
Germany	159

Production of Steel (in thousands of tons)

Year	United Kingdom	Germany	United States
1890	3,579	2,195	4,275
1896	4,133	4,745	5,282
1900	4,901	6,260	10,188
1908	5,300	10,900	—

Note: Figures in these tables are compiled from R. C. K. Ensor, *England 1870–1914*. (Oxford: Clarendon, 1941).

including industrialization and the creation of a western-type army and navy. Production figures for coal and steel show the way in which British industrial predominance, which had been one pillar of the Pax Britannica of the nineteenth century, was surpassed. Germany, by dumping and subsidies, built up her steel industry in peacetime to an uneconomic position. When war came, its huge capacity proved an invaluable military asset.

IMPERIAL RIVALRIES

The spread of the Industrial Revolution during the latter half of the nineteenth century meant increasing competition among European powers for raw materials and markets. Africa and China alone remained as major fields for exploitation. Discoveries in tropical medicine, and the invention of canning, made possible the opening up of the heart of Africa, partly by traders in search of gold, ivory, and tropical products, and partly by missionaries seeking to carry the Cross to the "heathen" Black Africans. River steamers and breech-loading rifles established European supremacy. The continent was divided among the European powers. China, narrowly escaping reduction to colonial status, was compelled to grant special rights to Western powers and was partitioned into spheres of interest. The scramble for Africa and China in the last quarter of the nineteenth century was accompanied by a revival of imperialism among European peoples, a sentiment which was intensified by the growth of democracy. The newly enfranchised industrial classes evinced a powerful desire to see the map painted with their own particular patriotic color. At the same time the building of empire provided jobs for soldiers, traders, and administrators and profitable investment for the capital of the middle and upper classes. The period was marked by the growth of emotional imperialism; in England this phenomenon was called jingoism, from the words of a music-hall ditty.

Competition for colonies increased the friction arising among the European powers on various other grounds. It has been suggested, particularly by socialist-pacificst analysts, that this friction was actively provoked by certain interests which desired to profit by the sale of munitions to rival governments. The iron and steel industry in Europe and America had been greatly stimulated by the Prussian and American wars. While in the United States the steel industry found outlets for its increased productive capacity in railway expansion and in steel construction for buildings, rivals in Europe found other markets. Certain armaments salesmen earned the suggestive epithet "merchants of death." Culpability of individuals and corporations in the deliberate kindling of war is difficult to prove. Undoubtedly some small wars and insurrections all over the world were deliberately provoked by unscrupulous arms salesmen. But responsibility for the great conflagrations of the twentieth century is another matter. Arms races certainly increased international tension; but the arms manufacturer could claim that he was performing a patriotic service by ensuring that his own country was adequately prepared against the danger of an attack which the unsettled times seemed to threaten. At the same time he was pushing his wares abroad.

Thus the nations of Europe prepared themselves for a major conflict. The French dreamed of *revanche* against Germany. Germany sought colonial equality and challenged Britain's naval hegemony. Russia and Austria were competitors for the spoils of the disintegrating Ottoman Empire. Britain, holding tightly to its economic and political hegemony of much of the world, clashed with almost every other power in some corner of the globe. Armed with the weapons furnished by the new industrialism, the massed forces of nations-in-arms were prepared for war by military leaders who had absorbed Clausewitz's theories. Total war had become both possible and likely. The Crimean War and the Austro-Prussian War were the last of the limited wars fought by great powers in Europe. The Franco-Prussian and Russo-Turkish wars (1878) showed signs of a new age, even at the tactical level. In the American Civil War and the German wars there was increased use of firepower and also dispersion and flexibility in the field. The use of entrenchments in America recurred both in France (1870) and at Plevna (1878), thus anticipating the Western Front of World War I.

16

TOTAL WARFARE BEGINS

THE ANTICIPATIONS
OF TOTAL WAR

B y the end of the nineteenth century, there were many signs of Armageddon. Perhaps the clearest were in the writings of Ivan S. Bloch, who in 1897–98 had published his six-volume study of war, in which, as we have seen, he had made that remarkably accurate forecast of the nature of the First World War. This work was translated into German in 1899. Bloch declared that in the event of a great war in Europe, the technical development of weapons, coupled with the increased possibilities of harnessing the political and economic organization of the state to war, would make a stalemate between the fighting forces of the contending nations inevitable. As a result, civil populations would be subjected to a fearful war of attrition. He declared that the victor would suffer as much as the vanquished and that the ultimate result would be the collapse of organized society.

We saw that Bloch's book was influential in persuading the czar to call the Hague Conference, actually because Russia could not afford to modernize her artillery. But Bloch's warnings went largely unheeded by military men, perhaps because he was not a professional soldier. Moreover, in his last volume, the only one to be published in English and French, he suggested that technical developments had made traditional war all but impossible; he thus diverted attention from the horrible truths about future wars which his brilliance had uncovered. Military men who might have found ways to prevent fulfillment of the prophecy failed to respond. Only the Germans, who had recently integrated their field and foot artillery, and who had increased the proportion of machine guns as a result of observing conditions at Port Arthur, were in any degree prepared for the kind of war that emerged after the Battle of the River Marne. The others modified their training, organization, and equipment only after siege war had occurred on a huge scale.

Yet there were other signs nearer at hand which were written in the kind of language the military understood best—namely, in actual warfare. Three wars at this time gave a foretaste, each in its own way, of what the century was to produce. These were the Spanish-American War, the South African War, and the Russo-Japanese War. At first sight they appeared to prove Bloch wrong, but this was because all were "limited wars" insofar as they were restricted in objective or in locality or by the inability of the contestants to engage the whole of their national strength. However, taken together, they made up a picture which portrayed the total warfare of the future, the kind that Bloch had forecast would occur when modern European nations went to war.

A significant lesson of the Spanish-American War was that the new "democratic" state was vastly more powerful in its relation to war than was any earlier political organism. The Spanish-American War might not have occurred if it had not been for the new "yellow press" which played upon the mixed feelings of liberalism and nationalism in the United States. Sensational reports of the cruelty of the Spaniards whipped up a fever which made war all but inevitable. This war showed that democracies were not immune from the fever of aggressive war. In the hands of autocracy, the press could obviously be used for even more nefarious purposes. The state had, in the cheap newspaper, an organ of propaganda which greatly increased its control over its people. Public opinion could be molded at will; and nations, both autocracies and democracies, could be inspired to endure hardships far beyond anything known in earlier centuries. Future wars would thus be long and bitter. But the democracies were as yet poorly prepared for them.

The South African War, fought by Britain against the two small Boer republics, was mainly a war of movement; but this was principally because relatively small forces were being employed in a huge theater of operations. The Boer War, as it was popularly called, taught some lessons which had already been made clear during the American Civil War. The Boers were hardy farmers who were excellent marksmen with their Mauser rifles. They were not hampered by traditional military concepts and methods. In their own country, which they knew intimately, they were a stubborn foe, even for a nation with the wealth and resources and military tradition of Great Britain. It was found that the magazine rifle and smokeless powder had made frontal attacks more costly for infantry and that cavalry attacks with swords (the traditional *arme blanche*) had become suicidal. However, as the Boers were mounted riflemen—a kind of soldier for whom European cavalrymen had hitherto felt nothing but contempt—an excuse was provided for the retention of mounted troops for a generation, long after they had outlived their usefulness. Trenches lined by infantry with breech-loading rifles and machine guns now came to dominate tactics.

A second lesson of the war in South Africa was taught by the bitterness of the struggle. Britain had begun the war badly, largely because of inferiority of numbers. In due course, she was able to put enough men in the field to overcome the organized forces of the enemy, but the Boers resorted to guerrilla warfare on a broad scale. Here was the concept of the nation-in-arms carried out to the limit. This wholehearted national resistance was not overcome until Kitchener had cut the country up by blockhouses and barbed wire, had gathered the women and children of the "burghers" into concentration camps, had burned the farmhouses which gave shelter and provided supplies for the partisans, and had swept each section of the country free of warriors. Inevitably, in the improvised conditions of the camps, many women and children died of disease. The revelations about this by Miss Emily Hobhouse, an English humanitarian, shocked Britain and led to a softer peace than the military victory portended.

The third war of this period, the Russo-Japanese War, was fought between two great powers, and gave even more indication of what was to come. Perhaps only the fact that the contestants were separated by a continent crossed by a single-line railroad prevented this from becoming the first total war of modern times. It had many features which were to reappear in Flanders a decade later. Both sides had modern rifles; the Japanese had Hotchkiss machine guns and eleven-inch howitzers; and the only defense against these weapons was to dig in. Entrenchments protected by barbed wire around Port Arthur resembled the Western Front of the future. Heavy concentrations of artillery dominated the field of battle and frontal attacks were unprofitable. Neither Russia nor Japan could afford a long war, and when the way to victory by turning a flank became impracticable, a stalemate resulted.

A final lesson to be drawn from these three little wars was that Mahan's recently expressed doctrines of sea power appeared to be vindicated. Although widespread and unwarranted popular fear of the bombardment of the east coast of the United States by a Spanish squadron led at first to faulty deployment of the American fleet, it was sea power that eventually ensured the defeat of Spain, both in Cuba and in the Philippines. In the South African War, British control of the oceans had prevented any of the European well-wishers of the Boers from intervening. And the Japanese naval victory of Tsushima destroyed Russia's naval power and her possibility of cutting the shorter Japanese supply lines. Thus the three wars which ushered in the twentieth century provided plenty of indications of the nature of the warfare of the future; and they also reinforced the lesson that, in a major war, the command of the sea might well be the decisive factor.

But perhaps the most significant lesson of the three small wars which preceded the First World War was that new powers had appeared to upset the old balance of power and therefore to threaten world peace. That Britain had such difficulty in coping with a small colonial people like the Boers suggested that her century-long hegemony over subject peoples was ended. Japan's defeat of a great European power forecast a serious challenge to white supremacy in Asia. The American defeat of Spain, which marked the "coming of age of the United States," led to the acquisition of Caribbean bases, of the Philippines, and of other Pacific possessions and was followed by the start of the Panama Canal in 1904. Popular agitation for a United States navy "second to none" indicated that domestic support was growing for a more aggressive foreign policy. Meanwhile, however, as a result of a traditional distaste for the antagonisms of the Old World, the United States held back from major involvement in world affairs.

GENERAL STAFFS

One direct effect of this preliminary period of small wars was that both of the two great English-speaking democracies geared their military machine more firmly to the state by the adoption of a variety of the system of military leadership already developed by Germany and France. In both countries there had long been strong opposition to the creation of a general staff, partly because both British and Americans had been satisfactorily successful against primitive peoples, and in Britain also because the Queen's cousin, George, duke of Cambridge, as commander in chief, had resisted it. On the other hand, in both countries there were strong supporters. In England, Spenser Wilkinson, an Oxford professor, wrote *The Brain of an Army* in 1890, which had great influence in bringing about the ultimate adoption of a general-staff system for the British Army. Civilian authority was preserved by the creation of an army council which included members of the government. The office of commander in chief was abolished, and many of his functions were taken over by a new chief of staff in 1906. A Committee of Imperial Defense, established in 1904, and the Imperial General Staff, as it was called in 1909, were used to coordinate the defense plans of the United Kingdom with those of the self-governing British dominions. However, the dominions retained independent control of their own armed forces.

The structure of the army was revised. Weaknesses in mobilizing adequate forces early in the Boer War had shown that if war with a great European power were to break out, the country would be virtually defenseless. Hence, the Cardwell system, created in the period 1868–74 to provide a garrison for the far-flung empire of the late nineteenth century, was replaced in 1905–6 by the Haldane system, a combination of regular and "territorial" (or reserve) forces, designed to provide an army for operations on the continent of Europe. As a result of Lord Haldane's work, when the First World War came, Britain was able to put fifteen divisions on the continent within a few months of the outbreak of hostilities.

The new British general staff was significantly different from its German counterpart. It was not a separate corps, and therefore it did not become distinct from the regiments. At the same time, it was subordinate to a civilian member of the cabinet and so was in no position to act independently in the making of policy. Even so, contemporary with the reorganization of the army and the creation of the general staff, successive foreign secretaries, the Marquess of Landsdowne and Viscount Grey of Talloden, authorized staff talks with the French without informing the whole cabinet. The result was that Britain was, in effect, committed to come to the aid of France in the event of a German invasion. At a time when European alliances were hardening into two bitterly hostile groups, this secret military arrangement, which was kept from the Cabinet from 1904 to 1912, and was never submitted to Parliament, was of great importance in making war certain.

Similarly, largely through the work of Elihu Root, secretary of war from 1899 to 1904, a general staff was created in the United States. Although consciously modeled on the German system, it was, like the British general staff, subordinate to civil authority. The constitutional powers of the president as commander in chief were left untouched, and Congress kept a tight hold on the purse strings. The new policy permitted a more cohesive organization of the United States army to band together the scattered units which, in the nineteenth century, had effectively policed the plains. Hence, when the United States became involved in a major war, the incorporation of the states' National Guard divisions, and a rapid general expansion of the army, were facilitated.

THE WAR BEGINS

The growth of national armies in the nineteenth century and of a competitive arms race, the Anglo-German naval rivalry described in a previous chapter, the development of techniques of mobilization which, once started, were difficult to stop, all pointed to the likelihood of a major European war in the event of some particularly critical "incident." The fact that many of the European powers were rivals for overseas empire, the existence of alliances with non-European powers like Japan, and the certainty that a long war would raise the vexing question of the freedom of the seas for neutral trade, made it likely that a major European war would encompass the world. The resources and armies which could be conscripted and committed forecast a conflict of unprecedented magnitude. Shortly after the war began, Germany had mobilized and deployed nearly 2,500,000 first-line troops and Austria about 1,500,000. On the Allied side Russia had sent approximately 3,000,000 men to the front and France nearly 2,000,000. Britain delayed conscription until 1916 but eventually mobilized 5,000,000 in her armed forces. Among the other great powers, the total forces raised were: Germany, 13,250,000; France, 8,000,000; Russia, 15,000,000; and the United States, 4,700,000. By the end of the war, 65,000,000 men of all nations had borne arms.

It was at first expected that the war would be short; both sides hoped for a quick victory. Germany, which took the initiative by violating Belgian

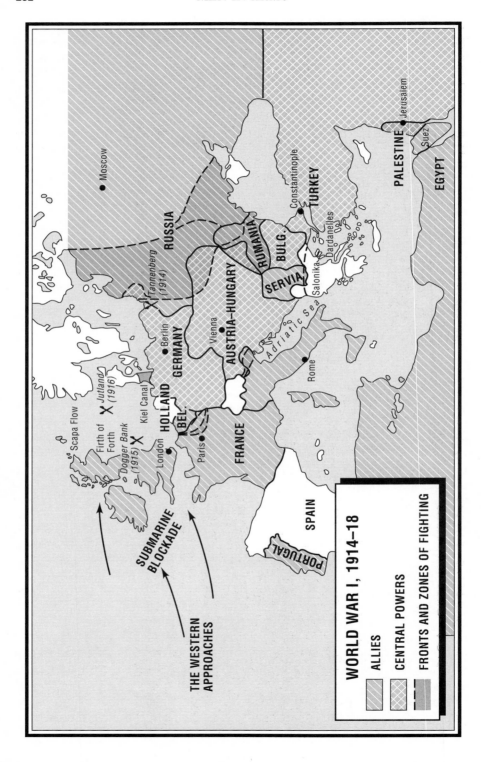

THE WESTERN
APPROACHES

SUBMARINE
BLOCKADE

Scapa Flow
Firth of
Forth
Dogger Bank
(1915)
London
Jutland
(1916)
Kiel Canal
HOLLAND
BEL.
Paris
FRANCE

SPAIN

PORTUGAL

Moscow

RUSSIA

Tannenberg
(1914)

Berlin
GERMANY

Vienna

AUSTRIA–HUNGARY

RUMANIA

SERVIA

BULG.

Adriatic Sea

Rome

Salonika

Constantinople

Dardanelles

TURKEY

Jerusalem

Suez

PALESTINE

EGYPT

WORLD WAR I, 1914–18

ALLIES

CENTRAL POWERS

FRONTS AND ZONES OF FIGHTING

neutrality to which the powers had agreed in 1839, relied upon a plan drawn up before 1914 by a former chief of staff, Count von Schlieffen. The purpose of the Schlieffen Plan was to avoid a long war on two fronts, which the Franco-Russian Entente seemed to threaten. As the vast area of Russia seemed to make a quick decision in the Eastern theater unlikely, Schlieffen argued that it was imperative that Germany should knock France out before Russian mobilization was completed. To achieve this end, he planned to invade France through Belgium, to weight the right of his armies for an overwhelming blow, and to lure the French to attack on the Rhine while the German armies on the right were sweeping around to accomplish an encirclement as effective as that at Cannae. However, by 1914 Schlieffen was dead and the control of the German military machine was in the hands of a nephew of the great Field Marshal von Moltke. He replaced a strategy of expedients by a hard plan. General Helmuth von Moltke (1848–1916), fearing that a French invasion in Lorraine might cut his communications and that the railways could not support the full attack on his right, strengthened his center. Furthermore, the rapid advance of the great German armies beyond their German railheads led to confusion and loss of contact which no staff planning had foreseen. An unexpectedly early Russian offensive caused von Moltke to draw divisions from the West for use against the Russians. Hence the Cannae on the Western Front was not achieved, the German invasion was halted at the battle of the River Marne in September 1914, and the invaders withdrew to the Aisne to dig in for a long war.

The failure of the adulterated Schlieffen Plan was the immediate cause of a great stalemate on the Western Front, which was to last for more than four years; but the real reason for the stalemate was the fact that developments in technology had led to a predominance of the defensive and had not yet been adapted to the offensive. This was most evident on the Western Front, where both sides could concentrate their biggest forces and where defeat would be decisive. The great railway network in this thickly populated industrial area, built with an eye to strategy as well as to commerce, had made it possible to rush huge armies to the frontier within a few hours of a general mobilization. In the later years of the war, the gasoline engine was increasingly used to haul supplies of ammunition and food from the railheads to maintain the huge armies deployed from Switzerland to the North Sea. Strategical considerations on the Western Front came to be subordinated to the tactical possibility of capturing German trenches.

TACTICS AND WEAPONS

Tactics on the Western Front were reduced to suicidal infantry assaults in formations which were theoretically skirmish order but which were sometimes so thick as to be almost the shoulder-to-shoulder lines of the eighteenth century, though without their drilled rigidity. The chief variations of

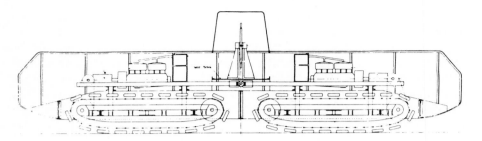

THE CATERPILLAR LANDSHIP, as submitted to the British Admiralty. 4 March, 1915. Leading particulars are as follows: total weight, 25 tons; over-all length, 36 feet; over-all width, 12 feet 6 inches; height to top of turret, 10 feet 6 inches; height to top of body of car, 7 feet 3 inches; turning radius, 65 feet; pressure on ground, 12 pounds per inch. The landship was powered by two 46-HP Rolls Royce engines. (Murray Sueter, *Evolution of the Tank,* London, 1937. p. 67)

these tactics attempted by the military leaders on both sides included preliminary artillery barrages to achieve surprise, creeping barrages behind which the infantry advanced, box barrages which isolated sections of enemy trenches, and saturation barrages which concentrated the fire of all available arms on a small area to obliterate it. In the battle of Verdun (1916), artillery dominated the battlefield and the infantry's function was only to take or to hold ground cleared by shells.

The ammunition required by the artillery for this kind of warfare was on a scale which no one previously had contemplated. Military leaders on the allied side were peculiarly slow to realize that vigorous measures must be taken to obtain it by gearing their great industrial plants to war production. When General Sir John French, the British commander in chief in France, was sending daily demands for more and more shells, the politicians had to overrule the soldiers at the War Office before the shells were produced. Then the Ministry of Munitions, under David Lloyd George, had to take control of war production away from the War Office.

Some attempts were made to achieve tactical surprise and the desired breakthrough by the introduction of new weapons. For instance, the Germans used poison gas shells in Poland in 1915, but with little success. On 22 April 1915, the Germans tore a great hole in the allied front at Ypres by the use of chlorine gas discharged from cylinders against French colonial troops; but they failed to take advantage of the breach and the Canadians quickly sealed it. Gas, although "improved" during the course of the war, failed to be a war-winning weapon because, surprise having been lost, protective measures were introduced; and the prevailing winds being westerly, the Allies had an advantage for retaliation. In the long run, the introduction of gas did the Germans more harm than good because Allied propagandists made effective use of this breach of the Hague convention. The Germans claimed that as cylinders were used, and not projectiles, it was not illegal.

A second example of the misuse of a new weapon which might have provided the tactical surprise necessary to breach the fortifications on the

Western Front is to be found in the story of the tank. Armored cars existed before the war began; but they were restricted in use by the necessity of staying on hard roads. The endless track was already available, being used on the great farms of North America. The employment of the gasoline engine in an armed and armored caterpillar tractor came just before the realization that new weapons had forced a stalemate. In 1916 the tractor and the armored car were effectively mated by Colonel E. D. Swinton with the blessing of Winston Churchill, the First Lord of the British Admiralty. Tanks were used for the first time on the Western Front on 15 September 1916, but not in the way their creator, Swinton, had advised. He had been overruled by Douglas Haig, the British commander in chief, and had been ordered to produce a small quantity for immediate use rather than build up a big tank force to achieve a knockout blow. Operating on unsuitable ground, and restricted in mobility, the tanks achieved only a temporary success. Skeptical commanders had failed to have infantry ready to exploit the breach. The great advantage of surprise was lost. However, it must be noted that the early tank was mechanically primitive and unreliable.

A third technological development must be mentioned in connection with efforts to achieve a tactical superiority on the Western Front. The gasoline-powered airplane, invented by the Wright brothers in 1903, developed with remarkable rapidity under war conditions. By 1916 and the Somme, virtually every future tactical use of the aircraft had been exploited operationally. The first function of the aircraft was reconnaissance; aircraft of the Royal Flying Corps played a significant role in the detection of the movements of Kluck's First German Army in 1914 and hence in the Marne counteroffensive. No longer could an army expect to move freely without detection; night movements of troops became the rule, and elaborate camouflage precautions had to be taken to attempt to disguise planned movements. From early 1915 aerial photographic reconnaissance assumed great importance, especially in preparing trench maps and in the detection of gun sites for counter-battery attack. At about the same time aircraft began to be employed to control and direct artillery fire. All these functions demanded protection. Therefore, specialized fighter aircraft mounting machine guns synchronized to fire through the propeller were developed. The Fokker single-seater fighter, which appeared on the Western Front in late 1915, gave the German air force immediate superiority. For the rest of the war, air superiority fluctuated as one side or the other developed better fighter aircraft. The spectacular mass dogfights of the air war were, however, less significant than the mundane activities of larger numbers of aircraft used more closely with ground operations. In the last two years of the war, these activities were extended to support of infantry and tanks, and to the ground strafing of enemy rear areas, particularly in the Battle of Amiens (1918).

Air bombing on the Western Front began as a tactical operation in support of ground forces. At Loos, in 1915, the Royal Flying Corps's tactical bombing program included strikes at German railway junctions and yards, bridges, and supply dumps. From the first, however, the Germans had shown an interest in the more distant, or strategic, possibilities of the air arm. They used Zeppelins (i.e., airships) and later giant Gotha bombers to raid military installations and cities in both England and France. These raids had more effect upon civilian morale and politics than the material damage warranted, and they caused a major diversion of British resources in order to build up a home defense organization. The German example was followed by the French and British air services; in 1918 a separate strategic bombing force, the Independent Air Force, was formed by the British to carry out raids upon industrial centers in Germany. Under its commander, General Hugh Trenchard, this force conducted a bombing offensive against Germany until the end of the war. He was planning to strike at Berlin in 1919 with giant Handley-Page aircraft capable of carrying a bomb load of 7,500 pounds. Thus, in the heat of operations, and long before the air-power prophecies of the interwar years, there had been conceived a strategic role for air power that was to bear bitter fruit in years to come.

MILITARY CONSERVATISM
AND
TACTICAL THEORIES

With these scattered exceptions, military leaders were far too conservative to attempt to introduce revolutionary weapons as a means of gaining victory. Where they did show some ingenuity as the war progressed was in varying their infantry attacks. They varied the time of the attacks, tried to conceal the accumulation of supplies and the movement of troops, and used different methods of coordinating infantry and artillery. However, these variations had two common features to which reference was made earlier: the employment of infantry in heavy formations in a direct charge and the use of artillery as an offensive weapon to clear the ground. The results were not impressive, except in casualties.

Military leaders on both sides were undeterred in their insistence that only through such attacks could the way to victory be found. In this determination they were largely conditioned by the lectures of one of their number, the French General Ferdinand Foch, who for some years before the war had taught at the French War College. France was wedded to an offensive doctrine to regain Alsace-Lorraine and also the departments occupied by the German invasion in 1914. The gist of Foch's teachings had been an emphasis on the importance of morale. He said, "The purpose of discipline is not obedience but to make men fight in spite of themselves," that there could be "no victory without fighting," that "to make war always means attacking," and that "a battle won is a battle in which one will not confess oneself beaten." Colonel de Grandmaison had developed this morale-raising teaching into a rigid doctrine of what he called the *offensive à outrance* (offensive beyond

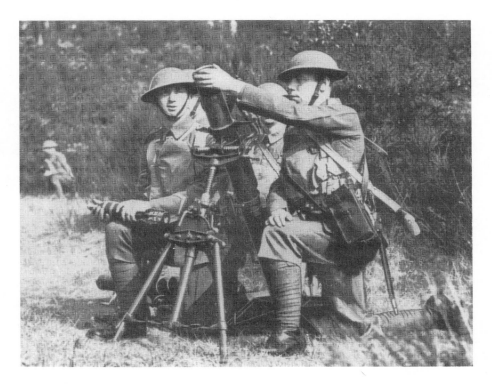

INFANTRY MORTAR, World War I. The mortar was originally an artillery siege piece or defensive weapon with a broad gauge and a high trajectory of fire. From this beginning was developed the howitzer, on the one hand, and the field mortar on the other. (Low, *Modern Armaments,* Gifford, London, 1934, p. 66)

anything else). This spirit had nearly brought a French defeat in 1914, when Plan XVII, a strategy of attack in the center, had failed to take account of German numbers and flexible strategy. However, as the war went on, the generals on both sides developed an oversimplified, and somewhat illogical, doctrine of morale and mass attack which stated, in effect, two contradictory things: "superior numbers will always win," and "superior morale will always win." From a combination of these they developed the principle that whatever the circumstances, it was correct to attack. This was almost their only answer to the great advantage which the rifle, the machine gun, and entrenchments had given to the defensive.

As the war proceeded, continuous entrenchments became stronger. Concrete supplemented earth and sandbags; and when the Germans retired to a new defensive line, the Hindenburg Line, in March 1917, it was hailed as impregnable. Its plan was based on the principle of defense in depth, using concrete pillboxes, which the Germans had developed from experience at Verdun and on the Somme.

Sounder offensive tactics developed slowly. The principle of infiltration in place of a general assault in line had been suggested early in the war by Captain Laffargue, a French officer; but the Allies made little attempt to model their offensives on it, relying on heavy artillery bombardment, later supplemented by the infantry tank. However, at Verdun, in 1916, the Germans used combat groups of about fifteen men, and then developed this idea further in March 1918, when the German chief of staff, Ludendorff, pushed small groups of infantry forward and through the line and switched the weight of his attack from place to place by moving troops laterally by motor bus on roads immediately behind the line. Thus flexibility crept painfully back into war. The great Allied counteroffensive which finally broke the German lines, and apparently justified Foch's doctrine of "attack," was based on a rapid series of attacks at different points, as well as on an overwhelming mass of men and matériel.

It must be noticed that throughout the war, although many cavalrymen were dismounted to fight as infantry, large numbers of cavalry were kept behind the lines of the Western Front for a breakthrough that never came. On other fronts, where the concentration of rifles and machine guns was thinner, the horsemen saw more action. But the value of cavalry on the Western Front was a negative one. Large numbers of men were kept idle, and forage for the horses entailed a tremendous strain on the supply lines. Overlooking the fact that horse-drawn transport was still vitally important and used even more fodder than the cavalry, Major General J. F. C. Fuller, the British exponent of tank doctrine, went so far as to allege that if the Allied leaders had not maintained a mistaken belief in cavalry, bread rationing might not have been necessary in England.

THE STRATEGY OF THE INDIRECT APPROACH

In addition to the tactical problem of breaking through on the Western Front, there was an interesting problem of grand strategy which led to strong differences of opinion. While the French and most of the British military leaders insisted the war would be won or lost on the Western Front, an important group of civilian leaders, among whom Winston Churchill and David Lloyd George were prominent, argued that the Allies should make use of their command of the sea to attack the Central Powers in the rear. Most professional military opinion declared that the "Easterners," as the school which advocated this strategy of indirect approach was called, were ignorant of the logistic problems involved in supplying large forces at a great distance from their principal bases. Efforts made elsewhere than on the Western Front were therefore usually inadequately supported, either on account of the problem of geography or because of the lukewarm interest of the British and French military commands. This "indirect approach," as the British military historian Basil Liddell Hart dubbed it, tied down millions of Allied troops in operations peripheral to the decisive Western Front. The Germans left those distant theaters to their Austrian and Turkish allies and husbanded their own troops for action on the Western and Russian fronts. Turkey, which had joined the Central Powers in November 1914, at first defended herself successfully against Russian, French, and British attacks. The attempt to force the Dardanelles, link with the Russians in the Black Sea, and knock the Ottomans out of the war by an Anglo-French naval bombardment in February 1915 was a costly failure. It was followed by the abortive landings on the Gallipoli peninsula. An Anglo-French landing at Salonika (which, in the fashion of total war, ignored the desire of Greece to remain neutral) was only a little more successful. Not until 1917 did the British, under General Edmund Allenby, make significant advances into Turkish territory, advancing from Egypt and capturing Jerusalem at Christmas of that year. It is perhaps some vindication for the Easterners' school of indirect approach that the crack-up of the Central Powers came first in the Near East and Balkans. On the other hand, the "Westerners" could still claim that the only place where the war could be, and was, won was on the Western Front.

THE PROBLEM OF COMMAND

This great strategic dispute draws attention to a vital problem of command. Under the German system of government, the military leaders, controlling the high command and operating through the general staff, had a centralized control over the war effort. The only political group which might have challenged that military hegemony was the Social Democrats; but upon the outbreak of war their German patriotism easily overcame their socialist doctrines, and they left the direction of the war effort to the traditional leaders of Prussia, the Junkers and the military hierarchy. After the dismissal of Falkenhayn in August, 1916, von Hindenburg had been placed at the helm, with Ludendorff as his chief of staff. This pair were masters, not merely of the military machine but of the economic side of the war effort too, and they continued to be in command until almost the end of the conflict.

In the democracies, the problem of command could not be settled so easily. The military leaders, who were appointed by the civil government and were subservient to it, were, in theory, left in complete control of all matters military. But in modern war the dividing line between military and civil matters is not easy to draw; and in any case, grand strategy at the summit is a responsibility of the civil government. In both France and Britain this problem led to bitter disputes. Control of the war effort could not be surrendered completely to the military leaders without endangering the principle of the sovereignty of the people. Hence civilians were drawn in to make decisions about military policy and to encroach on the traditional sphere of the professional soldier. In several major instances their instinct was sounder than the soldiers' expertise.

But their intrusion frequently met resistance from the soldiers. Thus Marshal Joseph Joffre refused permission for the Military Committee of the French Chamber of Deputies to visit the Western Front on the grounds that civilians had no place near the battle line. Not until October 1917 were the constitutional rights of the committee recognized by the commander in chief, and by that time Clemenceau had fully asserted civilian authority. War, he believed, was too important to be left to soldiers.

The disputes in Britain, between soldiers and statesmen over the supply of munitions, and over the grand strategy of the war, have already been mentioned. Lloyd George, who became prime minister in December 1916, insisted on taking control into civil hands; but he did not succeed in gaining the confidence of his professional advisers, nor did he entirely rely on their professional competence. The military men accused him of acting for purely political motives, of being concerned more with votes than with the safety of the Empire. On the other hand, it must be remembered that even when the prime minister had doubts about the competence of his military leaders, it was no easy matter to dismiss them without destroying the confidence in their war leadership that had been deliberately built up by propaganda. The civil-military dispute did not become so urgent in the case of the United States in this war, partly because American armies were operating further from their base, partly because of the constitutional position of the president as commander in chief, partly because of the personalities and abilities of War Secretary Newton D. Baker and General John J. Pershing, but most of all, perhaps, because the duration of American war effort, and particularly of effort in adversity, was not long enough to bring matters to a head.

One fertile source of trouble between civil and military leaders in Britain was the question of a unified Allied command. From early in the war, the politicians were convinced of the disadvantages arising from a divided command on the Western Front when it faced the centralized control exercised by the German High Command. Influenced by a belief in the military prowess of the French generals, and by some distrust of their own, Lloyd George and some of his colleagues, supported by General Sir Henry Wilson, were prepared to subordinate the British Army to the French commander in chief. "Wully" Robertson, the British chief of the Imperial General Staff, who had risen to eminent position from the ranks, and Haig, resisted this move grimly because they were doubtful about the competence of Nivelle and French motives. They used the argument that the British soldier would fight better under his own leaders. It is difficult to believe that there was not also an element of personal ambition, particularly as Haig was a strong advocate of the complete integration of the troops of the dominions, and even of American troops, into the British army or under British command.

When American forces began to arrive in France in 1917, both British and French military leaders pressed for the incorporation of American infantry into their own armies to fill the gaps caused by the awful casualties which they had suffered. They were astonished by the insistence of General

Pershing that American troops would fight better as a national army; and they yielded only reluctantly and partially to this argument. So, while unity of command at the highest political level had been achieved by the creation of a Supreme War Council, it was not matched by an effective unity of military command when, in March 1918, the great German offensive threatened to drive the armies apart, forcing the French back on Paris and the British on the Channel ports. At this point, Haig, seeing the danger, proposed the appointment of Foch as generalissimo to coordinate the action of all Allied armies on the Western Front. President Wilson ordered Pershing to accept Foch's directions in strategic, but not tactical, operations. This was a timely solution of that vexing problem of inter-Allied cooperation, which had nearly led to defeat.

THE ECONOMIC FRONT

But the war was much more than a military affair in which questions of strategy and tactics, and of military command, were all-important. Far more than ever before, the nations as a whole were harnessed to the war effort. Germany, ringed about by enemies and cut off by the blockade from her former sources of supply, had to accept state direction of the whole of her economic life. In the first few months of the war, Emil Rathenau, an industrialist, set up controls over the whole field of production. Later the General Staff sponsored the creation of a new *Kriegsamt,* or War Office, to organize war production and to produce *Ersatz,* or substitute, materials for those necessaries in short supply. Rationing was an inevitable part of this machinery. Other nations adopted measures of the same type, although to a lesser degree.

The impact on national life and on society was tremendous. People learned to submit to state control to an extent not before contemplated. Labor was directed, taxation was increased to phenomenal heights, women went into industry and into the armed forces, and in Russia even fought, and finance and industry came under state direction. The use of substitutes led to changes in the eating habits of the household and some of them were permanent; some new inventions, like artificial fertilizer that replaced imported phosphates in Germany came to stay and were important stimulants for future peacetime industrial developments. But the effect of such innovations on the life of the people was less important than the cumulative effect of state direction and control. While in the democracies the legal supremacy of parliament or assembly was carefully preserved during the war, the tendency to rule by order-in-council or executive order, and the willingness of the people to accept such rule, was to have far-reaching effects and to set precedents for a new kind of state control.

Only those states, whether autocratic or democratic, which were already highly developed industrially could fully reorganize themselves for this kind of war; and failure to do so spelled defeat. Imperial Russia was

the first of the major states to crack under the strain. At the beginning of
the war Russia, the "Colossus of the North," had been portrayed in Allied
propaganda as a steamroller; and it was confidently expected that when her
millions were armed, she would roll over the Central Powers. Instead, when
the Dardanelles-Gallipoli strategy failed to open the way for the arms she
needed, Russia became a liability. The war on the Eastern Front rolled
backward and forward much more freely than in the West, and the sufferings of the Russians were even more severe than those of the troops on
the Western Front. Russia and Austria, lacking adequate industries, proved
much less able to supply their armies than Germany, France, and Britain.
While Austria could be propped up by Germany, Russia was too remote
from her allies to be effectively sustained. Eventually her communications
system collapsed, and revolution broke out both among her industrial
workers and in the armed forces. The net result of failure on the military
and economic fronts was the communist revolution. This same phenomenon
was to reappear in Germany in 1918. The economic dislocation which the
war produced, thus helped to create the communist totalitarian state.

THE NAVAL WAR

Meanwhile, war on the economic front had included the traditional use of
sea blockade, but with new weapons, new methods, and a new ruthlessness.
When the war began the rival navies, unlike the armies, were already mechanized. By December, 1914, all German shipping except submarines had
been swept from the seas. In that month the Battle of the Falkland Islands
secured for Britain the decisive advantage of full use of the waters of the
Southern Hemisphere. Despite the fact that one major cause of the war was
the Anglo-German race in building battleships, only twice did the fleets
of capital ships meet, at the Dogger Bank action, 24 January 1915, and at
Jutland, 31 May 1916, an unexpected engagement in which the British lost
more heavily in ships but were left in control of the North Sea. After that
battle, the German fleet did not come out again to seek its opponents' main
fleets until it came to surrender. Even so, both before and after Jutland,
the necessity of watching and blockading the German navy tied up a large
proportion of the Royal Navy in the remote bases of the Firth of Forth and
Scapa Flow.

   The Germans thus made good use of the principle of the fleet-in-being,
while following the traditional policy of the weaker sea power, commerce
raiding. For this purpose they had a powerful new weapon, the submarine.
When Britain proclaimed the blockade of Germany and published a contraband list that included every kind of goods which might conceivably be used
in the war effort, the Germans, arguing that the remote blockade was not
legal, retaliated by proclaiming all the waters around the British Isles war
zones in which vessels would be subject to attack by German submarines.
Because merchantmen were being armed and the Allies were using Q-ships,

that is, armed vessels disguised as tramp steamers which uncovered their gun when the U-boat surfaced for the kill, the Germans sank merchant ships on sight without providing for the safety of the crew and passengers. In 1917, the rate of sinkings grew so high that Britain, dependent on imports for raw materials and food, found her economy seriously threatened. The Allies were forced to protect their trade by organizing convoys. But they were short of escort vessels because of the overconcentration in the North Sea. The German submarine campaign had come close to bringing Britain to her knees.

The most important result of the blockades imposed by both combatants was their effect on neutral nations. The United States, by far the most important neutral, naturally became the champion of neutral rights, following her traditional policy of the "freedom of the seas." Britain's insistence on comprehensive contraband lists, the black listing of firms which "traded with the enemy," the assertion of the right of search, and the forcing of neutral ships to report at certain British control stations aroused bitter resentment among American shipping interests. However, in the long run these resentments were overbalanced by the German policy of torpedoing neutral ships without warning. It was this latter practice which played the greatest part in influencing America to enter the war on the side of the Allies.

CONFLICT OF IDEOLOGIES

The submarine campaign was not the only factor in deciding American policy. American financial interests had lent money to both combatants but, influenced partly by a greater sympathy, and partly by the fact that the blockade made it easier for the Allies to operate in the United States, the loans to the Allies amounted to one and a half billion dollars against 27 million dollars to the Central Powers. Although many Americans of foreign extraction wanted the Central Powers to win, there was a far greater bond of sympathy for the Allies, especially for France, with whom the United States had had ties of friendship since the early days of the Republic, and for Britain, with whom she had a common language and a common inheritance of free institutions and of law. When the collapse of Czarist Russia made the war appear more clearly as a conflict between liberty and autocracy, the way to American participation was made easier.

At the same time, it is important to realize that this element of ideological sympathy was stimulated by deliberate propaganda which both sides undertook, but in which the Allies had a natural advantage because they were able to identify their cause with the cause of constitutionalism against autocracy, and therefore of liberty against tyranny. They also proved to be more adept in the arts of propaganda. After the war, some of the deceptions of the propagandists were to recoil upon their heads. In the 1920s many Americans became convinced that the United States had been dragged into war, not so much as a result of the submarine campaign, but as a result of the activities of Wall Street financiers and also of Allied propagandists who had exaggerated German atrocities in Belgium, and had invented such lies as the famous story of the "German corpse factory." This was alleged to have produced fats for industrial use. The manipulation of public opinion and the deliberate use of propaganda were as important elements in the new total warfare of the twentieth century as was genuine ideological sympathy. However, one of the most important causes of the ultimate Allied victory was the moral crusade preached by President Wilson. His enunciation of the Fourteen Points promising a new kind of world in which war would be banished became the basis for discussion of the terms upon which the Germans would surrender.

It would be doing less than justice to the armed forces, however, and to the material efforts of the workers on the Allied side if it were not made clear that by this time Germany had suffered severe defeat on the Western Front and was losing her allies. Undoubtedly what had saved the Allied cause was the timely appearance of huge fresh American armies backed by overwhelming material resources. By early 1918 both sides were war-weary. The German High Command had promised the German people that the United States would never be able to bring her armies across the Atlantic through the U-boat blockade. Although the Army of the United States was insignificant in size in 1917, by the middle of 1918 a quarter of a million American soldiers were landing each month in France. When this news reached the German soldier and when he saw the spirit and equipment of the American forces, the end beckoned.

17

TOTALITARIANISM
BETWEEN THE WARS

THE LEAGUE OF NATIONS

T he World War of 1914–18 had had a most serious effect on the society of the whole world. Its physical consequences alone were enough to slow the onward march of civilization and to destroy that general belief in the inevitability of human progress which had marked much of the philosophy of the nineteenth century. The material cost of the war, including property damage, has been estimated at twenty-eight billion dollars; and the number of killed and permanently disabled, military and civilian, at twenty million. As the vast majority of the latter were young males and potential fathers, the actual loss to the world's population was much greater. Dislocation caused by direct losses was only a small part of the story. The wartime redirection of channels of trade, the stimulation of uneconomic industries and of submarginal agriculture, the postwar rearrangement of political frontiers on lines of national "self-determination" cutting across well-established regional economies, the huge debts incurred by the belligerent governments, and the heavy reparations imposed on the vanquished, all placed a tremendous strain on national economies and on the balance of international payments. Currencies crashed, unemployment figures rose, unrest flourished, and moral standards declined.

For the first time in human history, the scourge of war came to be regarded by a large part of mankind as a primary evil. Some evidence of this may be found in the victors' attempts to punish "war criminals" and those alleged to have been guilty of causing the war. There was talk of bringing the Kaiser to trial and even of hanging him. While these plans came to nought, the same tendencies revealed themselves in an indictment of guilt written into the peace treaty with Germany, and in the large sums which were to be levied for reparations. The Allies were moved to a considerable extent by a desire for revenge; but these things were also indicative of a new attitude; namely, that those guilty of making war and of war crimes must by some means be brought to atone for their deeds.

Clearer evidence of this new attitude to war is found in the fact that the statesmen entrusted with the task of making the peace treaties were also charged with the planning of a system for the prevention of a recurrence of the blood-letting and devastation. There obviously was a widespread revulsion against war. Many people became outright pacifists, declaring that war was a far greater evil than any other evil which it might be used to remedy; others sought to distinguish between "just" and "unjust" wars and to demand machinery for preserving peace. Hence the Covenant of a new type of political organization, the League of Nations, was written into the treaties which put a legal end to hostilities. Many people believed that mankind was entering a new era in which war would be no more.

The basic principle of the League of Nations was the new concept of "collective security." The member nations solemnly promised to refrain from recourse to war to settle their grievances until three months after other methods had failed. While no clear and effective method was established for the prevention of aggression, it was hoped that the moral pressure of organized world opinion would be a powerful deterrent. If that failed, the Covenant provided that economic and military "sanctions" could be imposed.

From its birth, the League was weakened by troubles which were probably inevitable in a new and revolutionary political device for which the nations were not yet ready. It was quickly found that it was impossible to obtain resolute action to check incipient or actual aggression, especially when powerful nations were concerned. Furthermore, when attempts were made to define aggression and to tighten the machinery for preventing aggressive war, some nations were more anxious to limit such obligations as the Covenant had already imposed upon them; and many of those who were anxious to improve the machinery for preserving peace were at opposite poles in their ideas. Disappointed by their failure to obtain an Anglo-American guarantee in 1919, the French wanted an international army to restrain Germany; the British Commonwealth regarded the League primarily as a diplomatic meeting ground where differences could be resolved by customary British methods of discussion and compromise. As these were by far the strongest powers in the League in the twenties, little progress could be made toward a more effective organization.

Furthermore, a fundamental weakness from the beginning was that the League was not universal in its membership. The United States, which had sponsored the plan at Versailles, returned to its traditional policy of isolation from European entanglements and did not join. Russia, Germany, and Japan were, for long periods, nonmembers. In these circumstances, the League of Nations lacked the broad base which was important for its success. It is not surprising, then, that within a few years of the creation of this world organization to ensure peace, new defensive alliances were being forged (like that between France and the Little Entente of Czechoslovakia, Rumania, and Yugoslavia); and regional pacts were arranged to guarantee peace in especially dangerous areas (like the Locarno Treaties which dealt, among other regions, with the western border of Germany). These were but poor alternatives for the hopeful aspirations of the founders of the League.

The frustration of the idealists led to a new attack on the problem of war. From the United States came a proposal for a general agreement to outlaw war. By 1931, fifty-six states had signed the 1928 Kellogg-Briand Pact solemnly renouncing war as an instrument of policy. Without any provision for enforcement, such an undertaking among sovereign nations proved to be of little practical worth. It would be honored only as long as each of certain signatories was convinced that peace would serve its ends better than war.

DISARMAMENT

Contemporary with the schemes to check aggressive war and to outlaw war altogether, there was a long-drawn-out attempt to render war unlikely by a process of general disarmament or, more accurately, by a general agreement to limit national arms. This was based on the realization that the anti-German naval arms race had been one of the important causes of war in 1914, and also on the belief that a nation which possessed a great and expensive military establishment was likely to be tempted to make practical use of it. The arms limitation imposed on the Germans at Versailles had been accompanied in the Treaty by a vague statement about aspirations toward general disarmament; the League Covenant provided for a plan of multilateral arms reduction; and one of the first actions of the Council of the League was to instruct its Permanent Armaments Commission to explore the problem. When the First General Assembly met, it was argued by many delegates that the Council's committee, which was dominated by the military advisers of the great powers, would not really be interested in cutting the size of armies, and so the Assembly appointed a "Temporary Mixed Commission," which included civilians, to prepare a plan for disarmament.

But still progress was very slow. There was suspicion among the powers and therefore a reluctance to lay down arms; and the technical difficulties were great. The armed strength of a state and its requirements for security are affected by the degree of its industrialization, by the literacy and technical capacity of its people, by the circumstances of its geography, and by other immeasurable factors. Proposals to use the amount of money expended as the measure for scaling arms down were unacceptable to those nations in which the cost of manufacturing arms was high. No covenient formula could be found by which various kinds of technical troops could be equated with infantry and with one another. Hence it was 1931 before an agreement to call a

World Disarmament Conference was reached. When the conference met on 2 February 1932, the Nazi seizure of power in Germany was only a year away. It had met too late.

Earlier, in November 1921, as a result of the impatience of the United States at the slow progress being made toward disarmament, and more especially at the high cost of American naval programs, a conference to deal with naval disarmament had been called at Washington. There the chief naval powers, the United States, Great Britain, Japan, France, and Italy, agreed to a ten-year "holiday" in the building of battleships and aircraft carriers, to limitations on size, and to a ratio of 5:5:3:1⅔:1⅔ in capital ships. Agreement may have come only because some admirals believed that battleships were out-of-date.

But the Washington Conference achieved no comparable success in other categories of vessels: The cruiser question led to acrimonious debates between Britain and the United States about the kind and number of cruisers each should possess; the submarine problem led to trouble between France and Britain; France was reluctant to accept absolute equality with Italy since she felt that she had much greater responsibilities; and Japan accepted naval inferiority only as a temporary expedient. The reason for such success as was achieved was simply that the United States, which could so obviously outbuild all other powers if a naval building race began, wanted to cut the arms budget; and battleships were the biggest single item in that budget.

The failure to bring about general disarmament, on which great hopes for peace had been placed, was one of the chief causes of the failure of the League. The problems of collective security and disarmament were inextricably entwined. No nation could disarm until it felt reasonably secure; but no nation would make a contribution to collective security by surrendering its freedom of action and a vital part of its sovereignty while its potential enemies were strongly in arms. Every power protested that it had only the minimum armament needed for its defense. Each power saw its neighbor's arms as a potential threat. There appeared to be no way out. When the World Disarmament Conference met in 1932, the Germans claimed equality in arms on the ground that the general disarmament, toward which they alleged the limitations placed on Germany at Versailles had been only a first step, had not been achieved. The first signs of a new arms race had thus appeared. By late 1933, Adolf Hitler had recalled the German delegations from the conference, had announced the rearmament of Germany, and had pulled the Germans out of the League.

Actually, Germany had set out on the way to rearmament long before the rise of Hitler. Willingness to renounce war as an instrument of policy was, not unnaturally, less general among the vanquished than among those who had triumphed in 1918. War-weariness, of course, existed in Germany. But the militarists had long been all-powerful, and their defeat in 1918 simply left them convinced that only by the re-creation of Germany's armed strength could the verdict of the war be reversed. By the terms of the

Versailles Treaty, the German army was limited to 100,000 long-service troops for purposes of maintaining internal security; battleships, tanks, and military aircraft were forbidden; the Great General Staff was to be dissolved; and war industries were to be demolished. From the first, these conditions were evaded. The Treaty army was increased by 150,000 *Schutzpolitzei,* who were theoretically policemen but were trained and armed just like soldiers. The General Staff lived on under another name, the *Truppenamt,* or Troop Office. Skillful measures were taken to impede the Allied Control Commissions in their work of supervising the demolition of war industries and of limiting the army. Illegal private armies, the *Freikorps,* were regarded by the military authorities of the *Reichswehr* with paternal friendliness; and disciplined organizations of veterans, like the *Stahlhelm,* were encouraged. Private civil flying, glider schools, and airlines, aided by government money, prepared the way for the day when military planes would be available. Troops practiced for mobile warfare with cardboard tanks. After the Treaty of Rapallo, April 1922, Soviet Russia permitted training on Russian soil with tanks denied to Germany by Versailles. Under the designation "Development Projects in Experimental Motor Boat Technology," the German navy kept its hand in the field of U-boat design. Since some of these designs were actually built in Spanish yards for Turkey, testing was also carried out. The German army made up for its limitations in size by high standards of efficiency; and it consisted of highly trained specialists, every man a potential officer or NCO ready for the day when Germany once again would have a great national army. Furthermore, compelled by circumstances to seek every means to increase its power and effectiveness, the *Reichswehr* was more willing than most other armies to investigate new doctrines and methods of warfare.

While Germany was obliged by the Treaty of Versailles to abandon conscription, other countries took the same step voluntarily after 1919. Armies and military service were generally unpopular; economic difficulties meant that military budgets must be cut to the bone; the prospect of peace by way of collective security and disarmament seemed to make the maintenance of large military establishments unnecessary, and the experiences of World War I raised a cry of "never again." After the war Britain quickly repealed conscription, demobilized her armies, and returned not merely to the pre-1914 Haldane system but to something like the Cardwell army designed mainly for imperial garrison service. In the United States, universal compulsory service for a three-months' period was proposed but had to be abandoned in face of congressional opposition. Thus the United States also returned to the old system of a small long-service army.

INTERWAR MILITARY THEORIES

Advocates of a regular army argued that it was greatly superior in efficiency to a conscript army. General Sir E. B. Hamley, British military writer, professor of military history at Sandhurst and commandant of the Camberley Staff College, pronounced in *The Operations of War* (1867), a standard textbook, the doctrine that a "regular army is immeasurably superior to an armed population," and this position was echoed by military writers with a conviction that was probably more than mere rationalization. Professional soldiers, disgusted by the warfare on the Western Front, argued that too many soldiers were as dangerous as too few because of the impossibility of maneuvering a great mass army into position for an overwhelming assault. A small, efficient, professional army and a quick victory were believed to be the remedy for the trench-warfare stalemate.

In France, the debate about the nature of the postwar army followed the same lines as in Britain and the United States, but other factors influenced the French decision. In the first place, the Revolution of 1789 had left a tradition of universal service which was quite as strong as the contrary British and American fear that large armies were a danger to constitutional liberty. While the Channel and the Atlantic, and reliance on sea power, enabled Britain and the United States to avoid the implication that democracy entails the obligation of universal military service, France had no such security and so built the Maginot Line, manned by an *armée de couverture* (i.e., defensive covering troops) to shield an offensive force. This fostered a defensive mentality. The French debated the army question as an alternative between taxing for a highly trained, costly, professional, long-service army, and calling up young men for regular military service. There was opposition to both plans, but eventually the taxpayers outvoted the recruits. The victory at the polls of the Cartel des Gauches (Leftist Alliance) in 1924, the Leftists who, by tradition, supported a large national army of conscripts as against a small professional army won out and ensured that France would retain the system of the nation-in-arms.

But it was a victory for which a price had to be paid. The term of conscript service, which had been raised to three years in 1913 in face of German militance, was reduced to eighteen months in 1923 and to one year in 1928. This meant, in effect, that apart from her *armée de couverture* on the frontiers, France possessed no organized trained troops at all. Her other regulars were engaged in training conscripts; and the latter, as soon as they were trained, were released. Thus, while Germany managed to create a highly efficient army of a quarter of a million, France, which was generally believed to be the most powerful military nation in the world, had little more than a militia behind the armies in her frontier defenses.

A debate about the maintenance of the nation-in-arms as against the *armée de métier* (professional army) was paralleled by a related but distinct debate about mechanization. While professional soldiers generally, in their dislike of the armies and tactics of 1914–18, were almost universally agreed that there must be a return to the small regular army, not all were willing to agree that such an army must be highly mechanized. World War I had shown that the defensive had become dominant. Discussion after the war ranged about methods by which mobility and decision could be restored.

The great tank battle of Cambrai, the battle of Amiens, the campaigns which had brought about the defeat of the German army in the West, and Allenby's campaign in the desert were all studied intensively. Trench warfare was generally regarded as having come and gone during the last war; but there was no general agreement on the means by which it could be avoided.

Some French soldiers, following the main line of development of the war, believed that the emphasis must be put on firepower, with the infantry taking over ground in which all opposition had been eliminated by artillery. Slow-moving heavy tanks, operating with the infantry, would help to cut down casualties. Other military specialists like the Italian Douhet, the American Mitchell, and the White Russian Seversky, argued that aircraft would completely revolutionize war, that the great battles of the future would be aerial battles, that land and sea forces would at most be ancillaries to the dominant air arm, and that wars would be won by seizing control of the air and by aerial bombardment. Some writers like the German general Von Seeckt and the American historian R. M. Johnston (*First Light on the Campaign of 1918*) put all their faith in small, highly trained, ground armies for a quick victory. In England, General J. F. C. Fuller, who had planned the battle of Cambrai, advocated all-tank formations. Captain Liddell Hart, the military correspondent of the *Telegraph* and later of the *Times,* while also urging the use of tanks in large numbers, taught that armies must be balanced units of all arms using aircraft and tanks to restore mobility and motorized infantry to keep up with the speed of modern war. In France, Charles de Gaulle, one of Marshal Pétain's aides, who had lectured at the Ecole Supérieure de Guerre in 1925–26, opposed reliance on great conscript armies. In the United States, Brigadier General Adna Romanza Chaffee, Jr., (1884–1941) and a few tank enthusiasts explored ideas about their use in future wars.

But the conservatives in all armies, many of whom were in high positions and were chiefly concerned with justifying their World War tactics, were hard to convince. An editorial in the British *Army Quarterly* in July, 1921, defended the tactics of 1916 and 1917 with the argument that their results should be measured, not in terms of ground gained but in terms of the exhaustion of the enemy. It claimed that German losses in the battles in those years had made possible the victory of 1918 and it took no account of the fact that the Allies had also suffered crippling losses, nor of the fact that American manpower had turned the tide. British Field Service Regulations, in 1924, asserted that "infantry is the arm which in the end wins battles" and while admitting that "to enable it to do so, the cooperation of other troops is essential," implied that infantry was the arm around which all tactics should be built. In 1932, the tank enthusiast, General Fuller, was retired from the Army. Only 500 copies of his *Lectures on Field Service Regulations, III: Operations between Mechanized Forces* were distributed in Britain; but they were known in Germany and also in Russia. Liddell Hart was influential in Germany more than at home. Although Guderian and other German tank generals were to complain later that they had been thwarted by conservative opposition, they actually fared much better than their counterparts in the West. The French army was trained to misuse tanks by splitting them up among the infantry; and in Britain in 1935 twice as much money was spent on cavalry as on tanks and armored cars.

To some extent, the slow development of new weapons of war, like the tank and the military aircraft, was the result of parsimonious military budgets. When their appropriations were pared to a minimum, the military chiefs had good excuse to avoid experiment, to neglect costly development projects, and to adhere to well-understood tactics and arms. However, even after rearmament had begun in Britain in 1935, although the navy and the air force were planning new forms of warfare, the tank forces were still behind the times. When the British army began to motorize and mechanize its forces, conservatism held back the construction of tanks and put greater emphasis on the tracked infantry vehicle, the Bren carrier. All armies were slow to equip their tanks with large guns, preferring to concentrate first on the development of defensive armor. But it was only a tracked, armored vehicle mounting a heavy gun that could fully restore mobility to war. The tradition that firepower was defensive only, and that the bayonet and the *arme blanche* (the sword) were the proper weapons for the offensive, was dying hard.

AIR POWER

The controversy between the conservatives and the innovators was brought into sharper focus in the discussions about the use of air power. Basically the question was whether air power should be used as an ancillary to the existing ground and sea forces or should be granted an independent and co-equal status. The former would presumably increase the effectiveness of cooperation by the preservation of the traditional principle of unity of command; the latter would permit a fuller development of the potentialities of the new arm by men who specialized in it, and would make possible the implementation of air power in a revolutionary way, by striking at the enemy far behind his protecting military and naval forces.

Toward the end of World War I, Britain had accepted the principle of an independent air force. In order to maintain it in peacetime, its leaders, exaggerating the effect of Germany's Zeppelins and of the RAF's own bombing campaign against Germany, stressed its independent role. Between the wars, British airmen built up a doctrine of air power based on what was unfortunately miscalled "strategic" bombardment. Up to 1923 "strategic" bombardment was related to close support for armies and navies. After that date it referred to more remote targeting which it would have been more correct to name "grand strategy."

The theory was that the first duty of the air force was to destroy the enemy's air power. When that was achieved, the destruction of the enemy's economy could be undertaken. Cooperation with the army was to be the work

of special squadrons but was limited chiefly to aerial reconnaissance and artillery spotting. The use of Royal Air Force aircraft, independently of the
army, as garrison "police" forces in the deserts of the Middle East gave some
justification for theories of air bombardment. Conservatism and shortage of
money slowed development. The heavy bombers to carry out this distorted
role did not get beyond the drawing board until January, 1937. The deterrent
striking force was therefore not in existence when the war began. Meanwhile
the home-defense force had been neglected; and had it not been for a privately financed RAF entry in the Schneider Trophy contest for float plains
in 1931, Britain might have developed no modern fighting plane by 1939. The
Supermarine S6, which won the trophy, was adapted to become the Spitfire,
and in 1934 the air ministry belatedly accepted the monoplane design. In By
1939 the Hurricane had also been developed and the British scientist Robert
Watson-Watt had invented radar; but the heavy bomber force was not yet
equal to the tasks envisaged for it by British theorists of air power.

In 1921 in the United States, Brigadier General William Mitchell had
publicized the potential of air power by reducing an obsolete battleship, anchored and unmanned, to a smoking hulk by his aerial bombardment. Henceforth air enthusiasts in all countries argued that the capital ships of 1914–18
had become as obsolete as the mastodon. As this problem could hardly be
settled with certainty by a single experiment, the conservatives prevailed,
but mainly by use of the argument that as long as potential enemies built
battleships they had to be matched. The Royal Navy succeeded by 1924 in
establishing some degree of control over naval air power by the creation of
the Fleet Air Arm. But a complicated joint system of RAF and RN command
of air forces operating with the fleet continued until 1937. Hence, since there
was all too little money for the Royal Navy as a whole, it was not surprising
that the air service suffered. Even after the Admiralty took over completely
in 1939 a few months before the war, development in naval aircraft, and in
carriers, still lagged behind development in land aircraft and behind naval
air services in Japan and in the United States, where the naval air arm was
always directed and operated by the Navy, and where Mitchell's ideas had
had some influence even though he had been discredited.

Few other countries followed Britain's lead in creating an independent
air force. In France, the air force was subordinate to the army and thus was
inevitably weak. Even in Germany, when rearmament began, although the
patronage of Hitler's henchman Hermann Goering gave prestige to the *Luftwaffe,* it was subordinate to the *Reichswehr;* and, although it made rapid
strides in the development of good aircraft, its function was chiefly limited to
cooperation with the ground forces.

TOTALITARIANISM

By and large, it can be said that, during the peace which followed the first
total war, development in arms and in methods for using them fell behind
contemporary advances in science. This lag was especially true in the

democracies, where there were greater hopes for the creation of an effective system of collective security, and where the voters had a more direct influence on the spending of money for military purposes. But the axiom that weapon development always lags in peacetime operated even in the new totalitarian states. However, while the democracies were turning their backs on things military and were wrestling with the economic upheaval caused by the last war from which many of them had not yet recovered, Russia, Italy, and Germany in turn began to organize the whole structure of their state toward efficiency in war. The totalitarian state was fundamentally a war state.

World War I and its aftermath had brought the totalitarian political system into full bloom, but the principles on which it was based actually go back far beyond the war to several different sources. The concept of the nation-in-arms, coupled with the power generated by the Industrial Revolution, was its seedbed.

The communists of the mid-nineteenth century had grasped the significance of force in politics and had become convinced that their plans for the introduction of socialism could be achieved only by violence. They regarded the normal relation between classes as a state of war, and they studied tactics for use in armed revolution. In their political and economic jargon the communists reveal the extent of their military interest by the use of such military terms as "labor front," "battle of production," etc. While Marx and Engels, and their successors Lenin and Trotsky, spoke of internationalism and pacificism, they also believed that the revolution could be brought about by the disintegration of the state, and particularly of the army, in war, and that the revolution would have to be defended by the workers in arms. Pacifism was a weapon to be used for the purpose of rotting the armies of capitalist states. It was not an end in itself. As long as capitalism existed, there must be, according to communist theory, a state of war even when there was no actual fighting.

After the 1917 revolution in Russia, the communists had been faced with the necessity of defending themselves against a White counterrevolution and against both German and Allied intervention. They managed to buy off Germany by concessions at Brest-Litovsk, but they were compelled to organize a "Red Army" to replace the czarist army which their two years of propaganda had dissolved. Inevitably they had to use the experienced officers and NCOs of the old army; but for a long time they avoided the term "officers," using instead "technicians," "specialists," "instructors," "commanders," or "red commanders." At first, the practice of electing those officers prevailed, but that was soon abandoned. Communist theory continued to speak in terms of the nation-in-arms, or rather of the "armed workers," but in face of the danger from outside, the creation of a professional army was imperative; and this tendency toward a regular force was undoubtedly strengthened by the fact that the Communist regime was a minority regime dependent on force for its existence. Gradually the symbols of professional militarism returned: an officer class, a rigid rank structure, medals and decorations. While 75 percent of the Red Army in 1924 were militia troops, by 1939 it was a completely professional army.

Nevertheless, the Red Army and Soviet military doctrine retained important influences from communist revolutionary ideology. Political commissars, with authority to screen a commandant's military orders for political implications and with a duty to supervise the political indoctrination of the troops, became an important part of the Russian military organization. The Red Army developed what was called "Marxist military doctrine," which combined an emphasis on offensive strategy based on a belief in the superior spirit of the Red soldier, with the use of "political warfare"—that is to say, propaganda and subversive or partisan activity behind the enemy's lines. The defensive positional warfare of the 1914–18 war was spurned, and the writings of Fuller and Liddell Hart, the apostles of mobility, were intensively studied.

Realizing that the weakness of the czarist army had been to a large extent caused by the industrial weakness of the country and by the lack of social cohesion in the state, the communists introduced a five-year plan designed to build up the nation's heavy industry. The first five-year plan was followed by others, all aimed at strengthening Russia for war. In effect, in order to direct the economy from consumer goods to war production, the Soviet government imposed the kind of direction which other nations had found necessary during the last war.

Furthermore, because the regime feared a counterrevolution, it imposed a rigid control over the movements of its citizens and over their actions and utterances. Since the state had become the owner of virtually all property, it directed the whole foreign trade of the country, and it used that control to implement its policies. Finally, as all political activity except that of the Communist Party was suppressed and punishable by death or exile to Siberia, the Party had become identified with the state.

In all these respects, Russia went much further than any of the belligerent countries had gone during the war. These measures were taken by the Soviet regime because the Communists regarded themselves as being engaged in a war to the death with all capitalist states; some precedent for all their actions could be found among the policies of the belligerents in the First World War. The Soviet Union simply carried rigid wartime controls to their logical conclusion. Total war had produced the totalitarian state.

FASCISM

Like the communist revolution in Russia, the fascist revolutions in Italy and Germany were brought about by the war. In Italy, dissatisfaction caused by failure to gain the rewards promised by the Treaty of London in 1915, coupled with the industrial unrest in north Italy in the period of postwar readjustment, created the conditions in which Mussolini seized power and set up a Fascist dictatorship. Like the Communists, the Fascists adopted the monolithic one-party state and used many of the same political tricks and tactics. Mussolini, having established himself as Il Duce in Italy, proclaimed his determination to win back the Roman Imperium over the Mediterranean area. His followers, wearing black shirts as uniforms and adopting a Roman salute, were organized in "fascisti," literally little bundles of the rods carried by ancient Roman officials. The Fascist state, built on force, glorifying war, and regimenting the Italians in military fashion, showed how militaristic ideas could be imposed upon a democracy and could subvert it.

The German brand of fascism, called national socialism, was also born of war, since it was made possible by psychological factors and by the depression which was in part caused by the war. Writing in justification of his war leadership, Erich Ludendorff had advanced a theory of total war which argued that preparation for war must come before the fighting, that the military command must have unchallenged authority, that the war should be fought by the whole nation and not merely by the armed forces, that it should be fought over the whole area of the enemy territory, and that all methods of propaganda should be used to strengthen the home front, provided that the information distributed was based on truth. About 1920, a Munich professor, Dr. Karl Haushofer, discovered a lecture published in 1904 by Sir Halford Mackinder, a British geographer, which described Europe and Asia as the "heartland" around which the rest of the world was grouped. Mackinder also argued that if Russia and Germany ever united to take advantage of their interior lines of communication, "the empire of the world was in sight." From this concept, Haushofer built up the pseudoscience of "geopolitics" which in effect refuted Mahan and declared that the forces of geography had destined Germany for world leadership. Many young Germans, driven to desperation by the shame of defeat, embraced Haushofer's theories; he had close connections with the General Staff; and, most important, he was acquainted with prominent members of the National Sozialistische Deutsche Arbeiterpartei (the National Socialist Workers' Party, i.e., the Nazis), and with their leader, Adolf Hitler.

The Nazis, absorbing the cult of total war and the doctrines of conquest implicit in the theories of the geopoliticians, forced their way to public notice by screaming from the platform that Germany had not been defeated but had been "stabbed in the back," and by denouncing the "shame" of Versailles and the iniquity of Germany's indictment for war guilt. They used the familiar strong-arm methods of the communists. They adopted from the *Freikorps* and the military organizations of German veterans the idea of organizing bands of "storm troopers" (*Stürm Abteiling,* or SA) for street fighting; and they became masters of the art of political propaganda and of the technique of the "big lie."

When they obtained control of the government, the Nazis followed the normal totalitarian pattern. The one-party state was ruled by use of the machinegun, by control of the organs of propaganda, and by filling the concentration camps. Although the Nazis did not abolish capitalism (perhaps because they had been aided to power by financiers like the iron and steel magnate Thyssen), they exercised a rigorous control over the whole economy by methods similar to those used by Rathenau during the last war. Control of currency, production, and exports and imports, and the orientation of the economy toward making arms, techniques used by the belligerents during the war, were introduced by the Nazis in Germany in time of peace.

In part, the Nazi aim was to use state-directed rearmament to solve the unemployment problem, which had been the curse of postwar Germany; and there was an immediate economic revival. But uneconomic war production could have only one end when there was no alternative means of subsistence for the German people. Indeed, as Hitler made perfectly clear in his book *Mein Kampf,* the Nazi philosophy was that peace was merely a period in which preparation could be made for total war. Following the arguments of Clausewitz, Ludendorff, and Haushofer to their logical conclusion, the Nazis made the state into a war machine tuned for action. The distortion of the economy toward war production and the system of *autarchy,* or self-sufficiency, which they developed was made possible by depriving the people of consumer goods. Propaganda made the Germans accept "guns as well as butter" (some said "instead of" butter), and those who resisted were ruthlessly dealt with by the all-powerful state. An adverse balance of trade, which could not be avoided despite stringent controls, was repaired from time to time by sharp trading practices with weaker nations and by the seizure of the gold reserves of Austria and of Czechoslovakia by "peaceful" annexation. But such methods could not long continue. The German totalitarian state, born of total war, rushed on inevitably to a yet greater war.

THE ARMY AND THE NAZIS

The only element in Germany which might have been able to overthrow Hitler once he had seized power and had set up his Nazi regime was the *Reichswehr.* But the army leaders made no move either to defend the republic or to seize power for themselves. They watched the Nazi rise to power with sympathy because they believed that they could control Hitler, who was promising the restoration of their chief interest and concern, German military strength. But they deceived themselves. The former corporal took over the leadership of the High Command himself; and his chosen political storm troopers (*Schützstaffel,* or SS), given military arms of all types and called the Waffen S.S., later became the elite shock troops of the German army. Indoctrinated with Nazi creeds, they were, like the men of the Red Army, typical soldiers of total war.

The German repudiation of the Treaty of Versailles, rearmament, the reoccupation of the demilitarized zone in the Rhineland, and the Italian attack on Ethiopia were the first steps toward a new world conflict. In Spain, a civil war was fought in which fascist and communist totalitarianisms were found on opposite sides, but in which the real issue and ultimate result was the destruction of constitutional government. That war, although limited to the Iberian peninsula, was an ideological conflict of the total war variety and was a preview of the Second World War. In it communists and fascists tried out their new weapons and methods of warfare. Terror and frightfulness, like the aerial massacre of Guernica, were foretastes of what was in store. And the difficulty experienced in defining neutrality in an ideological war of this kind showed how far the concept of war had changed since the dynastic conflicts of the eighteenth century.

The military doctrine of the new German army—total war and the *Blitzkrieg,* or lightning attack with all arms—was in startling contrast to the pacifism of the Western democracies, where the cry of "No more blood-baths" still prevailed. Despite all the portents of the gathering storm, the western democracies continued to believe that they could avoid war. In the United States in 1934, books and articles exposing the methods of the armaments manufacturers led to an investigation under Senator Gerald Nye, and aided the passage of neutrality legislation to insulate America from European wars. About the same time, as a result of Nazi policies, Britain began to rearm, but under the leadership of Stanley Baldwin and Neville Chamberlain she also continued to work for the appeasement of the dictators. In June 1935, without consulting any other power, and seeking to avoid the naval rivalry that had preceded the last war, she made a naval agreement with Germany. This was a breach of the Treaty of Versailles and it in no way limited German building during the next decade, for German naval rearmament had a long way to catch up.

Appeasement, culminating in British and French concessions at Munich at the expense of Czechoslovakia in September 1938, may have delayed the war and may even, as some have claimed, have given the democracies time to make up some of their deficiencies in military strength, but it proved futile as a policy for preventing a war which the devastating effects of the First World War and the consequent rise of the totalitarian state had ensured. Hitler was a genius at propaganda and a charismatic leader with a remarkable sense of timing his aggressions. The German seizure, first of the Sudetenland and, then on 1 March 1939, of Prague, was followed by Anglo-French guarantees to Poland and Rumania. When Hitler attacked Poland, the Second World War began. Thus a long series of peacetime aggressions that started with the occupation of the Rhineland in 1936 culminated in war in 1939.

18

WORLD WAR II

ORIGINS OF THE WAR

The drift to modern total warfare had been evident from the beginning of the twentieth century. Looking back from our present point of vantage, we can see that everything was moving relentlessly in that direction. However, to contemporaries that trend was not always clear, and, when the nations came to grips in World War II, events served at first to obscure the fact that the conflict was total.

In the Orient, Japan had long been troubled by overpopulation and by a thirst for markets, resources, and empire. She had swallowed up Manchuria in 1931 and had soon moved across the Great Wall. By July 1937, her militarists had sent mobile columns to conquer China itself. The Japanese invasion sparked both Nationalist and Communist resistance. Until the Japanese threatened, Mao Tse-tung and his Communists cooperated with Chaing Kai-shek and the Kuomintang. The strategy and tactics of the offensives undertaken in the Sino-Japanese War are worthy of study as early examples of the new mobile warfare. But against feeble opposition the Japanese made remarkable headway that was, however, no precedent for future conflicts elsewhere. Nevertheless it soon became clear that the conquest of sprawling China was not going to be easy. China's vast spaces, and her almost inexhaustible supplies of expendable manpower, offset the Japanese superiority in mobility, armor, firepower, and material. This campaign therefore provided no clue to the future of warfare.

The initial phases of World War II proper also misled observers about the nature of the trends of modern warfare. At dawn on 1 September 1939,

having been assured by his Russian Treaty of 23 August that he would not
have to fight a major war on two fronts, Hitler ordered his troops into
Poland and, in a terrain peculiarly favorable to mobile operations and without any natural defenses, the *Reichswehr* carried through a classic campaign straight from the new manuals on armored warfare. Surprise aerial
bombardments destroyed the meager Polish Air Force before it could take
to the air; the gallant but ineffective Polish cavalry was swept aside; and
when serious resistance was met, as in the labyrinth of the city of Warsaw,
terror bombardment snuffed it out.

The technique of the *Blitzkrieg* was thus proved in practice. The Germans had about a million men mobilized, but the defeat of Poland was
achieved by only seven or eight armored *Panzer* divisions which encircled and
cut up a much greater number of Polish divisions. The total war potential of
Germany had not been used. Poland, attacked from the west by Germany and
from the east by Russia, was crushed in three weeks. Thus, although it led at
once to the disappearance of Poland from the map, the Second World War
began, in some senses, as a "limited" war. *Blitzkrieg,* in the Prussian tradition, had achieved as nearly perfect a limitation in time as could be desired
by the most optimistic planner.

A deceptive appearance of limitation was also produced by the attitudes and behaviors of the great powers. While German radio stations,
with unconsciously ironic accuracy, blared out frightful tales of "unheard
of (*ungehörte*) Polish atrocities," the propaganda war in the West was singularly restrained. On the eve of hostilities the British had published information about the torturing of Jews in Nazi concentration camps; but it was
prefaced by a virtual apology for the publication of hate propaganda. In
the West this tragic story was doubted; in Germany it made no impression
at all; and not until the curtain was torn back in 1945 was the full extent
of Nazi shame realized. Thus in 1939 that aspect of the totality of the war
was hidden.

While Poland was in her death throes, her two western allies stood
helplessly by and watched. They possessed neither the armored might to
draw off the attackers by invading Germany nor the air power to halt
aggression by the threat of bombing. In both countries governments and
peoples were not ready to accept the fact that all-out war must come. On
3 September 1939, up to the last minute, it was not known whether France
would declare war at all; and although England had always made it
perfectly clear that she would honor her pledges to Poland, the Chamberlain government was still in power and was unable to shake off the
lethargy of its former policy of appeasement. American newspapermen described the second stage of the war, following the conquest of Poland, as
"phony."

While Germany and Soviet Russia digested Poland, the western powers
dug themselves in behind the Maginot Line and along the Belgian frontier
and, in a somewhat leisurely fashion, prepared for a war which they professed

to believe would be short but which they vaguely felt might last a very long time. Conscription, an approach to the nation-in-arms, had been introduced in Britain in March 1939. In September, when war broke out, Chamberlain had hesitatingly consented to plans for a fifty-five-division army to be ready within the next two or three years. But mechanization of the army, despite the lessons of the Polish campaign, was very slow. One skeleton armored division was sent to France; but a second was not to be equipped until the second year of hostilities.

No one appears to have fully realized that modern war between great powers would entail *both* the full mobilization of the nation's manpower *and* full mechanization of its armies. The long debate between those who had urged the creation of armored professional armies and those who had defended the "armed horde" had obscured the possibility that in total war both might be needed at the same time. The course to be followed if *Blitzkrieg* failed to secure a quick victory had not been properly explored, either in the West or, for that matter, in Germany.

THE PHONY WAR

During the first winter of the "twilight" or "phony" war, the main weapons used against Germany were blockade and propaganda. Instead of bombs, the RAF dropped leaflets: it was incapable of a bombing offensive, and German retaliation was feared. However, if the Germans could not be talked out of their sins, it was hoped that economic pressure would bring them to their knees. Even here the war was incomplete and therefore limited. The 1914–18 blockade had been reintroduced in an improved form with a Navicert system by which neutral vessels carried certificates to show they had no contraband. The point of control had thus been carried back to the port of lading. Blacklists of neutral firms who traded with the enemy had been drawn up. Yet, although the blockade of the seas worked smoothly and efficiently, the situation was actually very different from 1914. The whole of East Europe was open for German trade; and *Ersatz* products and stockpiles of essential commodities were in much greater supply. Propaganda and the economic weapon were parts of an apparatus of total warfare, but used by themselves, and in this incomplete fashion, they were practically impotent.

During this same period, Hitler also relied on economic warfare. Following the precedent of 1914–18, he proclaimed a submarine blockade of the British Isles and backed it by dropping a "secret weapon," new magnetic mines, in the shallow approaches to the ports. German aircraft sorties for this purpose were the nearest approach to those great aerial attacks on enemy territory which the prophets had long declared would herald a new war. Otherwise, in this first winter of the war, the Nazi aerial onslaught in the West was strictly limited to reconnaissance.

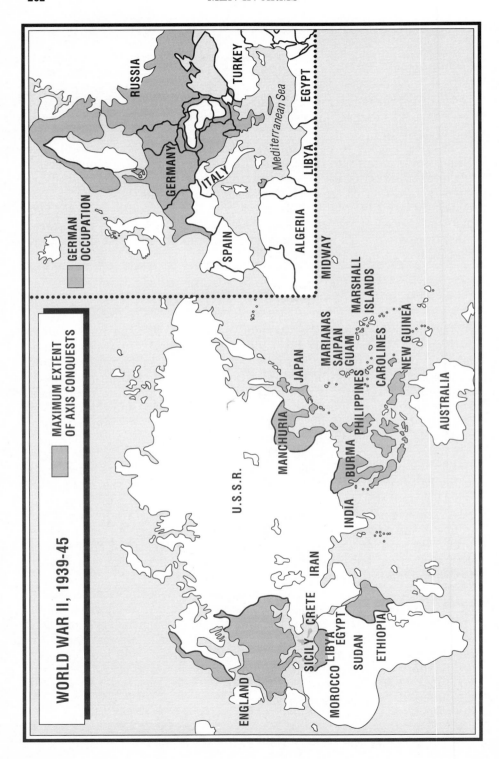

WORLD WAR II, 1939-45

MAXIMUM EXTENT
OF AXIS CONQUESTS

GERMAN
OCCUPATION

BLITZKRIEG

The third stage of the war, the renewal of *Blitzkrieg* with attacks on Denmark, Norway, Holland, Luxembourg, Belgium, and France in the spring of 1940, was again, like the attack on Poland, limited in time. Denmark fell without a blow, Luxembourg in a day, Holland in five days, Belgium in less than three weeks, and France in six. Furthermore, in the Battle of France, although the Germans had used a hundred divisions, a few armored columns won the victory. *Blitzkrieg* had again shown itself to be a method of restricting the full impact of war by the very speed of its onslaught.

The German victories in the West were won by the use of new tactics and weapons against armies which were inadequately prepared for war and whose strategy and tactics were faulty. The Nazis used specially trained parachute troops to clear difficult obstacles (for instance, a few were dropped on the Belgian fort Eben Emael), Stuka dive bombers to give close support, and tanks to probe, pierce, and fan out behind the Allied lines. They were superior in training, and they exploited surprise as a psychological weapon to give them an advantage in morale. For these armies the Maginot Line, which had never been properly extended along the Belgian border to the sea, presented no great difficulties. General Maurice Gamelin (under whose command, as a result of the lessons of 1918, the British expeditionary forces had been placed) relied entirely on a static and linear defense. When the Low Countries were attacked, the Allied armies on the left flank were ordered forward to their aid, despite the fact that, because the Belgians and Dutch had sought to protect their neutrality, plans for cooperation in defense had not been worked out. Before a front could be formed, the Allied armies were pouring back in retreat. The Nazis had originally planned to repeat the strategy of the Schlieffen Plan, a thrust on the right. Instead, acting on a plan developed by General von Manstein which turned the Schlieffen Plan inside out, they thrust their armored columns through the Ardennes, which were regarded by the French as unsuitable tank country and were therefore only lightly defended. French tanks were dissipated along the whole front; and when their front was breached, the Allies possessed no "mass of maneuver" to throw against the bulge. The Allied air forces were unable to prevent the *Luftwaffe* from dominating the field of battle; and at the crisis of the Battle of France, the British refused to transfer their metropolitan air force from Britain to bases in France. It was a hard decision; but the RAF could not have turned the tide. The planes and pilots were thus preserved for the Battle of Britain.

Although the war had been limited thus far, there were ominous signs that it was moving into a new phase of intensity. Five of the victims in 1940 were neutral states which had vainly striven to keep out of the conflict but had been struck down without warning. Neutrality had been ignored because it stood in the way of the aggressor. In the First World War, when neutrality had been infringed upon, it was only after formal warnings. Now it was breached in the night.

The aggressor was fully abetted by "fifth columnists" (the phrase was coined during the Spanish Civil War to indicate those in enemy territory who were working to soften the defenders). Major Vidkun Quisling, a Norwegian Nazi and former war minister, by betraying his country to the Germans, gave another new term to the dictionaries of many languages. "Quislings," traitors who for ideological reasons were ready to sell their country, although not a new phenomenon in history, were now much more numerous because of the growth of social and economic cleavages within the state. Class warfare, whether real or imagined, had brought ideological disunity and had loosened some of the cement which held nations together. Total victory, leading to the complete disintegration of a defeated state, had thus become much more likely.

The internal weaknesses of Belgium and France had been cunningly exploited by the Nazis. The small amount of sabotage and subversion carried on by quislings and parachute troops was deliberately exaggerated to destroy the morale of the defenders and to soften them before the *Blitzkrieg*. Terror was spread further by the machine-gunning of refugees on the highways. Such methods greatly intensified the impact of *Blitzkrieg*. Neutralism in Belgium and defeatism in France had sapped the strength of the defenders and had laid those countries wide open to the psychological propaganda of total warfare.

When the fall of Norway put an end to complacency in Britain, Winston Churchill replaced Chamberlain as prime minister. The fall of France had a yet greater result. It convinced President Roosevelt and many other Americans that the United States could not stand idly by, watching the overthrow of Britain. Already in November 1939 Congress had revised the neutrality legislation to permit the belligerents to buy arms in the United States on a "cash and carry" basis. This new legislation favored the democracies, because they could use the ocean to get the arms they needed; but America was still shielded from being dragged once again into war through financial entanglements or through the loss of her ships. When the Nazis broke through to the Channel and the Atlantic Coast and threatened to overrun Britain too, rifles, machine guns, and ammunition were hurriedly rushed across the Atlantic; in August 1940, President Roosevelt met Canadian Prime Minister Mackenzie King at Ogdensburg, New York, and arranged to set up a Permanent Joint Board of Defense for the north half of the Western Hemisphere; on 3 September he announced the exchange of overage American destroyers for British bases in American waters. The United States was passing from benevolent neutrality into a state of passive belligerency.

Inside the United States, a vast arms program and the passage on 3 September of the selective service law, the first "draft" in the peacetime history of the country, showed an increasing public awareness that geographical isolation was no longer a secure defense. In campaign speeches both presidential candidates, Roosevelt and Willkie, while promising to keep the nation out of "foreign wars," pledged themselves, if elected, to aid those countries which were resisting aggression. After the election President Roosevelt declared that while Britain was the spearhead of resistance to world conquest, the United States must become "the great arsenal of democracy," and he proposed a scheme to make possible the continuation of aid although Britain's dollar credits were nearly exhausted. On 11 March 1941, Congress enacted what Winston Churchill later called the "most altruistic act in history," legislation by which the president could lend or lease arms and supplies to any country whose defense he regarded as vital to the safety of the United States, with repayment to be made in kind or in any other way deemed satisfactory. (Canada did not receive lend-lease aid but in the Hyde Park Agreement of 20 April 1941, arranged to dovetail her economy for war production with that of the United States. About the same time, she granted to Britain a billion dollars in credit.) Thus, to achieve an integration of defense production for total war, traditional concepts of international finance were being swept aside while, as yet, the United States was still technically neutral. More than anything else, this fundamental change in American policy away from strict isolation and neutrality showed that the peoples of the democracies were coming to a realization of the danger which totalitarian aggression spelled for their way of life and to believe that, to save themselves, they must revolutionize their traditional ways and attitudes. When the United States entered the war, techniques were worked out for close coordination with Britain in industrial output, shipping, and food production, far beyond any degree of integration previously achieved by allied sovereign states at war.

Meanwhile, on 10 July 1940, the Germans had begun daylight aerial attacks, the Battle of Britain. By 7 September however, the *Luftwaffe* had been beaten by the Royal Air Force. A handful of Spitfire and Hurricane pilots flew desperate sorties round the clock. Radar gave them early warning. Goering was compelled to resort to night attacks. These soon degenerated into an indiscriminate bombardment of London. A defensive air victory of enormous significance had been won. On 17 September Hitler postponed the invasion, Operation Sea Lion, indefinitely.

Thus began the next stage of the war, which was to last until June 1941, when the Germans invaded Russia. Although in this period the war seemed once more to be in many ways limited warfare, this was actually the time when both sides were organizing themselves for a total fight to the finish. In Europe, the victims of the Nazi conquest were ground down under German rule, and the world saw what totalitarian victory meant. The Gestapo ruled by torture; and the subject states were pillaged for the benefit of the conqueror. The total eclipse of historic nations was a sign that war had become unlimited in its objective.

Halted by the Channel, the Nazis were forced, by the very dynamics of the domination which they had created, to seek further conquests. For a time they had talked with Franco about a march to Gibraltar to close the Mediterranean at the western end. Then, in March 1941, they found it necessary to come to the relief of Mussolini in North Africa, where the Italians, who had boasted that they would capture Suez, were hard-pressed.

The vital strategic significance of the Middle East had been recognized by Churchill when he sent an armored brigade to Egypt in 1940 at the time of Britain's greatest peril. In campaigns fought between 1940 and 1942 on the borders of Egypt and Libya the use of mechanized and armored forces by both sides led to a new degree of mobility in modern warfare. The desert, for the most part flat and trackless, was eminently suited to tank actions which resembled operations at sea. With both sides using armor, fronts became fluid. In desert campaigns in earlier wars, logistical problems had severely limited cavalry actions. They still created great difficulties and often brought promising offensives to a halt; but the essentials of mechanized warfare—gasoline for tanks and water and supplies for the troops—could now be moved much further and faster, and in greater quantity, and cached longer than fodder and water for horses. Troops from Australia, New Zealand, India, Palestine, and South Africa as well as British and Free French forces stopped the Italians and Nazis from cutting the Suez and so penetrating to the Indian Ocean and to the threshold of southern Asia. By February 1943, they had driven Rommel's Afrika Korps back to Tunisia.

HITLER ATTACKS RUSSIA

A campaign against Yugoslavia and Greece in the Balkans delayed Hitler's real objective, a long-planned invasion of Russia, because he regarded a firm Balkan flank as a prerequisite for such an invasion and also because he wished to deny the British the chance to bomb the vital Ploesti oilfields in Rumania. From the time he had put off the idea of invading England, it had been clear to him that one day he must seek a reckoning with Stalin. He could not hope to maintain his domination of Europe with great potential enemies in both West and East. The inevitable result of totalitarian victory was that the conqueror had to destroy all possible enemies.

The Nazi invasion of the Soviet Union in 1941 seemed at first likely to knock Russia out at one blow. Great encircling pincer movements cut off large pockets of the Red Army and forced them to surrender. Within five months the Nazis had reached the outskirts of Moscow, had enveloped Leningrad, and had penetrated the Crimean peninsula and the Don Valley. But this three-pronged attack, the strategy of which was dictated by Hitler himself, failed to achieve its object, the destruction of Soviet ability to resist. By 7 December it had become clear that a quick victory was not possible. Hitler announced the end of the campaign for the season, only to be taken by surprise by a Russian winter offensive for which his troops were ill prepared.

By mid-August 1942, concentrating on the southern front in order to seize desperately needed oil fields, the Germans reached the foothills of the Caucasus and, further north, the great industrial city of Stalingrad. This was the high water mark of German conquest in the East. By the end of the year a Nazi army was surrounded near Stalingrad, and Leningrad had been relieved. The tide had been turned by stubborn Russian courage, by the building up of war production in factories carried back to safety across the Urals, and by British and American supplies sent in perilous convoys through Arctic seas or by the long route through Iran.

The war was clearly becoming total in objective, in method, and in its impact on peoples. Under the whip of totalitarianism, the whole economy of Germany was harnessed for war. Drawing forced labor from the subject territories, the Nazi war machine was able to increase its potential effort vastly. In the Todt Organization at least five million men and women labored on immense war projects, such as constructing defenses (normally a military function), repairing bomb damage, and building airfields. After the defeat at Stalingrad in January 1943, the Nazis decreed total mobilization. Even the children were drafted for work in factories and fields.

Total mobilization for war on the home front was also adopted by the United Nations, as the opponents of fascism now called themselves. Indeed, after the war it was discovered that Britain was harnessed to the war effort to a degree that was greater in certain respects than was the case in Germany. For instance, British married women were compulsorily employed in war work on a much greater scale than the German *Hausfrauen.* But in all countries the same techniques were followed. Great national armies, navies, and air forces were conscripted and trained; labor was registered, drafted, and directed; property was requisitioned; industry and production were controlled; consumer goods were severely cut and rationed; and travel and transportation were limited. This arbitrary government was rendered inevitable by the emergency. It became an offense to spread alarm and despondency or false rumors; and state-directed organs of propaganda sought to condition the thoughts of the people. All these were the tools of totalitarianism. Total war had driven the democracies to adopt some of the methods of their adversaries. But it must be remembered that all of these had first been forged in the previous total conflict. For the Soviet Union, which was already a totalitarian state, total war measures were more easily assimilated.

Despite the great mobilization of manpower for war industry, the size of armies continued to increase. In the early years of the war relatively small but highly powerful armored forces had won great victories, but when peoples were fighting for their national existence, great mass armies were thrown into the field. For the invasion of Russia in 1941 the Nazis had used 160 divisions, of which no fewer than 20 were armored, and in a very short time there were 9,000,000 men struggling on the Eastern Front. Even so, the war did not become static as it had on the Western Front in 1914–18. Great armored offensives encircled huge pockets of resistance, defense was organized in depth, fronts were fluid, and the battle raged over great areas. As the German juggernaut rolled on toward Moscow, the retreating forces ruthlessly scorched the earth they left behind. Red Army units that were overrun either fought in pockets until all further resistance was useless or went underground to organize guerrilla warfare against the invader.

In those parts of the country like the Ukraine, where non-Russian peoples were to be found, many Soviet citizens and deserters welcomed the invaders, and some even joined the Germans in the fight. Total war had engulfed the civilian population as well as the soldiery. Strangely enough, while the war on the Eastern Front was apparently one of fascism versus communism, and while large numbers on both sides accordingly deserted to the enemy, Stalin at the same time found it expedient to proclaim a patriotic war for the defense of "Mother Russia." Any and every ideology was recruited for the conflict.

Fascist and communist ideologies played an important part in conditioning the armies locked in combat. The Red Army had from its birth been politically indoctrinated, and although its commissars were abolished for a short period during the war, they soon returned. The German army also included divisions of political troops, the *Waffen* S.S., heirs of the storm troopers who had violently opened the way for Hitler to seize power in Germany. The *Waffen* S.S. were first formed to be militarized police to relieve the army of its occupation duties in the conquered territories, but they soon became the elite shock troops of the German forces. Ultimately they were increased to twenty divisions, some of them armored (the *Panzer* S.S.), having the pick of all equipment and personnel.

A GLOBAL WAR

The war became global at the end of 1941, when Japan embarked upon the conquest of the "Greater East Asia Co-Prosperity Sphere" and involved the United States by attacking Pearl Harbor. Even before Pearl Harbor, the United States had in fact been cobelligerent, giving much aid to Britain, with whom Roosevelt had made common cause in the Atlantic Charter, and also to Russia. The enormous American industrial power had already been thrown behind the democratic cause by the Lend-Lease Act; the United States had established air bases in the Danish colony of Greenland and had also relieved Britain of the occupation of Iceland. Virtually abandoning the former decision to keep American ships out of belligerent waters, the United States Navy had taken over the task of preventing enemy submarines from acting in American waters, which were now considered to extend to within 700 miles of the British Isles. In September 1941, after several submarine attacks on American naval vessels and merchant ships, the United States Navy had been given orders to shoot at any hostile submarine encountered in the neutrality zone. By November a shooting war had begun; one American destroyer had been sunk by U-boats and another damaged. This was stretching neutrality to the limits.

A most significant American contribution to the winning of the war was the enormous material aid which was put into the war against the fascist powers. In the vital shipbuilding industry alone, American shipyards built over five thousand ships to maintain the sea communications between the industrial plants of the New World and the fighting fronts in the Old. In addition, millions of rifles, thousands of guns, and equipment and weapons of all kinds were produced for the fighting forces of the other United Nations as well as for those of the United States. American industrial power, in this war even more than in the last, made victory possible.

For all the major contestants, this war was a war of matériel. Only those nations which had great industrial plants or could build them could claim to be major military powers. Hence it was the expansion of her war industries, as much as her military contributions, that raised Canada to the status of a middle power on the Allied side. In Russia, the transference of a great part of the industrial machine beyond the Urals and out of reach of Nazi attack was a significant step toward the defeat of the invader. Even in relatively backward China, the creation of a cottage industry for war production was an essential part of the war effort. The technology front was as important as the fighting fronts.

The planning for the conversion of this great industrial effort of the nations fighting fascism into effective military power led to important discussions among the Allied leaders on strategy and grand strategy that revealed serious differences of opinion, but Roosevelt agreed with the British to concentrate upon the defeat of Germany before turning to Japan. He was anxious that American troops should be involved in the all-out fight against Hitler within a few months of the United States's entrance into the war. Although wishing to open a second front on the continent of Europe, which the Russians had been demanding ever since they were attacked, he agreed to the invasion of French North Africa as a first step. Churchill had persuaded him that lack of shipping and of other material made a major attack on Europe impossible for the present and that the preliminary invasion of French North Africa would help clear the Axis from that continent and eliminate the serious threat to the Middle East. Some of the president's advisers wanted to fight the Germans by a more direct attack in overwhelming strength which, although costly in lives in the early stages, could conceivably shorten the war. They were suspicious of Churchill's "Mediterranean strategy" of attacking the "soft under-belly of Europe" through Sicily, Italy, and the Balkans, which seemed to them like playing the old imperialist game and which might antagonize the Russians. The British were anxious to avoid heavy initial losses which might have a dangerous effect on popular morale; and they also paid more attention to the political implications of strategy, and especially to the results that might follow the defeat of Germany if the Russians advanced too far into Europe.

THE WAR AT SEA

The strategy of invading Europe depended upon success in the war at sea. While the war on land had remained "phony" for seven months, at sea the conflict was intense from the first hour. On the day after Britain entered the war, a German submarine torpedoed the passenger liner *Athenia* without warning, in disregard of international law. The sinking had not been planned, and the submarine captain was reprimanded by Hitler; but in fact the Germans took up again the policy of unrestricted submarine warfare almost where they had dropped it in 1918. As in the First World War, German naval strategy was aimed at destroying the seaborne commerce upon which Britain depended; the submarine attack was therefore especially heavy in the western approaches to the British Isles from the Atlantic Ocean. At the critical time of the German invasion of Norway, British submarine commanders were also authorized to sink vessels on sight, but only in a narrow strip of Norwegian coastal water; and similar zones of unrestricted submarine attack were declared later in the year near the coasts of Libya and Italy. Ruthlessness of this kind was not new in submarine warfare. It was merely carried over from the First World War when it had been adopted by the Germans because the submarine lost most of its effectiveness if it adhered to international protocol.

While Nazi totalitarianism was not responsible for changing the nature of the war at sea, it contributed to the ruthless use of new weapons. These, of themselves, brought a new dimension of effectiveness into war. They included magnetic mines dropped from aircraft early in the war in the shallow coastal waters of the British Isles. This was quickly countered by the organization of a huge research and computing program by British scientists who arranged for the degaussing (demagnetizing) of ships' hulls by electronic belts. But in September 1940, British and Allied losses were higher than in any similar period at the height of the submarine campaign in World War I.

The Battle of the Atlantic, as the submarine attack on communications between America and Europe is called, is usually said to have begun on 6 February 1941, when Hitler issued a directive stressing the importance of attacking ships bound for Britain. As submarines and aircraft became more efficient, the battle became three-dimensional, fought under and over the ocean as well as on the surface. Countermeasures also improved and in 1942 the U-boat threat seemed to be coming under control when the greater involvement of the United States gave Nazi commanders rich pickings on the eastern seaboard of North America and also in the Caribbean. Thereafter, the operation of submarines in wolf packs assisted by aerial reconnaissance, greater ranges and speeds, and eventually the invention of the *Schnorkel* tube which enabled a submarine to stay below the water while it recharged its batteries, made the *guerre de course* the most serious that Britain had had to face in any war.

However, the techniques and weapons of anti-submarine warfare also improved. Convoys and escort services had been organized early in the war. Aerial reconnaissance was undertaken from land bases, and "jeep carriers" of 10,000 tons were improvised to cover the mid-Atlantic gap that could not be reached from the shore. Asdic (sonar) sound detection was improved. Radar was adapted to naval use. Anti-submarine weapons, depth charges and "hedge-hog" launchers came into greater use, and "killer groups" were

THE BATTLE OF THE ATLANTIC—A gunlayer on a merchant ship in convoy, an uncomfortable, lonely job. (Directorate of History, Department of National Defence, Canada)

deployed against the wolf packs. Long after the war was over, it was learned that the British had benefited from knowledge of German naval codes to an extent that was kept a very strict secret, even from Allies. When Poland fell, a Polish cavalry officer had escaped with a stolen German cipher machine and brought it to England. The British had already broken the codes, but this machine, known as Ultra, enabled them to decipher messages instantaneously. Nevertheless, the information, which was shared with Allies, was not of itself enough to defeat the submarines. By March 1943 the Battle of the Atlantic reached a crisis, but the rate of allied replacement of ships lost had begun to surpass losses. Even so, the German skippers, ably directed by Admiral Doenitz, fought on to the bitter end. In 1945, new long-range craft were about to appear that might have changed the course of the war had it not been ended by victory ashore. But when defeat came on land, the submarine crews lost their spirit. Ordered to surrender, they did so without protest. The *Schnorkel* and other devices had come too late to affect the decision in the war.

On the surface, German naval power was too weak to bid for mastery of the sea. The Nazis avoided the possibility of being trapped to fight a major sea battle like Jutland by the simple device of never concentrating their small battle fleet. Instead, they used their heavy ships as commerce raiders and compelled the defenders to use as convoy escorts either ships of equal strength or submarines. At different times between 1939 and 1941 the pocket battleships *Deutschland (Lützow), Graf Spee,* and *Admiral Scheer,* the battle-cruisers *Scharnhorst* and *Gneisenau,* the battleship *Bismarck,* the heavy cruisers *Admiral Hipper* and *Prinz Eugen,* and several armed merchant cruisers were loosed in the sea lanes to work havoc until they were either sunk or driven back to base. When this proved too costly for the Nazis, the new battleship *Tirpitz* was stationed in Norway as a threat to the convoys bound through Arctic waters to Russia. Despite the severe losses inflicted by these ships, their efforts only served to prove that surface commerce raiding, although it imposed a severe strain on British naval resources, was not, in itself, a war-winning weapon.

Through lack of adequate surface sea power, the amphibious operations which the Germans could attempt were severely limited. While they captured Norway by a combined sea and air assault aided by trickery, and futile British attempts to interfere confirmed that surface ships alone could not face strong land-based air power, a Nazi invasion of England continued to be impossible because Hitler lacked sea power. In the Mediterranean, Crete was seized by glider and parachute landings in face of Allied superiority on the sea; but that operation proved that an invasion dependent on aircraft acting alone could not be carried out without suffering enormous losses.

In narrow waters air power seriously altered the effectiveness of navies. After Italy entered the war in 1940, her fleet showed timidity in seeking action, was crippled in harbor at Taranto by an air strike, and was defeated in the night action called Matapan. However, the Royal Navy lost control of the Mediterranean because enemy air superiority made it impossible to operate through the narrow channels of that sea. The Royal Navy's shortage of radar until mid-1941 and its lack of the carriers and adequate aircraft to provide air cover for fleet operations severely limited its effectiveness until late 1942. Supplies for the British Eighth Army had to be carried around the Cape of Good Hope. Malta was beleaguered and, although never invaded, could be relieved only by submarines, from time to time, and by desperate convoys pushed through from each end of the Mediterranean. On the other hand British submarines, surface vessels, and aircraft took a heavy toll of shipping bound from Italy to Libya. From all these fierce encounters new tactics were being constantly evolved, and new theories of the relation of air and sea power had to be worked out.

THE UNITED NATIONS STRIKES BACK

When the Allies had built up their strength to strike back, the Anglo-American landings in North Africa and in Sicily and Italy, where Canadian and other Allied forces were also employed, were only possible because the United Nations possessed overwhelming naval as well as aerial strength. They were now able to fulfill the requirements of successful amphibious operations—namely, to land strong forces in any one of a number of different places, to prevent enemy interference by sea, to give close support from the guns of the fleet both on the beaches and further inland, and to maintain communications between home bases and the invading forces. Yet the battle for the mountainous terrain of the Italian peninsula, begun by landings at Taranto, Reggio, and Salerno in September 1943, was fought the hard way. With the exception of the Anzio landing in January 1944, because of high-level disputes over the allocation of naval resources, no attempt was made to turn German positions by the use of sea power. However, when, in June 1944, the time came to attack the West Wall, there was no such parsimony. Operation Neptune, the overture to Overlord, the Normandy campaign, employed 702 warships (excluding the minesweepers) and over 9,000 craft of all kinds.

Technical ingenuity made important contributions to the success of the invasion. Frogmen reconnoitered the beaches; specially designed landing craft ferried men and tanks ashore; others gave close rocket artillery support on the beaches; tanks were modified with a device to clear paths through minefields; more efficient systems of communications enabled the large naval, military, and air forces involved to be controlled in a single operation; artificial harbors called Mulberries compensated for the lack of a major port at the outset; and oil for the land attack was carried across the Channel by Operation Pluto, pipelines laid under the ocean. The amphibious landings in Africa, Europe, and the Pacific proved that a dominant sea power had not lost its old advantage of being able to strike virtually where it liked and to exploit surprise to the full.

The Anglo-American drive from the beaches of Normandy to the heart of Germany from June 1944 to May 1945 (which, like the sea war, was facilitated by superior air and naval support) was matched by the even greater effort of the Red Army which had stopped the last great German offensive on the Eastern Front in 1943. For four years of war the Russian front always occupied at least two-thirds of the German ground forces and a large part of the *Luftwaffe,* an indication of the power of the Russian military effort. Attacked from east and west, the Nazis were everywhere driven back into Fortress Germany and, when that fell, were compelled to surrender unconditionally on all fronts. These great Allied attacks were the product of the industrial resources of Britain, Russia, and the United States, as well as of the rest of the free world, which, when fully mobilized, outweighed Germany's productive capacity even when that was supplemented by the enslaved states of Europe. Victory in Europe was won by the exploitation of superior economic power, superior numbers, superiority in the air and on the sea, and a tactical training on the ground in which the Germans no longer possessed a decided advantage. It was won by the development of close cooperation between the various forces of the Allies, civil and military.

THE PACIFIC WAR

In the course of the Pacific War, begun by the surprise air attack on Pearl Harbor which had caught the defenders off guard, these same factors, the elements of total war, were also clearly revealed. The initial attack, without a declaration of war, was itself in line with the practices of total warfare and was in the Japanese military tradition; for they had begun the Russo-Japanese War in exactly the same way, by a surprise attack on the Russian fleet at Port Arthur. Japan had come to believe that the European War had so weakened the Allies that she could pick up an East Asian empire at will, make herself economically self-sufficient, and form a defensive cordon powerful enough to discourage her enemies from attempting to dislodge her. She planned a line of "unsinkable aircraft carriers," islands stretching from Rabaul in the Bismarck Archipelago to the Kuriles, north of Japan. Behind this protective screen she intended to swallow and digest the possessions of Great Britain, France, Holland, and the United States while also finishing off her Chinese meal. The Pearl Harbor attack was intended to knock the United States off balance and so prevent the possibility of interference by American fleets while Southeast Asia was being overrun. Instead, as Samuel Eliot Morison has pointed out, it was a "strategic imbecility," for it ensured that the American people and government would enter the war with their full power without long discussion; and also it made certain that they would be satisfied with nothing less than total victory to atone for that "day of infamy," 7 December 1941.

Meanwhile, the Japanese conquest of Southeast Asia showed once more what could be attempted with superior sea and air strength. The Pearl Harbor attack itself revealed the great offensive power of a carrier task force. In 1939 the Japanese navy was the only one which gave the carrier a place in its fleet ahead of, or equal to, that of the battleship. The war was to justify such prescience from the very beginning. In November of 1941 a Japanese fleet had slipped across the Pacific, refueling en route and making good use of the cover of weather fronts to hide its movements. Off Hawaii on 7 December it had launched 300 aircraft, which sank most of the battleships of the United States Pacific fleet in water considered by the Americans to be too shallow for the use of aerial torpedoes. Shortly afterward, the sinking by Japanese land-based airplanes of the British battleship *Prince of Wales* and the battlecruiser

Repulse seemed further proof of the vulnerability of surface vessels to aerial attack. In view of these demonstrations of the importance of air power in naval warfare at the outset of the war in the Pacific, the accidental absence at sea of all the United States carriers at the time of the attack on Pearl Harbor, which preserved them for future battles, must be regarded as an extraordinary stroke of fate.

Within a few months of entering the war, Japan had achieved her main territorial objectives in the "southern resources area" of Southeast Asia. Her warships had penetrated to the Indian Ocean and seemed likely to link with Hitler at Suez. The new Japanese Empire stretched from the Home Islands to Sumatra. Elated by these early successes, Admiral Yamamoto, the commander in chief of the Combined Fleet, succeeded in persuading his superiors to expand the objectives to include Midway, the Aleutians, and the Solomons. This expansion stretched dangerously the sea communications on which Japan's retention of her conquests obviously depended. Winston Churchill said later that the turning point in the war against the Axis came when Squadron-Leader Len Burchall, a Canadian flying with the RAF, spotted a Japanese task force steaming toward Ceylon.

Already in an action in the Coral Sea (4–8 May 1942), the first action in which surface ships did not exchange a single shot, a Japanese amphibious attack upon Port Moresby had been foiled, compelling the invaders to undertake instead the crossing of the difficult Owen Stanley mountain range in the interior of New Guinea. Then, a month later, the unsuccessful attack on Midway brought on a naval battle which cost Yamamoto all four assault carriers of his striking force. Thus less than six months after Pearl Harbor, and while Japanese conquests went on, one of the most important battles in history marked the turning point of the war against Japan. At Midway the Japanese lost 30 percent of their carrier pilots and two-thirds of their big carriers, the vessels which henceforward were to have at least an equal right with battleships to be classed as capital ships. The battle was described by Admiral Nimitz as "essentially a victory of intelligence" (because it resulted from the breaking of the Japanese code). It was also a triumph for the courage of American carrier pilots and for superior technology, since the Americans had the advantage in radar.

In the Pacific, carrier-borne planes were used so much that the war is sometimes seen as one fought mainly in the air. In normal weather aircraft and their bombs and torpedoes were the weapons used in fleet actions instead of the big guns of old. In the vast reaches of the Pacific, it was the aircraft of the fleet carriers, and not shore-based planes, that dominated naval warfare. The old belief that carriers were unusually vulnerable ships was proved untrue. Though the United States lost four in the first year of war, thereafter no big ones were lost. Similarly, of sixty carriers commissioned by the Royal Navy during the war, only seven were sunk by enemy action and only one after 1942. At Midway the Japanese had lost heavily in carriers through lack of adequate air cover.

Big ships, carriers and orthodox surface vessels as well, with their heavy armament, remained a basic ingredient of the fleet largely because of the

WAR IN THE PACIFIC. A kamikaze attack on *U.S.S. Bunker Hill.* (Directorate of History, Department of National Defence, Canada)

extra power of defense which radar and the proximity fuse had given. Since radar had added to their offensive power by making possible effective shooting without visible sighting, surface vessels still had an important role. At the Coral Sea it had been surface ships in the background which had barred the way of the Japanese invading force. Off Guadalcanal in late 1942 the United States more than held its own in a series of cruiser battles that were in part responsible for the Japanese decision to withdraw from the island. At Leyte Gulf, when the Americans were recapturing the Philippines, every kind of naval vessel took part and the action included both carrier strikes and gun battles. This battle stands beside Tsushima and Trafalgar among the most crushing naval engagements of all time; only Tsushima had been won with as little loss to the victors and in that engagement the Russian fleet which the Japanese defeated had sailed half round the world and was laden for an ocean voyage. But at Leyte Gulf both fleets were cleared for action. It was a most remarkable victory: Japanese losses were the greatest ever endured in naval warfare in so short a time. They lost three battleships (one of them, the *Musashi,* with nine 18-inch guns, was one of the largest warships afloat), four carriers, ten cruisers, and nine destroyers sunk, while they accounted for only three small American carriers, two destroyers, and one escort vessel. The most vital part of these statistics was the figures of carriers sunk because, although Japan still had eight or nine built or rebuilding, she had lost at Leyte Gulf all the rest of her experienced carrier pilots, and new ones could not be trained in a day. She therefore turned to a desperate expedient to stave off the irresistible advance of her foes, Kamikaze, or suicide, attacks upon the ships of the United Nations. That was total warfare in its most absolute form.

In the Pacific war amphibious landings were as essential steps to victory as in the German war, but their relation to the campaign as a whole differed significantly. Whereas the invasions in Europe depended upon the success of a few big landings on hostile shores, and especially upon Operation Neptune, the Pacific campaign was a series of hops from island to island, bypassing strong points which could be safely left because they were isolated by superior sea power. Amphibious operations in this theater were different from those in Europe in two significant ways: they were undertaken at much greater sea distances from base, and they were directed against targets of limited depth where all the shore battles were usually fought within range of the guns of the fleet. Fleets made up of units of all kinds were used. These "task forces" of World War II took the place of the battle fleets of ships-of-the-line which had remained unchallenged from the seventeenth century to World War I. Naval operations over the great distances of the Pacific were made possible by the development of the technique of refueling at sea. The Americans exploited sea power over greater distances than ever before.

The Japanese garrisons attacked in this way lived up to the traditions of their warrior code by fighting to the last. Thus, at Iwo Jima a garrison of 20,000 was attacked by 60,000 Americans, and only 200 Japanese lived to be taken prisoner of war. These figures, matched during other assault landings, show that the war was approaching absolute totality. But with all their fanaticism, the Japanese troops did not emulate this suicidal conduct when the heat of battle was absent. The Japanese High Command had planned that island garrisons left stranded by the leapfrog American advance would resist to the death. Instead, the Japanese Army insisted that isolated garrisons which had been left to starve must be withdrawn by surface vessels, or even by submarines, a diversion which further weakened the shrinking Japanese fleet.

The chief reason for the defeat of Japan's cordon defense was economic. Japan could not maintain the power with which she began the war. Even in the first year of war she had a net loss of warships. Up to the end of 1944 she lost 275 combat ships, excluding escort vessels, and she replaced only 162 of these. During the same period the United States lost only 128 and added 1,005 warships by new construction. Of the latter, more than 200 were submarines: the American underseas fleet made a major contribution to Japan's defeat.

Japan's industrial output was in large part dependent upon supplies from overseas; in fact, the war had been precipitated by the imposition of an American embargo on exports of aircraft, machine tools, chemicals, strategic metals, and gasoline to Japan. The war brought economic collapse. Japan's overseas trade was cut off; as soon as the United States Navy recovered from Pearl Harbor, preparations were made for attacks upon Japan's sea communications which were to prove decisive; 5,000,000 tons of her shipping were sunk; her harbors were mined; many of her industrial cities were bombed; her manpower was drained away in the campaigns in China and Southeast Asia. Japan lacked the industrial power to defeat the United States, let alone the whole United Nations, especially after the end of the German war allowed them to turn all their resources against her.

THE SIGNIFICANCE OF AIR POWER

So, the most important aspect of the naval warfare in World War II was the meteoric rise of American sea power. This was based upon the unprecedented ability of both navy and industry to expand enormously without loss of efficiency and upon the United States Navy's rapid appreciation of the revolutionary effect of naval air power. The Americans quickly adopted the strategy, tactics, and techniques in war at sea which air power imposed. With her vast resources, in the course of only four years of war, the United States became by far the strongest naval power in the world.

But the most important military and naval development in World War II was the inception of air power as a major factor. It is interesting to examine how far it realized the ideas of the theorists of the twenties. By and large it is possible to say that at least until the development of the atom bomb, while air power did not become so dominant as its extreme advocates had prophesied, it did revolutionize the strategy and tactics of war.

Before 1939 the various major powers had developed their air forces on lines dictated by circumstances of geography and politics. Britain, which alone had an independent air force, had been compelled through fear of a German aerial attack and by considerations of economy to concentrate on fighters. As a result, she had not been able to build up the strategic bomber force in which many of her airmen believed. Germany, thinking in terms of *Blitzkrieg* and a short war, and Russia, whose greatest asset was her tremendous manpower, had concentrated on the use of air power to support land armies. France, bedeviled by politics, had allowed her air force to decline in quality. The United States, aware that no enemy would be likely to land on her shores, had turned attention to war at a distance and had concentrated on a long-range bombing offensive.

All these characteristics and strategies were to influence the course of the war. Nazi dive bombers, the Stukas, acted like artillery in blasting a path for the columns which overran Poland in 1939 and France in 1940. Because of their slow speed, they later proved to be sitting ducks when faced with superior aircraft. The Germans had not grasped the fact that air power cannot be decisive until air superiority has been gained. On the Eastern Front the Russians had produced a superior dive bomber, the Stormovik (IL-2), which played an important part in the Russian winter offensive of 1941–42. Later rocket-equipped versions were famed as tank destroyers. In the West, although there was great pressure from the public for the development of dive bombers (an invention of the United States Marine Corps) to match the Stukas, for tactical purposes the air forces eventually used aircraft of a more orthodox type, which quite properly relied on speed.

Having reached the Channel in 1940, the Nazis had failed to win the aerial Battle of Britain because the British had radar, the Hurricane, and the Spitfire. Defeated in the air, and not properly prepared for an amphibious operation, Hitler had had to call off his Operation Sea Lion to invade Britain. As a result the war settled down to a stalemate just as in the First World War, but without the same bloody conflict in the trenches. At this stage Winston Churchill said that he could see no way to victory except by an "absolutely devastating attack by very heavy bombers from this country upon the Nazi homeland," a comment apparently indicating acceptance of the theories of strategic bombing, but quite unrealistic in terms of the existing strength of the RAF and actually not in keeping with Churchill's own belief in the effects of aerial bombardment. The British Commonwealth Air Training Plan had been set up in December 1939 to train air crews from the United Kingdom and other British dominions in Canada for the attack on Germany; but it took time to furnish the large number of crews required, and the four-engined bombers capable of reaching the far parts of the Reich were hardly off the drawing board. Bitter experience had already taught the Royal Air Force that it could operate with reasonable economy only at night, when accurate attacks on industrial targets were unfortunately not possible. Hence in 1941 Churchill informed the chief of the air staff that while the policy of strategic bombing would continue to receive full support, a decision in the war could be reached only by a simultaneous armored assault on the ground.

The entry of Russia and of the United States into the war made that armored assault possible but, reinforced by the United States Army Air Forces (USAAF), strategic aerial attack continued as well. The Royal Air Force (RAF), because of the difficulty of locating and hitting smaller industrial targets, resorted to heavy night attacks on industrial centers, beginning with thousand-plane raids on Cologne and Essen in May and June of 1942. Only in 1944, when the British had new navigational aids, did they return to attacks on individual factories. Under pressure of war, tremendous improvements had been made by the Royal Air Force in speed, range, bomb load, and armament. The Lancaster and Halifax four-engined bombers were a far cry from the Whitleys and Wellingtons of 1939; the blockbuster, of ten tons, made earlier bombs seem trivial.

The U.S. Army Air Force concentrated on daylight bombing of industrial targets and at first suffered heavy losses. Precision bombing having been found impracticable, it had to develop the technique of pattern bombing. Even so, only 20 percent of the bombs fell in the target areas—that is, within a radius of 1,000 feet of the aiming point of attack. The renowned B-17 Flying Fortress, introduced in 1935, was completely overshadowed by the B-29 Superfortress, which appeared at the end of the war.

The prosecution of this policy of strategic bombing, especially by daylight, had been heavily challenged by the *Luftwaffe* and, as has been seen, the Allies had been compelled to vary their tactics. In this regard, whereas the American Air Force had hoped to protect its bombers by building up their firepower, it soon found that it was also necessary to protect them with fighter escorts. Until long-range fighters were available, the daylight bombardment of the remoter parts of enemy territory was not possible. By early 1944, however, fighters, especially the Mustang and the Mosquito night fighter, were able to accompany bombing raids to all parts of enemy territory in Europe, and they were also ordered to seek out Nazi fighters and destroy them. Enemy losses greatly increased, and the air battle of Germany was won.

The defeat of the *Luftwaffe* in the air was the result not merely of German errors in production and training policy, but also of Allied bombing. Through overconfidence, the Nazis, failing to appreciate the lessons of the Battle of Britain, had not utilized the full capacity of their aircraft industry until the initiation of the combined Anglo-American bomber offensive in June 1943 forced them to desperate measures to produce fighters. A year before this time, as a result of a shortage of aviation gasoline (despite increased synthetic output), the *Luftwaffe* had cut down the length of flight training by two-thirds. The inevitable result was a serious decline in the quality of its personnel. During 1943 and early 1944, continuous Allied attacks on airframe factories contributed significantly to Allied aerial advantage in the ensuing months. By a policy of strategic bombing directed at aircraft and gasoline production, and also by defeating the enemy concentrations sent aloft to ward off the strategic attack, the Allies were able to seize command of the air and so to prepare the way for a combined assault of all arms.

The winning of the air battle had made the landings possible. German air generals have stated in interrogation that on D-Day only eighty fighter planes were operational to oppose the landings in Normandy, and there were certainly no more than 120. But it is important to note that only 28 percent of all the bombs that were dropped on Germany during the war had fallen before 1 July 1944. Three-quarters of the full weight of the air attack came after the ground assault on the heart of the Nazi Empire had been launched in the West. Thus heavier assault was also made possible by the winning of supremacy in the air.

Air power gave invaluable aid to the Anglo-American land forces. Shortly before the invasion, the weight of the aerial attack had been transferred to transportation; and the communications behind the Atlantic Wall were severely impeded. This was only a part, but perhaps the most important part, of the contribution made by strategic bombing to the success of the landings. As the invasion forces rolled on into Germany, they continued to receive invaluable support from the tactical air forces which had been set up. Spitfires, Tempests, Mustangs (P-51), Thunderbolts (P-47), and Typhoons, which were literally flying gun platforms, and medium bombers armed with rockets were used with effect against infantry and armor; and on occasion, the American and British "heavies" were called in to blast a path through strong German defenses, though sometimes with debatable results. However, despite serious deficiencies in land-air cooperation throughout the campaign, air power continued to assist the land advance and contributed to the eventual annihilation of the German armies as organized fighting units.

GERMAN 88-mm ANTIAIRCRAFT GUN. This gun was also used with devastating effect as an anti-tank weapon in North Africa and Europe. (RMC Museum, Royal Military College of Canada)

Meanwhile the strategic bombardment continued with an ever-increasing ferocity that sometimes seemed unnecessary, as in Coventry, Hamburg, and Dresden. It is not clear how much the bombing contributed to the defeat of Germany. *The United States Strategic Bombing Survey,* which studied the effects of the bombardment immediately after the war, came to the conclusion that "the German experience suggests that even a first-class military power . . . cannot live long under full-scale and free exploitation of air weapons over the heart of its territories. By the beginning of 1945, before the invasion of the homeland itself, Germany was reaching a state of helplessness. . . . Her armies were still in the field. But with the impending collapse of the supporting economy, the indications are convincing that they would have had to cease fighting—any effective fighting—within a few months. Germany was mortally wounded."

At the same time the *Survey* pointed out that the German economy showed tremendous powers of recuperation and that the effects of bombing were far less than the Allied air forces claimed. Despite heavy air attacks, German production actually reached its wartime peak in late 1944. The *Survey*'s report showed that even though businessmen and economists had

THREE WORLD WAR II ESCORT FIGHTER PLANES fly in formation over England. From top to bottom they are the P-38 Lightning, the P-51 Mustang, and the P-47 Thunderbolt. (U.S. Army, *The War Against Germany: Europe and Adjacent Areas,* Office of the Chief of Military History, Washington, 1951, p. 9)

cooperated in working out the strategic bombing plan, there were serious errors in the selection of targets, mainly caused by inadequate intelligence or wrong judgments. It also showed that industries could continue working after raids which appeared to have destroyed them, and that the civilian population in a police state showed "surprising" resistance to the terror and hardships of repeated air attacks. One might add that the population of London had also shown that terror bombing did not have the immediate results which its prewar advocates had claimed. Finally, estimates of the effectiveness of the Allied bombing must not neglect the fact that the

Germans had been fighting a major war on their Eastern Front for months before the invasion, and the major Allied bombing effort, began. The Russian campaign had occupied the bulk of the German armies and a considerable part of their air forces.

PROBLEMS OF COMMAND

The appearance of a third armed force, air power, with which the older services must always operate closely but which could also act independently, complicated the problem of interservice, or "joint," command. For a variety of other reasons also, questions arising out of the exercise of command over the fighting forces became far more important in the Second World War than in any previous conflict. Further, because communications had improved and distances had shrunk and because the war was worldwide and fought by great coalitions, the problem of "combined," or international, commands had to be settled. Finally, because war had come to involve the whole national life, the age-old problem of the relations of civil and military leadership had been greatly aggravated.

It so happened that in this war there was a greater need for amphibious operations. In Europe a landing on the continent had to be forced, and in the Pacific the route to Japan stepped from island to island. The operations which necessitated joint command were therefore far more numerous than ever before. But amphibious operations in the past had often suffered from the inability of coequal land and sea commanders to work with one another. When one or the other had been given primacy, he frequently had not sufficiently understood the problems of the other service. The appearance of air power as a third force made coequal command yet more difficult and perhaps more dangerous; for two services might conceivably outvote the third and compel it to an action which it knew from its experience to be unfeasible.

The British had had a longer experience than Americans of amphibious operations and had usually, although not always with success, appointed coequal commanders. In their Chiefs of Staff Committee, they had evolved a form of command which seemed to be alien to the basic principle of unity but which usually worked, perhaps because it was suited to their temperament. Through it they had achieved a useful tradition of successful interservice cooperation. However, since Gallipoli, they had been less enthusiastic about amphibious operations, which seemed to them to be too difficult in the conditions of modern war. On the other hand, between the wars, the United States through the Joint Board worked out techniques for amphibious operations in preparation for the possibility of having to strike at Japan across the Central Pacific; and in 1935 the principle had been established that when joint operations of the United States army and navy or marines were undertaken a single officer would be put in command.

When the war came, the need for an organization to represent the American services in the Anglo-American Combined Chiefs of Staff Committee (to be described later) led to formation, for the first time in United States history, of an American Joint Chiefs of Staff. Thus the United States conformed at this level to the British system of coordinating the efforts of the services by a committee. But at lower levels, where forces of all arms of both countries acted together in theaters of war, the American practice of appointing a single theater-commander from one of the services was followed. While these solutions for interservice command did not always work without a hitch, they were, on the whole, remarkably successful. In this war, more than in any previous war, the necessity for full cooperation between the various arms was apparent; and by and large it was achieved.

On the German side of the war the problem of international, or combined, command raised fewer problems. As far as Nazi-Fascist cooperation went, Germany was so much more powerful than Italy that she invariably took control in joint operations. Otherwise the three Axis powers merely fought the war on parallel lines. Japan and Germany did not work in tandem. In the United Nations, however, where the military strength of the three leading powers was more equal, but there were several distinct major theaters, including West Europe, Russia, the Pacific, and the Middle East, no easy solution could be adopted. Military coalitions in the past had always found the problem of international command difficult. In this war also, when there was great disparity of outlook, as between the Western democracies and Communist Russia, it proved impossible. While the Russians demanded full information about Allied plans, they never revealed their own secrets to the Allies. The British and the Americans found agreement with the Soviet leaders on general strategy—for instance, on the timing of the Second Front—difficult to achieve. Even where the political ideals and systems of government of the Allies were similar, as between Great Britain and the United States, the essential cooperation under a united command was not easy. And it was found that the civil heads of states were often able to come to agreement with their opposite numbers more easily than the leaders of the armed forces, especially those at the operational level.

Nevertheless, Britain and the United States, despite certain difficulties, achieved remarkable degrees of cooperation and of coordinated command. The lessons of the First World War had been learned. For the overall direction of the war an Anglo-American Combined Chiefs of Staff was created with headquarters in Washington. Regional commands in various theaters of war were given to an officer of one nation, usually that which was contributing the larger forces. Harmony between officers of different countries at combined headquarters was a new feature of warfare. Undoubtedly the most outstanding example of it was at General Eisenhower's headquarters in North Africa and Western Europe, where unity by combined command was really achieved. It must be added, however, that the Combined Chiefs of Staff system was never fully extended to include the staffs of the smaller Allied powers.

A noticeable feature of the command at the highest level in the Second World War was that it was exercised by civilians who concerned themselves with many matters that were military in nature. While Churchill, Roosevelt, and Stalin, and Hitler and Mussolini to a lesser degree, might all be said to have had some earlier military office and experience, it was rather limited, and they were, in fact, civilians. The Japanese leaders had a great deal of experience. Yet all of these political heads of state were called upon to make decisions about military strategy; and each one of them paid great attention to matters which in earlier times would have been regarded as being outside the concern of a mere civilian. The top leaders supervised the formulation of grand strategy but also, aided by modern methods of communication, kept a close watch on the conduct of operations. Churchill, as prime minister and minister of defense, directed Britain's war effort. On military questions, he was advised by the Chiefs of Staff Committee. His volumes on the war showed how much he interested himself in even the minutiae of service matters. Roosevelt, the constitutional commander in chief of the United States forces, concerned himself less with military details, but nevertheless he also supervised major strategic policy. Stalin became a marshal of the Soviet Union and generalissimo. Hitler went further still and took upon himself the supreme military command. In 1938 he had created a new organization, the Oberkommando der Wehrmacht (OKW) through which he could command the armed forces; and in December 1941, he took over personal command of the army as well.

While it was imperative that the supreme civil authority in both democratic and totalitarian states should have ultimate control of military affairs, it was obvious that lack of professional military experience could be a handicap. The successful exercise of this high command by a civilian necessitated great trust in, and dependence on, the professional soldiers who commanded the armed forces. That trust was notably lacking in Germany. In certain spheres, especially those in which military decisions were affected by political affairs, the statesmen were often the best judges of the issues. British actions to support Belgium in 1940 and Greece in 1941 were militarily unsound but politically important. Hitler has been blamed for his insistence on unyielding defense on the Eastern Front and for his repeated orders to his armies to stand and fight. There were times when Hitler was militarily correct, but his policy ultimately led to the destruction of his armies; the strategic retirements contemplated by his generals, which would have prolonged the war, would probably have led to so great a loss of prestige that his regime might not have been able to survive.

The relations between statesmen and soldiers in the democracies, where strategy was drawn up by military chiefs under the supervision of the head of the state and on lines indicated by him, and where the operations were, to a great extent, left in the hands of professional military men, were more successful. But the line between military and political decisions is very hard to draw, and it is quite impossible to leave all military decisions to the soldiers and all political decisions to the statesmen. As we have seen, when the United Nations debated whether to invade Southern Europe through France or through the Balkans, Churchill's desire to strike at Central Europe to forestall the Russians had seemed to the Americans to savor of imperialism. It also conflicted with American anxiety to ensure a quick military decision. In fact, however, many such decisions, like those which stopped the advancing Anglo-American forces short of Prague and Berlin, should never have been made on purely military grounds. At that level no decision can be nonpolitical.

One reason that problems of command and of politico-military relationships had become so much more important in the Second World War was that war had become total and universal. It had come to permeate thoroughly the national life. As a result of the air attacks, citizens were often as active participants as the soldiers. In the occupied countries they were organized in partisan or underground armies to fight behind the lines for liberation or to commit acts of sabotage. Even where they were not active as fighters or in civil defense, their lives were affected by restrictive regulations, by the rationing of their essential needs, and by the restriction of their luxuries and comforts and pleasures. Government controls instituted for these purposes had to be tied closely into the national military effort and required careful coordination with military policy.

SCIENCE AND TECHNOLOGY

The unleashing of atomic power dramatized another revolutionary development of the Second World War, employment of science and civilian scientists in military effort. Modern science, which had had its birth in the early Renaissance, had been very slow to mature. The scientific renaissance had not fully come until the latter part of the seventeenth century; the technological developments of the Industrial Revolution in the eighteenth had owed little to its inspiration; the marriage of science and technology had been a later development. While soldiers had made use of certain aspects of scientific knowledge in the eighteenth century, especially in the sphere of siegecraft, the full application of science to war had been delayed until chemistry, physics, electronics, and other branches of science had been developed as distinct fields of research.

Thus it was the First World War, in which the technology of the Industrial Revolution had first been fully exploited, which had also seen the beginnings of scientific warfare. Mass-produced electrical and mechanically operated devices of great accuracy and intricacy began to take over various tasks in which the soldier or sailor had formerly made an empiric judgment. As machines are not subject to the emotional stresses and strains of conflict, and as they could perform tasks beyond the capacity of men, the place of the human element in fighting was bound to be affected although, of course, it still remained an essential ingredient. In these circumstances, scientists and laboratories had been absorbed into the war effort, with the result that, by 1918, almost all the principles used in scientific warfare in the Second World War (with the notable exception of atomic power) had made their appearance and had been tried out in practice. It remained only for this greater conflict to put them to their fullest use.

However, after 1918 science and technology had been restricted mainly to developments for peaceful application and their potentialities in war had been little exploited. This was true even in the warlike totalitarian states. On racial grounds, Hitler had exiled many of his most brilliant scientists who might otherwise have contributed much to his war effort. Elsewhere, wishful thinking that great wars were things of the past, the one-sided concentration of scientists on the works of peace, and the inability of military men to comprehend the trends of modern science had brought about a serious lag in the further development of the techniques which scientists had produced during the last war. Some few military laboratories existed, but the scientists were relegated to an inferior position by the dominant military direction; and the fundamental principle, which industry had long endorsed, that the scientist must be given freedom to develop his ideas and must be allowed to engage in pure scientific research, was completely alien to the thinking of most senior service officers. Hence, while industry in this period had produced the "most bizarre gadgetry" the world had yet seen, many in the armed forces tended to think of war in terms which were out of date. In a few military fields, notably in the development of radar, scientific advances had been made; but, in the opinion of Vannevar Bush, the head of the American scientific war effort, "the world slept."

It must be noted that in every nation, and especially in the United States, the extraordinary advance of scientific knowledge and of industrial techniques had produced a reservoir of military power in the form of industrial plants, techniques, skills, and "know-how" which, while developed for purposes of peace, were available for, and were used in, modern war. When the war came, a tremendous change took place which can only be regarded as revolutionary. The competition between the scientists on opposing sides in the race to produce new weapons and methods of warfare, and to checkmate those already developed, was as intense as the conflict between the armed forces. In the United States alone, 30,000 scientists and engineers were employed on problems connected with new weapons and new machines. Organizations like the National Research Council, founded during World War I, and the National Defense Research Committee and the Office of Scientific Research and Development, established in the Second, were to some extent paralleled by the British Scientific Advisory Committee to the Cabinet, by Canada's National Research Council, and by her postwar Defense Research Board. For the first time, scientists became full and responsible partners in the conduct of war, and as the war progressed the military leaders came to have ever greater confidence in their work. Scientists in or out of uniform were to be found using their peculiar skills and knowledge beside the front-line soldier, in aircraft flying over enemy territory, and in the battle at sea. Through "operational research," the application of scientific techniques to the study of special problems posed by military weapons and operations, science entered into warfare in a new fashion.

No detailed account can be given here of the great number of scientific developments which the war produced, but some mention must be made of a few. Radar, developed at first to give early warning of the approach of enemy aircraft, came to have a multitude of other uses, from directing anti-aircraft fire and night-fighters to aiding navigation and guiding lost aircraft. The proximity fuse, described by J. P. Baxter in *Scientists Against Time* as one of the four or five inventions that helped win the war, was developed by American scientists in response to a British request. It was used against aircraft and also against ground targets. RDX, an explosive nearly twice as powerful as TNT, greatly increased the effectiveness of depth charges, "blockbuster" bombs, and torpedoes. But these are just a very few of the contributions made by the scientists to the war effort.

It is worthy of note that many developments were the result of inter-Allied scientific effort, a new feature in war. While the United States was still neutral, a British Scientific Mission headed by Sir Henry Tizard had taken to Washington and Ottawa information about British secret devices and equipment and a list of urgent requirements. From this visit had come a degree of cooperation between the United States, the United Kingdom, and Canada which had greatly increased productivity. Some groups in both England and America had at first opposed the interchange of information on the ground that their own country would be making the greater contribution. It happened that they were both thinking of the same field, radar, in which they both believed, wrongly, that they had a complete monopoly! It is reasonable to say that inter-Allied cooperation in scientific effort was achieved because the scientists saw with greater clarity than the statesmen and soldiers what its results would be, and because they were successful in their pleas for its introduction. The greatest scientific feat of the war, the releasing of atomic energy, was a product of inter-Allied cooperation, although its full development was left by agreement to the United States. Penicillin, perhaps the greatest medical achievement of the war, was discovered in Britain but made available by a joint effort.

Discoveries of this kind, which could so obviously have a greater value for peace than for war, but which were developed speedily only because war needs were urgent and funds were therefore made available, are evidence that weapon development progresses faster under the stress of war; and they also suggest the theory that war itself has been a stimulant of progress. This theory, which gained a wide public credence at the end of the war, supports those like Sombart who would attempt to defend war as a creative institution. Against this historians like Nef have argued that war has always destroyed far more than it has created. War conditions intensify development and investigation though they may stifle pure research.

Although the United Nations may be said to have won the battle of science, it must be emphasized that no single invention was decisive, not even in the war against Japan. The Japanese had already been defeated by conventional weapons before the atomic bomb was dropped at Hiroshima. The Germans were not far behind in the contest in the laboratories. They produced, among other things, the V-1 flying missile, the V-2 rocket, the first operational jet plane, the Messerschmitt 262, the magnetic mine, the snorkel, and the true submarine, one capable of extended operation beneath the surface. This fact is significant because it casts doubt on the old belief that conditions of freedom are necessary for scientific progress. The scientists of the Western powers may have had an edge in the contest but, under totalitarian direction, Nazi scientists were foes to be feared.

TOTAL WAR

The use of the atomic bomb was not only a demonstration of the military power of science but also raised important ethical problems which threw light on the declining standards of conduct in modern war. The decision to use the bomb was in line with that trend toward moral nihilism which had marked warfare in the last hundred years. In this it had been preceded in kind by practices like the indiscriminate bombardment of industrial cities and unrestricted submarine warfare. The Hamburg and Dresden fire storms killed in the same order of magnitude as the atomic bomb dropped by the *Enola Gay*. The Nazis, Japanese, and Russians, partly in retaliation but partly on their own initiative, had perpetrated atrocities on a scale not known before in the history of civilized man. Races marked for extermination were slaughtered wholesale with cold-blooded efficiency, much as the Turks had dealt with the Armenians shortly before World War I. Spies, political police, "thought-police," terror, the concentration camp, and forced labor, types of oppression used in the totalitarian states in peacetime, were used in war on a much greater scale to cow subject peoples. These crimes were usually committed, especially in the west of Europe, by political soldiers like the S.S. and the Gestapo rather than by the regulars; but regular troops were not innocent of them.

The United Nations forces were restrained by moral considerations to a greater extent than the totalitarian states, but even their decisions were ruled, in the end, by expediency, as was the case in mutual abstention from the use of poison gas in World War II. Moral considerations were important only when they were also expedient; the doctrine of reprisal was used to justify actions which would otherwise have been ruled out; and the ultimate criterion was whether a contemplated policy, operation, or mode of conduct would help to win the war. Any weapon, however frightful, could be justified on these grounds. It is sometimes said that gas was not used because it has aroused the moral conscience of the world in the previous war. Actually

A GERMAN TIGER TANK, captured in North Africa in 1942. With its heavy armor and 88-mm gun, it sacrificed mobility for protection and firepower. (U.S. Army, *The War Against Germany and Italy: Mediterranean and Adjacent Areas,* Office of the Chief of Military History, Washington, 1951, p. 41)

gas was not used because of fear of retaliation in kind and even more because it is not a decisive weapon.

Thus the total war brought a decline in moral standards. Although offset to some extent by individual deeds of heroism performed through belief in a cause, this decline resulted in a general lowering of traditional standards of behavior. Postwar crime waves and increased juvenile delinquency were but minor symptoms of a disease which threatened civilization. Its most serious result would be the failure of moral standards to influence men, parties, and nations. Those lusting for power would then feel free to use any weapon and any practice to gain their ends. With the new weapons which science had developed during the war it was clear that man's control over nature was outrunning his capacity to create a stable social and political system in the world. Total war might lead to total barbarism.

Thus, from being at first a war which was limited in very many ways, World War II had reached a climax in a new degree of totality. It had been fought by great armies of millions, mechanized to an extent beyond the dreams of all but a few peacetime visionaries. Scientists were, for the first time, fully engaged in the war effort; and this war saw technology applied at an astonishing rate to fashion weapons which greatly altered the traditional methods of conflict. The great social and political developments of the twentieth century, the development of democracy on the one hand and of new "popular" autocracies on the other, had also contributed to the revolution in the conduct of war. The harnessing of the whole national life to war had ended, as it had to, in the complete destruction of the vanquished state. For the chief enemy, Germany, defeat, like war, was total. The huge

THE SOVIET T-62 TANK, one of the most successful armored vehicles of World War II. (U.S. Army, Center of Military History)

losses of manpower that had marked World War I had been surpassed by the total on the battle fronts. In World War II, though, Britain and France actually had fewer military losses. The wholesale liquidation of civilians (very much greater than in the earlier war) brought the sum total of war deaths to a figure (estimated at up to fifty millions) very much higher than the total number of deaths in World War I. Realization that yet more horrible weapons of mass slaughter were being produced as rapidly as laboratories and factories could turn them out renewed once more the determination which had gripped men everywhere in 1918. Once more the cry was raised that war must be abolished.

19

COLD WAR BEGINS

"**C**old War" was a term coined to describe the state of tension that developed between the nations of the West and those of the communist world shortly after the end of the Second World War. Technically, this was not war, but a political confrontation that could lead to war. International tension is no new thing; in fact, it could be described as a normal condition of relations among states. In the past, when tensions reached a level intolerable to one nation or another, war was frequently a means of resolving them. The Cold War was a distinctive phase in the history of international relations not only because of the fundamental cleavage in ideology and objectives between East and West, but even more because the revolution brought about by the development of nuclear weapons and by the perfection of new means to deliver them had in some respects restricted the use of war as a feasible method of pursuing political ends, but without eliminating the possibility of its occurrence.

If the problem of the Cold War had been simply that of finding a political accommodation between the Soviet and Western blocs in order to lessen tensions and reduce the danger of war, the difficulties would have been formidable enough. But relations between the two power blocs were infinitely complicated by contemporary developments that had nothing to do, in origin, with the Cold War, yet became elements in it.

Probably the most significant of these developments was the collapse of the European colonial empires in Asia and Africa, and the emergence of many new sovereign states. In the nineteenth century European military powers armed with magazine rifles and machine guns had been able to use railways and roads to maintain rigorous control of vast areas that had formerly been subject only to a vague suzerainty. They had spread the political, economical, and cultural characteristics of Western Europe, including the idea of democracy, though usually not its practice, across the globe. But despite the drastic disruption of ancient societies which European penetration brought about, there had been relatively little partisan resistance against it, perhaps because the Western regimes had made changes that were acceptable either to local ruling classes or to the majority of the people. However, imperial authority was spread thin, and it collapsed when it was seen to conflict with the principles of Western liberalism and when empires were weakened by World War II. Political and military resistance movements now made Western imperial rule impossible almost everywhere. But they proved less effective against communist totalitarian imperialism when that in its turn infiltrated areas lost to Western control but unable to function democratically. The disturbances that frequently accompanied the passage from colonial status to independence, and the economic and social problems with which the new states were confronted after the achievement of independence, created grave instabilities in the world state system. This fostered the spread of the Cold War, and occasioned a series of minor conflicts in which the danger of a major collision between the great power blocs was always present.

THE UNITED NATIONS ORGANIZATION

Before the Second World War ended, the United Nations took measures to erect barriers against the outbreak of a third. At Dumbarton Oaks in 1944 and at San Francisco in 1945, the wartime alliance was converted into a permanent organization to keep the peace. The United Nations Organization, like the League before it, attempted to reconcile the conflicting concepts of national sovereignty and collective security but, as a result of experience, certain changes in form were introduced to attempt to make the new world body more effective. The charter outlawed private war; and an international military force was planned. The most powerful body in the United Nations was the Security Council, in which the five great powers had permanent seats. Because it was believed that the new plan for collective security would not work unless there was harmony among those great states which would, in the event of a major police action, bear the brunt of the fighting and the responsibility, the veto power (which every state had virtually possessed in the League of Nations as a result of the unanimity rule) was retained only by each of the five great powers in the Security Council.

The Military Staff Committee of the Security Council, composed of the chiefs of staff of the great powers, was charged with the task of planning an international army and was also expected to discuss the limitation of armaments. The atom bomb had not yet been used with the U.N. Charter was drawn up. When it came up for discussion, it was so different in kind from ordinary weapons that it was considered an entirely distinct problem. In order to devote special attention to the abolition of atomic warfare, a separate U.N. committee on atomic power was created. Plans to create a U.N. army and to control national armaments were moves toward the limitation of national sovereignty.

WAR CRIMES TRIBUNALS

During the course of the Second World War, the United Nations had prepared plans to bring war criminals and aggressors to justice. The War Crimes Trials, held at Nuremberg and Tokyo between 1945 and 1948, were widely hailed as precedents by virtue of which individuals and groups could be made to answer for crimes against international law. If this could have been achieved, then a new chapter in history would be opened in which old concepts of national sovereignty would be radically altered in a time when wars between nations would give place to international police actions.

From the first, however, there was doubt whether society had yet reached that stage of development. War crimes are breaches of international agreements, such as the Geneva Conventions, which govern the conduct of military operations and to which most civilized nations have subscribed; and it was arguable that aggression became a crime when private war was outlawed by the nations signing the Kellogg Pact of 1928; accordingly, charges laid under these headings had some claim to be legal. But some critics pointed out that a third crime brought before the international tribunals, that of genocide, or the wiping out of races, had not formerly been proscribed by international law and that such action thus savored of retroactive justice. Many people believed that the proceedings against German and Japanese statesmen and military men merely represented the vengeance of the victors. They noted that no parallel investigation was made of war crimes alleged to have been committed by members of the Allied forces, and that the judges were all drawn from the victorious nations. Some Allied military leaders objected to the trial and conviction of German generals on charges of committing aggressive war and other offenses. They contended that a soldier must carry out his orders and that inflicting punishment after a war for offenses committed on the orders of superiors would tend to dissolve all military discipline. There was thus no consensus that the War Crimes Trials were inaugurating a new and revolutionary stage in the history of warfare and society.

THE ATOM BOMB

Even without counting the atom bomb, technological progress in war had made such strides during World War II that the urgency of methods to prevent a third conflagration was augmented by the stark knowledge that another war would lead to yet greater devastation and have an even greater

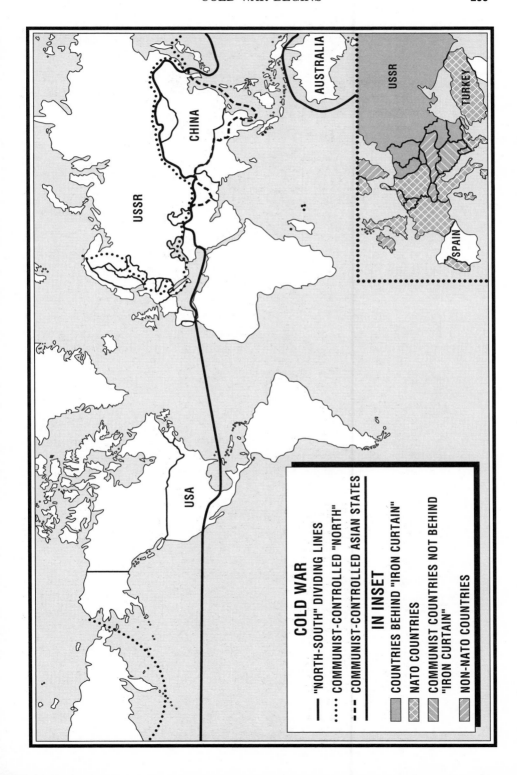

COLD WAR

"NORTH-SOUTH" DIVIDING LINES

········· COMMUNIST-CONTROLLED "NORTH"

— — — COMMUNIST-CONTROLLED ASIAN STATES

IN INSET

COUNTRIES BEHIND "IRON CURTAIN"

NATO COUNTRIES

COMMUNIST COUNTRIES NOT BEHIND "IRON CURTAIN"

NON-NATO COUNTRIES

impact on civilian populations. Recoilless guns, simple to produce and to operate, now gave infantry more protection against the tanks which had brought mobility back into war. Guided missiles, which Hitler's V-1 and V-2 had foreshadowed, were capable of spreading death and destruction far behind the firing line; and no defense against the V-2 had been found. With the prospect of supersonic, intercontinental ballistic missiles fitted with thermonuclear warheads, "push-button" warfare would soon enable nations to destroy one another from afar. Deadly nerve gases and bacteriological cultures that could spread disease far and wide were believed to be already available. The atom bomb, in the small crude form used against Japan, was equal to the effect of thousands of tons of TNT. But it also spread a radioactive effect that would bring immense harm to people. It appeared to introduce a new dimension into warfare. Men began to ask whether the human race—or, at any rate, civilization as they knew it—could survive another war.

When Western scientists designed the atom bomb during the war, they had feared that Germany might anticipate the democracies. The decision to use the bomb on Japan, made at the highest political level, had been defended by the argument that, although Japan was tottering, she still had armies in the field that would make necessary a bloody and costly invasion of the Japanese islands. (There were 2,000,000 Japanese troops, 9,000 Kamikaze planes, and ample supplies in the home islands.) The atom bomb was expected to save hundreds of thousands of Allied lives. Later analysis was to suggest, however, that the United States Navy had in fact already so fatally crippled Japan's access to essential foreign resources that victory was already assured.

During the war, when consideration was given to the awful possibilities of the new warfare, some engineers and scientists had begun to have qualms about prostituting their knowledge and experience in work that they feared might eventually lead to universal destruction. Man's technological capacity seemed to have outreached his power of social and political organization. Some scientists and engineers declared that, in face of the weapons of the future, the old law that every new weapon could be countered by new defenses could fail to operate. Led by Albert Einstein, whose theoretical work nearly half a century before had prepared the way for the splitting of the atom, and whose personal approach to Roosevelt had expedited development of the bomb, some of the country's foremost scientists insisted that man must now abolish war and stop the further development of thermonuclear weapons; others took it upon themselves to reveal the secrets of the bombs to foreign powers, especially to the U.S.S.R., which had transported many German missile experts to the East to counter those recruited by the United States.

In 1946 and 1947, following a plan devised by a veteran statesman, Bernard Baruch, the United States proposed that control and inspection of the peaceful use of atomic energy be entrusted to an international agency. This was rejected by the Soviet Union, ostensibly because of suspicions that the Americans intended it as a means of monopolizing the industrial development of atomic power. We now believe that when Truman hinted to Stalin at Potsdam in 1945 that the United States had developed a revolutionary weapon, the Communist leader's enigmatic response concealed the fact that Russian scientists were also well informed about atomic research and development. Then, allegedly because Klaus Fuchs, a German refugee scientist in Britain, and others in Canada, had given secret information to the Communists, the United States dissolved the international partnership with Britain and Canada which had rushed the bomb into production during the war. The Americans now apparently thought that they would enjoy a long monopoly of the use of atomic power for both military and industrial purposes.

THE IRON CURTAIN AND THE COLD WAR

By this time the uneasy wartime cooperation of the U.S.S.R. with its Western allies had ended. In 1945 at Yalta, Stalin had insisted on securing Eastern Europe as a Soviet sphere of influence and had established Communist governments in all the countries which he had "liberated." The Western powers were compelled to accept reluctantly Russia's denial of the claims of the "Lublin" Poles, a government in exile which had fought as an ally against the Nazis. As Churchill said when he received an honorary degree at Fulton, Missouri, from Truman's alma mater, an "Iron Curtain" had fallen across Europe.

Occupied Germany had been divided, at first into four zones of occupation controlled by each of the four major allies, the United States, Britain, France, and the Soviet Union. Then the Western powers amalgamated their three zones. Berlin, the historic German capital, administered jointly by all four occupying powers, lay deep in the Eastern section, linked on the ground to the West only by rights of passage along a "corridor" highway through East German and Russian barriers. This was the beginning of what came to be called the Cold War.

RIVALRY IN MISSILES AND SPACE

The American monopoly of the atomic bomb was broken by the Soviet Union in 1949; the first explosion of a hydrogen bomb by the United States came in 1952 and was followed, the next year, by the Soviet production of an H-bomb. Great Britain joined the "nuclear club" by exploding its first atomic bomb in 1952 and its first hydrogen bomb in 1957; then, in 1960, France tested an atomic device in the Sahara. In both the United States and Russia, German scientists who had been engaged during the Second World War in Hitler's V-1 and V-2 programs assisted in harnessing nuclear weapons to missiles. The rivalry between the United States and the Soviet Union in the production of missiles, especially the intercontinental ballistic missile (ICBM), was intense and enormously expensive. Experimentation in rocketry led to the probing of space. On 4 October 1957, the Soviet Union put its Sputnik into orbit around the earth, an achievement matched by the United States on 31 January 1958. Both nations bent their efforts not only to the launching of satellites but also to the perfection of the technology for manned space flight. Again the Soviet Union led the way, in the person of Major Yuri Gagarin, whose pioneer exploit in orbiting the earth was followed by Major Gherman Titov's flight, in which he circled the earth seventeen times. That the United States was still in the running in the space race was soon shown by the nonorbital flights of Commander Alan B. Shepard and Captain Virgil I. Grissom, Jr. Then, on 20 February 1962, Lieutenant Colonel John H. Glenn successfully rocketed three orbits from Cape Canaveral. The potential use of space vehicles for military purposes—whether for reconnaissance, surveillance, communications, or delivery of nuclear weapons—had doubtless spurred both powers to their remarkable endeavors in this field. But space technology had important connotations also for civilian use, especially for communications. It therefore led to intensive state-supported rivalry in which several other countries participated. The Russian successes in lifting heavier loads into space, and the spectacular nature of their space exploits, were of immense propaganda value in the Cold War. Yet the American launchings, conducted in the full glare of publicity instead of in secret, were an impressive demonstration of confidence and technical efficiency that served to counteract the effect of the earlier Soviet successes.

BIPOLARITY

The nature of the new weapons had a revolutionary effect upon the world state system. In the past, smaller states, if they possessed superior fighting qualities, more effective military organization, or favorable geographical conditions, had always had some chance of victory in wars against greater powers. Now, the cost of the newest arms was such that only the largest and wealthiest states could afford them; and the rate at which arms become obsolescent meant that smaller states were inevitably equipped with out-of-date weapons. The ability to feed and equip armies without resort to foreign aid had become an even greater factor in the reckoning of military strength. For these reasons, the relative military power of middle-sized and small states had greatly declined. There were now only two powers of the first magnitude, the United States and Soviet Russia. Britain, shorn of her empire, was not comparable with these. France's claim to great-power status in the United Nations was historic rather than realistic. No other power could pose a serious challenge except China. But China's strength, based on vast manpower and resources, was still potential rather than actual, since she would be occupied for some time to come with the immense problems arising from the communizing of an economy that was as yet technologically undeveloped.

The reduction of the number of first powers to two made the balancing of power much more difficult than it was in the days when there were more

U.S.S. AMERICA. The angled deck of this post-World War II United States aircraft carrier allowed simultaneous launching and landing operations and reduced the risk of accidents. (*Soviet Military Power: Prospects for Change, 1989,* Department of Defense, Washington, 1989, p. 130)

states in the equation. It so happened that the two leading states, the United States and the U.S.S.R., were fundamentally opposed in ideology, in social and economic structure, and in political organization. While both claimed to be "democratic," they had vitally different concepts of the meaning and nature of democracy. The political liberalism that forms the basis of American democracy, and the remnants of her traditional isolationist policy might have made it possible for the United States after 1945 to contemplate the continuance of a divided world if a greater unification could not be achieved; but communist doctrine made it more difficult for Russia to accept a world organization in which her sovereignty would be restricted and her influence outweighed. In addition, continual agitation by emigré minorities in the West, the implacable hostility of the American military concerned about security, and of what came to be called the "military industrial complex" concerned about profits, were as much a source of unease as of protection. On their part, the Soviets could not forget that the Western powers had tried to overthrow the Bolshevik Revolution by military means in 1919. The reduction of the number of powerful states, and the ideological antagonism between the two leading states, created great concern and threatened peace. As President Dwight Eisenhower put it in a speech to the United Nations Assembly, "Two atomic powers eye each other malevolently across a defenseless world."

In response to this situation, some of the lesser powers, and also groups in some of the middle states, talked in terms of "neutralism" or of creating a "third force." But a neutral policy was practical only for those states which, like India, felt relatively remote from the area of likely conflict or, like Sweden, considered themselves to be too precariously placed between East and West. For others, as the polarization of the world became clearer, there seemed no alternative to a policy of affiliation with one side or the other. By refusing to withdraw her occupation troops from Eastern Europe after the war because of her fear of Germany and her suspicion of Western intentions, Russia had quickly forced many of the East European countries into satellite status and had imposed communism upon them. This Soviet policy culminated in 1948 in the coup d'état in Czechoslovakia, a little power that had been attempting to make the best of both worlds.

SOVIET MILITARY POLICY

Czechoslovakia had not fallen to a Russian military assault: the influence of the Red Army encamped along her borders was decisive. Even before the Czech coup, the Soviet use of the veto in the U.N. and obstruction of attempts to limit arms, to control atomic energy, and to create an international army had shown that Russia would not accept the establishment of an effective world organization with full power to preserve the peace. The rape of Czechoslovakia brought the rest of the world to suspect that the Soviet Union had not abandoned the Marxist dream of world revolution; and there was reason to believe that that doctrine was now inextricably mingled with Russian imperialistic ambitions for world domination. A Russian note to Turkey in 1945 demanding a share in the control of the Bosporus was in the tradition of the policy bequeathed by Czar Peter the Great to his successors.

Soviet policy immediately after the Second World War would have been better understood if the West had then known more about Russia and especially about Soviet military doctrine; but unfortunately Russia was still a "mystery wrapped in an enigma." After the Bolshevik Revolution, the Red Army had consciously striven to develop a new "Marxist" military doctrine and military organization. Although the military theory that evolved was not greatly different in its broad outlines from that of nonrevolutionary states, a few important features appeared. The difference from Western military thought that was most important was the Soviet belief that war and peace are actually indistinguishable in a world in which capitalism continues to exist. This idea applied in two different dimensions. As was seen during the Second World War, modern conflict is not limited to the operations of armies, but is carried on behind the front by partisans and guerrillas, and by subversion and sabotage. In this respect, of course, Soviet military doctrine conformed closely to the practice of modern total war.

But the Russians carried the concept of total war into a second dimension. Although they denounced Clausewitzian thought in certain respects, they in fact developed his doctrine that "war is nothing else than a continuation of political transactions intermingled with different means" to an inverted, but quite logical, conclusion. They asserted that peace is only "a continuance of war by other means." Conflict (though not necessarily armed conflict) between socialist states and capitalism was regarded as inevitable. Although it was considered preferable to gain objectives without war if possible, Soviet statecraft had turned in peacetime to what are, in effect, lesser forms of war—namely, subversion, sabotage, encouragement of colonial rebellion, and satellite aggression. The Russians thus became masters in combining and operating various nonmilitary forms of war—political, economic, and psychological. They supported Communist parties if it suited their purposes, they fomented disaffection against "imperialist capitalist overlords," they encouraged their satellites to fight the Communist battle by aggression, and they could even—as in the Soviet-Japanese fighting on the Manchuria-Mongolia frontier between 1934 and 1939—engage in armed conflict themselves without considering that they were totally involved. All this kind of activity, falling short of total war, shaped part of what came to be called the Cold War.

The most dramatic evidence of the strategy of Cold War was found in the widespread espionage nets that were uncovered in the West, especially the one revealed by the desertion of a Soviet cipher clerk, Igor Gouzenko, in Canada. Communist ideology cut across national affiliations. Spy rings could easily be built up by the Soviet Union among the citizens of democratic states. Communist Party members and fellow travelers were potential or active agents of Russia. And the techniques of espionage, not uncommon in the old order of things, were multiplied and intensified. Inevitably, they forced Western countries to follow suit. The United States, for instance, created the Counter-Intelligence Agency (CIA).

ECONOMIC DISLOCATION

In the immediate postwar period the upheaval and dislocation caused by the Second World War and its aftermath provided a fertile soil for this new form of warfare. Cities and industries had been destroyed wholesale; populations had been rooted from their homeland and driven, as "displaced persons," to make new homes elsewhere; the Iron Curtain erected by Russia across Central Europe cut off industries from their markets and peoples from their food supplies; a great shortage of consumer goods endangered the stability of many national economies; and when industries had to be reconverted to peacetime production in order to restore prosperity, the machinery was not available.

The measures taken to restore shattered economies did not always contribute to a general revival. Every country in Europe, believing that unemployment was a source of unrest, strove to maintain full employment by state action. Nations were driven to continue wartime controls and restrictions, the same devices they had criticized when Hitler introduced them in time of peace. Import and export licenses and currency controls, while they protected national economies, hampered the revival of trade among countries that were becoming increasingly interdependent. Most important, the dollar shortage created by an imbalance of trade between the United States and the rest of the world made it virtually impossible for other countries to free their economies quickly. They did not have, and could not earn, the dollars they needed to buy food, manufactured goods, and machinery.

In these circumstances communism made gains in several countries in Western Europe, particularly Italy. So, even before the time of the Czechoslovak coup in 1948, it had already become amply clear that, in addition to the fear of a renewal of total war with all the frightful weapons science could devise, there was a nearer danger that the free civilization in the West might fall from within without a blow being struck.

The defense against such a threat was not weapons, but economic aid—not the long-term economic aid contemplated in the United Nations Charter through the work of the Economic and Social Council and other agencies, but direct and immediate financial assistance. American Lend-Lease, having served its original purpose, had been cut off soon after the end of the war. In 1947, it was, in effect, revived in the form of Marshall Plan aid—that is, dollar grants to countries for the purpose of reviving industries to restore the ability of nations to produce for themselves more of the goods they needed. The Marshall Plan was offered to the whole world, including Russia. The Soviets rejected the offer, however; and when their new satellite Czechoslovakia showed signs of accepting American aid, a word of warning from

Moscow soon compelled a change of mind. The Marshall Plan thus linked the nations of the non-Communist world. It was also an important weapon in the Cold War.

CONTAINMENT

At the same time, on the military level, the Western nations had given up hope of the creation of a United Nations force and had been compelled to devise a plan to counter Russian moves in Europe and elsewhere. The choice appeared to lie between appeasing the Russians or containing them. When Britain discovered that she could no longer bear the cost of maintaining forces in Greece, which had long been troubled by Communist partisans aided by the neighboring Russian satellites (Albania, Yugoslavia, and Bulgaria), President Harry Truman, on 12 March 1947, promised American aid to any country threatened by aggression. In July 1947, an anonymous article in *Foreign Affairs* called for a long-term, patient, containment of Soviet expansionist tendencies. It later became known that this was written by Ambassador George Kennan, a senior American diplomat and scholar of Soviet affairs who had served as ambassador in Moscow, and that it was based on a State Department paper. American military aid to Greece and Turkey was the beginning of what came to be called the Truman Doctrine, a policy of containment which was acceptable to many Americans largely because it contrasted with the appeasement policy that had led to war in 1939. It was a fundamental step away from the isolationism that had traditionally marked American foreign policy. Kennan later stated that he had not intended to advocate an American military deployment to surround the U.S.S.R. But the Soviets and many less involved observers naturally regarded it as a challenge.

In furtherance of the policy of containment, the United States worked to achieve the full cooperation of the free states of Western Europe to resist further Soviet aggression. American aid under the European Recovery Program was based on the premise that the countries of Western Europe could increase their trade with one another by reducing the barriers that hindered it.

REGIONAL ALLIANCES

The United States also encouraged plans to create a military union. A lead was given in this direction in March 1947, by the Dunkirk Treaty, by which the United Kingdom and France promised to aid one another if attacked by Germany. A year later this alliance was extended by the Brussels Treaty to include the Netherlands, Luxembourg, and Belgium. The new pact was to become operative in the event of attack by any power. Growing sentiment for a wider European union resulted in the creation in 1949 of the Council of Europe, which was believed by its supporters to be the first step towards a United States of Europe. While the reluctance of Britain to commit herself on the Continent chilled the hopes of a general European political union within the immediate future, some progress toward a "functional" union was made by the establishment of the European Coal and Steel Community and by a proposal for a European defense community. Advocates of "functionalism" believed that through various agencies European union would eventually be achieved and with it the strength, both military and economic, to resist the further encroachment of Soviet Communism. In the economic sphere, at least, such optimism was rewarded.

In 1957 the members of the European Coal and Steel Community agreed to establish the European Economic Community, or European Common Market ("the Six"), and also the European Atomic Energy Community (Euratom). The aim of the European Common Market was the progressive elimination of tariffs among member states, eventual free movement of labor and capital, wage standardization, and a common investment fund. By 1961, despite the complications that would result in her relations with the Commonwealth, Great Britain, after temporizing with a Free Trade Area of other European states ("the Outer Seven"), was to begin negotiations for entry into the Common Market. But these moves for European unity were then frustrated by President Charles de Gaulle, who feared Anglo-Saxon dominance in Europe and who sought to revive France's lost *gloire*.

Meanwhile, the Americans, underestimating the difficulties caused by deep-seated national antagonisms, inclined to be impatient with the slow progress of steps toward defense cooperation which European union would have facilitated. Their own attitude toward foreign commitments had undergone a sharp change. In 1948, an attempt by the Russians to blockade Berlin, inside the Russian zone of Germany, had been thwarted by a joint airlift. Then, in 1949, at the suggestions of the Canadian prime minister, Louis St. Laurent, nine "like-minded" countries in Western Europe and North America had set up the North Atlantic Treaty Organization (NATO) to give each other mutual guarantees against aggression. The United States, Canada, and Britain had thereby accepted military obligations on the Continent, and in 1950–52 a NATO army was organized, with an international general staff: Supreme Headquarters, Allied Powers in Europe (SHAPE). Countries like Norway and Belgium had abandoned their traditional neutralities, which were no longer an effective shield. In 1954, when France refused to ratify the treaties setting up a European Defense Community because of fear that it would be dominated by West Germany, a wider military organization which included Canada, Britain, and the United States had been quickly substituted.

Regional alliances of this kind had been provided for in the U.N. Charter; NATO supporters claimed that it, and other regional alliances, were an attempt to bolster the U.N. NATO differed from military alliances of the old type since it was based on similarity of ideology, had a permanent headquarters and secretariat, and attempted to include other forms of cooperation in addition to the strictly military. The Southeast Asia Treaty Organization (SEATO), formed in 1954 by the United States, Great Britain, France, Australia, New Zealand, Pakistan, Thailand, and the Philippines, seemed to be similar to NATO in that it includes nonmilitary modes of cooperation and had a secretariat. But it was not comparable with NATO in military significance because its members were not obliged to take military action, no supreme headquarters or permanently allocated forces were established, and some of the most important of the noncommunist countries in that part of Asia were not members. The danger of communist aggression in the Middle East prompted an attempt to form still another regional defensive alliance in 1955, composed of Great Britain, Turkey, Pakistan, Iran, and Iraq and known as the Middle East Treaty Organization (METO). METO was weakened in 1958 by the withdrawal of Iraq after a nationalist revolt in that country; but in 1959, with the approval of the United States, it was reconstructed as the Central Treaty Organization (CENTO). SEATO, METO, and CENTO were all short-lived. The concept that a system of regional alliances could be forged to keep the peace proved fallacious because they were thinly disguised military alliances without sufficient community of interest. However, in Europe, the alliance system was consolidated. It hardened the Cold War confrontation there beneath the dark shadow of a nuclear weapons confrontation. This was made evident when, after criticizing METO as a pact for aggression and a breach of the United Nations Charter, the Soviets fashioned a counteralliance, the Warsaw Pact. In 1955 the armed forces of the East European Communist bloc were put under a Russian commander, Marshal Ivan Konev, an indication that Russian troops were already stationed in satellite countries. Peace in Europe was henceforward to be preserved uneasily by a confrontation of nuclear superpowers, a new strategic concept that must be treated separately.

Foiled in Europe, Soviet and Communist planners turned to Asia and other parts of the globe. There, breaches of the peace occurred and conventional wars could still be fought. Those must be discussed in a separate chapter. Finally, in the uncertain state of world affairs, new forms of what was in effect a form of warfare, known as "low-intensity conflict," proliferated. The final chapters of this book will deal with these developments separately.

20

NUCLEAR STRATEGY AND
ARMS CONTROL

NEW TECHNOLOGIES

The acceleration of technical change stimulated by World War II continued unabated in the postwar decades. Before and during the war microwave radar, liquid-fueled rockets, and jet propulsion were developed, and in the ensuing decades there followed a vast array of other new technologies including improved plastics, solid-fueled rockets, microcomputers, lasers, fiber optics, solid-state physics, supersonic aerodynamics, inertial guidance, space satellites, and stellar guidance and navigation. These brought vast changes in everyday life, affecting industrial organization, work, daily subsistence and routine, and leisure. Many of these things led also to radical weapons development in the second half of the twentieth century. The pace was unprecedented. All of this inevitably affected both the structure of armed forces and planning for war.

But in advance of most of these scientific and technical innovations one was unique in its impact, namely, the nuclear revolution. The atom bombs used at Hiroshima and Nagasaki in 1945 were primitive forerunners of others that affected many aspects of war. Like technical development generally, their evolution was spurred by intense international competition and friction. However, in one respect this early atomic innovation from the first seemed to many people to be quite different from all the other changes. If it could not be restrained, it might annihilate civilization and perhaps all human life. Hence there ensued an intense debate about nuclear strategy and arms control that has not yet been finally resolved.

At the root of the problem was the general awareness that nuclear fission had vastly increased the damage that war could cause. In 1945 a world that was appalled by the extent to which bomber fleets had devastated cities with high explosives learned with horror that Hiroshima and Nagasaki were each destroyed by a single bomb. When fusion followed within seven years, it was realized that the destructive capacity of a hydrogen bomb was theoretically infinite, and that its cost was much less than the fission weapon. Subject only to limitations imposed by the size of the vehicle or the projectile involved, the H-bomb could devastate a small country with one strike, or could inflict crippling damage on a larger one. Missiles were soon to become available to travel through space to convey the bomb to distant continents. There was no defense. Loss of life and physical damage would be heightened by radioactive materials spread far and wide. We have seen that even before 1945 some scientists working on the bomb had argued that nuclear technology had so revolutionized the use of force in international relations that the bomb must be abolished or controlled. Nearly half a century later scientists speculating about the details of the process of destruction coined the phrase "nuclear winter," by which they meant that the planet, already suffering severely from man-made pollution, could be darkened by a globe-encircling radioactive cloud that would blot out the sun for a long period and could be fatal to all life, at least in the northern hemisphere.

THE NUCLEAR STRATEGY DEBATE

In earlier days a nation that wished to dominate its neighbors, or to protect itself against domination, could do so by force or threat of arms. Wise statesmen therefore always made ample preparation for war in time of peace. Furthermore, by so doing, even a small nation could often withstand a larger one's military potential, or at least curb its impact. Now a non-nuclear power would be helpless against the bomb. Many military men argued that the bomb was just another weapon against which, in due time as always in the past, an effective defense would be found. These held that the proper solution for the nuclear dilemma (that war could now be mutually and terminally destructive) was to prepare to fight a nuclear war. Only by so doing, it was said, could the danger of submitting to nuclear threats, or of bringing on annihilation, be avoided.

There were several corollaries to the opposing arguments in this nuclear strategy debate. One of these was that a high level of preparedness might lead to an accidental resort to nuclear weapons. A second was that in the event of a nuclear exchange, both parties might suffer what would be to any reasonable mind an unacceptable degree of damage. Third, a nuclear war would inevitably affect neutral nations. In the decades following World War II, as the debate about nuclear strategy continued, the strategists wrestled unconvincingly with these and similar pessimistic concepts.

At first only two states, the United States and the Soviet Union, could bear the cost of developing nuclear weapons. Added to their greater economic and conventional military power, their possession of the bomb led them to be popularly described as "superpowers." The superpowers, far from accepting the logic that nuclear war was irrational because it would lead to unacceptable destruction, and being afraid of either enemy attack or forced tribute, embarked on the most intensive arms race that the world had yet seen. All this suggested to some Westerners that as warfare now seriously threatened the continued existence of Western civilization, a sound approach to nuclear strategy was needed to protect it.

A strategy based on the destructive power of nuclear weapons is not, as some have said, "a simple and stable strategy." Nuclear power does not have the simple, direct relation to political power that conventional military strength once had. It has been suggested that this fact was first demonstrated during the four years, 1945–49, when the United States had a nuclear monopoly but did not automatically gain a monopoly of influence. This shortfall cannot be entirely explained by American war weariness; it may, however, be attributed to a commendable restraint; to have threatened a preemptive attack would have run counter to the canons on which democratic civilization was built. Civilized restraint is indeed suggested by the fact that soon after the war ended, the United States government took control of nuclear development away from the army, which had pioneered it in the Manhattan Project, and gave it to a new civilian Atomic Energy Commission. President Truman then called for a proposal for its future international development. In 1946 the Baruch Plan, which the United States presented at the United Nations General Assembly, offered to surrender the control and inspection of atomic power to an international agency, provided that all other states would forego national development of atomic weapons, and also provided that there was no veto to negate efforts to prevent abuse. But as we have seen, the Soviet Union saw this only as a ploy by which the United States could retain its current lead, threaten Russian security, and thwart its efforts to develop its own nuclear capability. It therefore replied by proposing that disarmament should come first. "General and complete disarmament," the Soviet position henceforward, sounded a high moral note but did not offer a practical solution for the atomic dilemma. Instead, it raised American suspicions about Soviet motives.

Before the invention of the nuclear bomb, the very high degree of mutual suspicion and hostility generated by the Cold War would probably have led to war within a relatively short time. Instead, as the French philosopher, Raymond Aron, noticed as early as 1954, nuclear technology appeared to have brought the substitution of crises to replace wars between the superpowers. It therefore led to the development of a new speciality in political science called "crisis management" to study the means by which crises in international relations may be settled without resort to arms.

The question of how nuclear power could be exploited to achieve political ends, especially national security, and at the same time made safe, gave rise to an ongoing debate among military and civilian specialists on strategy and politics, both in the United States and in other countries where freedom of opinion was permitted. It may be assumed that a somewhat equivalent debate also occurred behind closed doors beyond the Iron Curtain.

This debate, which was still going on forty years later, was of the greatest importance for the future of Western civilization. It compares in some ways with the centuries-long discussions about faith and reason that marked the end of the Middle Ages and the beginnings of modern society. Toward the end of that earlier debate, Machiavelli had enunciated that war was a legitimate political tool of princes. His analysis did much simplistically to explain international conduct in and after his day, a form of conduct that frequently led nations to resort to war. The current debate endeavors to restate that war-politics relationship, but it is in one sense more important, because resorting to war may now result in the end of Western civilization, or even of the human race. On the Western side, the nuclear-arms policies that emerge from this debate are primarily made by civilians. In the Soviet Union the military seems to have had predominance in their formulation.

Among the specialists in the West who publicly or privately debate the nuclear dilemma—that is, how nuclear power can be safely deployed or employed but at the same time controlled—are private advisers and consultants to governments, as well as military leaders and members of legislatures. Others work in universities, or in institutes especially set up to study politics and nuclear strategy. Some write for the press. The nuclear strategists have devised a lexicon that includes technical terms, the names of new weapons, acronyms, and words and phrases coined or borrowed to express theories, strategies, and opinions. Some of the participants in strategic theorizing are inclined to believe that the arguments which they present (and which at times sound like ritualistic casuistry) are beyond the understanding of ordinary people. But words and phrases from their lexicon, like "containment," "massive retaliation," "mutual assured destruction," "missile gap," and "window of vulnerability," which are used to encapsule complex ideas, have passed into general use. These were brought to public attention by the media which, as usual, often fastened on the sensational and, by so doing, too often provoked more alarm, despondency, or anger, than enlightenment.

NUCLEAR WEAPONS AND ARMS CONTROL

By and large the discussants in the United States belong to one or the other of two main camps or schools of thought. The first school seeks a strategy for the deployment of nuclear weapons, and even for their use if necessary. These have been called "nuclear strategists." They believe that the only language the Soviets understand is power, so they seek to challenge from strength. They want the United States to have at least parity or, better still, superiority. The second school hopes to make the world safe by an agreement with the Soviets to control, and eventually eliminate, nuclear weapons. These, known as "arms controllers," think that pursuing strategic deployment by itself is too dangerous.

A few ideologists used to advocate unilateral renunciation of nuclear weapons. Thus Lord Bertrand Russell, the distinguished British philosopher, campaigned for the abolition of all Western nuclear arms with the argument that the U.S.S.R. would have to follow suit. When Britain decided to get nuclear arms, there was widespread agitation for it to renounce them in order to head a "non-nuclear club" of nations that opposed proliferation. John Strachey, a leftist British writer and a onetime Labor secretary of war, declared that it is too optimistic to believe that war will abolish itself through the accumulation of thermonuclear abundance. He said that all that nuclear weapons have done is give man a choice between abolishing the bomb or being abolished by it.

After that time unilateralism declined. Few people believed that turning the other cheek would suffice. But unease remained. Most arms controllers, knowing that control by treaty must rest on mutual confidence between the powers, accept that, until confidence has been achieved, nuclear weapons must remain to provide control by deterrence. So they seek binding, verifiable control by treaty. They believed that their policy must eventually succeed because it is in the best interests of the Soviet Union, as well as of all other nations.

Nuclear strategists, who also wanted to avoid nuclear conflict, retorted that prohibition of nuclear weapons will not be safe because nations would cheat. Therefore, they said, the only safe path is to arm in order to avoid war by deterring aggression by the threat of retaliation. Deterrence was the main theme in American foreign policy and nuclear strategy. Arms control was secondary.

Michael Kreppon, in a Council for Foreign Relations book, *Strategic Stalemate, Nuclear Weapons and Arms Control in American Policies* (which has approving forewords by Lieutenant-General Brent Scowcroft, a former member of the National Security Council, and later assistant on national security to Presidents Reagan and Bush, and Paul C. Warnke, a U.S. arms control negotiator) said that within each of these two camps there were wings that derived their stance, at least in part, from ideology on the one hand or operational feasibility on the other. He added that individuals in each camp have often moved from one camp wing to the other, or even from one camp to the other, in response to developments in weaponry, to changes in the administration, to international events (especially Soviet pronouncements or actions), and, when the debate became particularly intense, to the logic of rhetoric.

Kreppon also said that, as a result, from time to time some individuals appeared to shift their ground, or to lack consistency. He instanced Senator Stuart Symington of Missouri, who switched from supporting the Pentagon's strategic weapons program to opposing them, and Donald Brennan, director of national security studies of the Hudson Institute and a consultant of the

Department of Defense, who reversed his stand in favoring arms control after the SALT I accords were signed. George Kennan, who as we will see, moved from containment to disengagement, also switched from reluctantly accepting the need for deterrence by a nuclear build-up to proposing instead a 50 percent reduction in nuclear arsenals to lead up to complete nuclear disarmament. Kreppon said, finally, that although Henry Kissinger has made "important contributions to arms control, . . . his positions on specific issues have hardly been a model of consistency over time." He added, "the public postures on arms control of some individuals like Paul Nitze [one of the architects of containment] suggest a correlation between positions held in and out of government." Kreppon's charges are cited here, not the question to motives or intellectual honesty of the individuals named, but to demonstrate that strategy in the nuclear age is a very complex and elusive science.

The public's reaction to what it knows of the course of events, or to what it has learned from the specialists' arguments, has often been influential, especially with politicians. The debate has been protracted partly because of the depth of Soviet–American antipathies, but even more because there seemed to be no satisfactory solution for the nuclear dilemma. Warnke suggested that the inability of the American side in the negotiations to find a compromise or consensus between nuclear strategies on the one hand and arms controllers on the other was the major obstacle to a reduction of the risk of nuclear war. He appeared to mean "greater even than Soviet intransigence."

While all the details of this fluctuating debate cannot be followed in the short space available here, the chief stages in its development must be synthesized. After the United States, Canada, and Britain had agreed under the NATO pact to station conventional forces on the continent of Europe, and even later after West Germany had been admitted to the Alliance, the Soviet and Warsaw Pact's conventional deployment in Europe (not counting Soviet forces in the Far East) was known to be still quantitatively greater than that of NATO. Although there was some question about relative quality and reliability, especially of the Warsaw satellites, Soviet tanks were not only more numerous than NATO's but were also more powerfully armed and armored. It therefore seemed that, by one swift attack, the Warsaw Pact's conventional armies could occupy all Western Europe to the Atlantic. Because NATO members refused to make the financial commitment necessary to match the conventional forces of the Warsaw Pact, the United States turned to a policy of nuclear retaliation. It hoped that the mere threat of such a response would be enough to prevent conventional war.

THE BOMB AND KOREA

The Korean War, which will be discussed later, had been a conventional war that had done nothing to provide an answer to the nuclear dilemma. There is, however, a report, also referred to later in this book, that toward the end of the war, Eisenhower's readiness to use the bomb was conveyed to the

Chinese through diplomatic channels, and that that explains their willingness to settle on the lines of the status quo ante bellum, i.e., the situation before fighting began. Bernard Brodie, looking back on it in 1959 in his *Strategy in the Nuclear Age,* said that one reason the bomb was not used in Korea was that the American Joint Chiefs of Staff suspected that the invasion there was a feint or diversion from the main area of confrontation, Europe. They did not want to be enticed by it to reveal the size of their nuclear arsenal (which at that time was very small). Brodie added that there were no suitable targets in Korea for a nuclear attack, that Britain and the other allies were opposed to one, and that it was feared that the Soviets might already have a usable bomb and might strike back with it, for instance, at American headquarters in Japan, an eventuality that would have been very embarrassing.

DETERRENCE OR DISENGAGEMENT

As Korea provided no practical experience on which to base a nuclear policy in Europe, or later in Asia itself, discussion of the problem continued to be theoretical and speculative. It was argued that to be effective as a deterrent, the proposed use of nuclear power should be not only overwhelming but also credible in the sense that the enemy had to be thoroughly convinced that it would be used. Many people believed, however, that any conventional war into which nuclear weapons were introduced would inevitably escalate quickly. This lead to a discussion of the effect of the use of nuclear weapons in heavily populated Western Europe. For many, this was unthinkable. As early as 1954 the influential British military historian and critic, Basil Liddell Hart, was preaching that massive retaliation, as this policy of an immediate devastating nuclear response was called, was unsound. But under Eisenhower, who had been elected to put an end to the war in Korea, the Pentagon, as a means of controlling defense spending, adopted a "New-Look" defense policy to exploit superior American technology, including nuclear capability.

The Korean War had brought a determination that there should be no more wars of that kind. Eisenhower's New-Look strategy seemed to offer a way out. His secretary of state, John Foster Dulles, brandished the bomb in Asia in 1954 when he had to face communist aggression in Indochina, and again when the communists threatened Quemoy and Matsu, offshore islands still in the possession of the Chinese Nationalists in Taiwan (formerly Formosa) from which the mainland could be bombarded. Dulles talked of "standing firm to the brink of war." There began to be talk of the possibility that one side or the other might launch a "first-strike preemptive war" to forestall an imminent attack, or even a "preventive war" to anticipate a war that seemed inevitable. From the middle of the 1950s each side began to accuse the other of preparing for such early use of the bomb, and there was also an increasing fear that nuclear war might be started by accident

NUCLEAR ARTILLERY. On 25 May 1953, the first live nuclear artillery test, Shot Grable, fired a 88-mm shell with a yield of 15 kilotons at the Nevada test site. (*For the Record: A History of the Nuclear Personnel Review Program, 1978–1986,* Defense Nuclear Agency, Washington, 1986, p. 99)

despite the fail-safe techniques that had been organized or, alternatively, by a fanatic with his finger on the button. In 1956 Henry Kissinger, then a Harvard professor and secretary of a Council on Foreign Relations study of nuclear weapons and foreign policy, in a book with that title, stated that it would be possible to fight a nuclear war in Europe with tactical weapons and keep it limited by not resorting to strategic arms. This gave little consolation. Most people believed that any war in so sensitive an area would spread everywhere. This was dramatically illustrated in the same year when Nikita Khrushchev "rattled rockets" over the British–Israeli attack in Suez.

It was at this time that George Kennan, author of the containment policy, came forward with a radically different proposal. In a talk to the BBC in 1957, he proposed "disengagement" from areas of confrontation with the Soviet Union (though he did not actually use that word). This terrified European NATO leaders who had steeled themselves to brave the prospect of nuclear war if deterrence failed. Paul-Henri Spaak, the Belgian secretary-general of the Alliance, pointed out that NATO's defense plans rested on depth for maneuver, and that in Western Europe that depth was already too shallow. Disengagement would lead to the disruption of NATO, which was the first goal of Soviet foreign policy.

One shortcoming of the strategy of massive retaliation as a means of deterring an enemy was that its application would be restricted to major threats. An enemy could still nibble away at minor areas where the opposition was softer. There were many such places not protected by NATO, by the North American Air Defense pact (NORAD), or by Dulles's "stand to the brink" in Asia. When (as we saw above) a British White Paper, a statement of government policy, adopted massive retaliation with nuclear weapons in 1957, critics immediately assailed it as being quite unsuitable for the protection of Britain's worldwide imperial responsibilities. Although those interests might now be written off in some circles as relics of an outmoded imperialism, the West still had yet other widespread concerns, responsibilities, and interests in the Third World that were vulnerable. Dulles had therefore built the network of alliances to endeavor to contain communism. As seen earlier, these had included the Southeast Treaty Organization (SEATO), and the Middle East Treaty Organization (METO, later called the Central Treaty Organization, or CENTO, in order to add Iran). But while these alliances had reaffirmed America's abandonment of isolation, they did not have the strength and durability of NATO, or of the Australia, New Zealand, United States (ANZUS) Treaty that had been negotiated in 1951, treaties that tied together peoples who shared the cultural and political traditions of Western European civilizations. The states allied in the new treaties had no similar sense of community. As we have seen, the new regional treaties did not last.

THE "BOMBER GAP" AND PREEMPTIVE STRIKES

Meanwhile, the Eisenhower administration, clinging to its faith in massive retaliation, still believed that if the threat could be made convincingly awesome and certain, nuclear weapons would not have to be used. The United States Strategic Air Command (SAC), in particular, continued to press for an expansion of the bomber attack force to ensure that the policy was effective. In the mid-fifties SAC began to imply that Soviet bombers represented a real danger, the greater because there had come to be what was called a "bomber gap" in Soviet favor.

The bomber-gap threat originated from American intelligence reports which the Russians had fed with misleading information. After the Soviets exploded their first fusion device, and when they introduced the Bison jet bomber, they staged multiple fly-pasts for a May Day parade in Moscow, giving the impression that there was more than just one squadron on show, thereby deliberately exaggerating the number that they possessed. The Bison, however, was only a light four-engine bomber that was inferior to the American B-47. The Bison's range would not have been sufficient to make a return trip on a bombing raid to North America. Even when the larger Bear bomber came into Soviet service, there were no more than 300 aircraft that could make the round trip to the United States without in-flight refueling, a technique they had not yet fully developed. So there was indeed a bomber gap, but it was actually in American favor by a ratio of two to one. Eventually that ratio became five to one.

SAC's answer to the supposed bomber gap was to push its strategic bombers forward to bases in Europe (with a host country's consent). The alleged bomber gap was also an incentive to the United States to build a fleet of new B-52 jets that could carry six nuclear bombs to the Soviet Union and return. On its part, the U.S. Army obtained so-called "tactical" nuclear missiles—Sergeant, Corporal, Honest John, and Lacrosse—some of which were more a reinforcement for the massive retaliation strategy than a preparation for nuclear land battles. In the mid-fifties the U.S. Air Force and Army also began to experiment with strategic missiles that could traverse the outer atmosphere and space.

Some nuclear strategy theorists on both sides of the Iron Curtain now began to talk more freely of the need to develop a preemptive strike capability. One month after the Soviet launching of *Sputnik* (the first man-made satellite) in 1957, the top-secret Gaither Report was presented to the U.S. National Security Council. It expressed concern about the vulnerability of SAC bases in the event of a surprise attack. The report called for active and passive civil-defense measures, as well as more effective area defenses against bombers and intercontinental ballistic missiles (ICBMs). It held that security could be achieved by the mid-1960s, but it forecast, erroneously as it turned out, that by 1960 there would be a very critical period of danger to SAC from Russian missiles. It said that during the current temporary period of American advantage, the United States should negotiate vigorously from a position of strength.

After the death of Stalin, when Nikita Khrushchev had emerged from a struggle with Nikolai Bulganin to become the unchallenged leader of the Soviet Union in 1956, he at first talked of accommodation with the West despite his Suez threat. A year later, after hinting that the Soviets had a lead in nuclear devices, he announced a moratorium on testing. This was quickly joined by the United States. These steps were followed in 1958 by the commencement of negotiations to work out the terms of a treaty. These negotiations toward banning the testing of nuclear devices seemed a hopeful step toward the eventual elimination of the bomb. But they had actually been brought about largely by a universal fear of radioactive fallout from testing. Radiation from earlier testing in the atmosphere had been experienced worldwide. Moreover, the United States had participated in the talks confidently because it had secret information about the state of Soviet bomber and ICBM development and deployment. Ultra-high-level reconnaissance planes (the U-2s) had photographed Soviet territory, looking for ICBM sites and bomber bases, and they had found that fewer were operational than was formerly believed. Khrushchev had been boasting.

Neither the nuclear strategists nor the arms controllers, lacking precise and certain intelligence information about Soviet intentions, and also fuller information about Soviet capabilities, could offer a convincing solution for the nuclear problem. Although the U.S. Air Force's Strategic Air Command (SAC) continued to urge an increase in the bomber strength to ensure deterrence, some studies by the RAND (Research ANd Development)

Corporation, which the air force had set up to study technical and other problems, questioned reliance on massive retaliation. They pointed to a need to limit the potential impact of a nuclear war. A British publication revealed that in January 1959, senior American officers, testifying before a Senate investigating committee, showed that there was in fact considerable disagreement among the military. Army General Maxwell Taylor had said that the United States now had more atomic weapons than were needed to deter. But General Thomas S. Power of the air force claimed that it was necessary to build in order to preserve the capacity to strike first because, "if one does not have that capability, an aggressor would be able to take the world away from us piece by piece," knowing that "as long as they do not strike us, we could never do anything about it." That sounded like a call for a first-strike capability. General Thomas D. White and Secretary of Defense Neil McElroy told the senators that, if the United States had the initiative, the targets would be enemy missile bases and airfields. This was called a "counterforce strategy." But they added that if the enemy delivered a surprise attack, retaliation would be against the Soviet Union "as a whole," that is to say against its cities, because once the enemy attack was launched, there would be no Soviet military target left that was worth hitting.

TWO DIFFERENT STRATEGIES

Hanson Baldwin, the *New York Times* military critic, defined two different kinds of deterrence. He called one "minimum," or "finite," deterrence; this targeted cities. "Infinite," or counterforce, deterrence was designed to destroy or substantially blunt the enemy's nuclear delivery capability. He said that the United States had a great overkill, much more than was needed for minimum deterrence, but probably had insufficient resources for infinite deterrence because that had to be geared to the size of the enemy's force, always an unknown quantity. Baldwin argued that finite deterrence could be just as credible as infinite deterrence; but he added that all deterrent forces must be invulnerable to an enemy attack. So, leaving aside the question of the morality of depending upon city destruction, rather than on counterforce, as the basis for deterrence, the fundamental question was whether the deterrent forces were secure.

There was uncertainty about the balance between the two opposing forces. In the Senate hearings Air Force General Thomas P. White had said that the Soviet Union did not have a great ICBM capability; and Secretary of State McElroy had reported that the American liquid-fueled Atlas and Titan ICBMs were either operational now or soon would be. Furthermore, the United States was rapidly developing a second generation of rockets with solid fuels that could be launched more quickly, notably Minuteman. The *Polaris* submarine, which could fire ICBMs from under water, would join the fleet in 1960. All this sounded optimistic. But an unattributed leak from a confidential session of the Senate Foreign Relations Committee alleged that McElroy had testified that by 1960 the Soviet Union would have 100 more ICBMs than the United States. He was said to have described this situation as a "missile gap."

THE "MISSILE GAP"

The story of the "missile gap" closely resembles that of the earlier bomber gap. By about 1957 some American intelligence analysts had begun to suspect that the Soviets might be outdistancing the United States in the production and deployment of ICBMs. They had then forecast that in four or five years the disparity would be tenfold in the Soviet favor. So the United States committed itself to build 1,000 Minutemen at a time when in fact the Soviets had fewer than a hundred similar weapons. Moreover, this commitment for a thousand had actually been a compromise. Military hawks had wanted from two to ten thousand, but U-2 reconnaissance had countered this by suggesting that Soviet claims of missile superiority were inflated. It was on this kind of information that Eisenhower had earlier been able to resist some of the demands from what he called "the military industrial complex" and had agreed to the arms limitations talks with the Soviets. The extreme demands of the hawks were now again similarly resisted.

Then in May 1960 Gary Powers, flying a U-2 plane, was shot down over the Soviet Union on the eve of a summit meeting between Eisenhower and Khrushchev. Eisenhower at first denied that there had been any aerial surveillance. But he then admitted it. He thus lost credibility because his self-correction seemed to suggest that he was not aware of what his administration was doing. American-Soviet relations soured. The summit was canceled. The arms limitation talks were suspended.

Meanwhile the death of John Foster Dulles in 1958 had removed from the scene the chief proponent of massive retaliation; and the change in the administration from Eisenhower to John F. Kennedy seemed likely to facilitate a revision of basic nuclear strategy to accommodate the need for a response more flexible than Dulles's. When Kennedy was elected, it was still widely suspected that Khrushchev really had the missile superiority that he had boasted. The United States, in contrast, still relied mainly on the manned bomber for deterrence. The bomber was ideal for the deterrent function because it could be recalled if dispatched in error or by accident. But rockets were different. They would make the nuclear threat much more frightening. When talk of a missile gap was followed by the Soviet launching of the first man-made satellite, *Sputnik I,* and then by Yuri Gagurin's space ride around the earth in 1961, both feats seemed more impressive than the American equivalents which followed. It could therefore be argued that Russian technology was not as inferior as had long been believed. Missiles, rather than bombers, seemed to be the weapons of the future. The public did not know what the U-2s had learned. The idea of a missile gap became quite disturbing. The hawks were encouraged.

MASSIVE RETALIATION

So, although there were still those who advocated "graduated deterrence," influential American thinking now turned to a different kind of strategy for avoiding a nuclear war. This was that the aim should be to threaten massive retaliation, not merely to blunt an attack but to ensure that any aggression, even if it brought widespread devastation, would be met with similarly devastating responses. This was called Mutual Assured Destruction. Cynics noted that the acronym for it was MAD. It was also called a "Balance of Terror." This theory held that if both sides were assured of being destroyed in a nuclear war, there would not be one. A capacity to survive an enemy's first strike, and so be able to retaliate, would make the first strike not worth delivering.

However, plans to deter aggression by threatening mutual assured destruction raised new concerns. For the threat to be convincing, the enemy had to believe that it would be fulfilled and would cause more damage than he was willing to accept. He must think that that would be the case even after he had delivered a successful first-strike surprise attack. A power that proposed to deter by being able to fight a nuclear war must therefore make full preparation in advance in peacetime. But too strong a deterrent force, or too blatant a preparation for nuclear war, might suggest to a potential enemy that the program for deterrence was in fact a cover for an intended first strike to obliterate his nuclear capability. That might provoke him to resort to his own forestalling preemptive, or preventive, first strike, and so might precipitate the nuclear war that MAD intended to avoid. This consideration supplemented the earlier fear that there might be an accidental discharge of nuclear missiles, or that desperate or fanatical men might resort to nuclear war to stave off economic collapse or political overthrow.

A second consideration also had to be taken into account. This was the need to be able to fight on after a preliminary nuclear exchange had brought casualties, damage, and dislocation on a scale never before experienced. Herman Kahn, in his *On Thermonuclear War,* published in 1960, had presented his calculation of the probable consequences of a nuclear exchange in terms of lives lost, of injuries, of diseases and epidemics, and of the dislocation of essential services such as food, water, power, transportation, and medical systems, as well as of civil administration. Although he was horrified by his own findings, and commented that "even a bad world government" might seem preferable to a nuclear holocaust, he nevertheless thought that a limited nuclear war could be survived. This brought up the question of how war could continue when all essential services had been destroyed. For some time there had, in fact, been discussion of what was called "broken-backed warfare," but there was no clear understanding what that meant or how preparations could be made for it.

SEA POWER

Another problem for the nuclear strategists was the function of sea power in the nuclear age. From World War I to World War II, dominance at sea had passed from battleship fleets to aircraft carriers. In the vastnesses of the Pacific, sea battles had become aerial contests between fleets which could be hundreds of miles apart. The introduction of long-range strategic missiles that could be fired by submerged nuclear submarines effected an important revolution in the potential role of navies in war. For the first time in history ships had the ability to remain hidden and at the same time to project their power far into the interior of land masses; they might exercise a decisive influence on world politics. This was far beyond anything that had ever been dreamed of by Alfred Thayer Mahan, whose *Influence of Sea Power on History* had been the bible of naval strategies through the first half of the twentieth century. It seemed as if H. J. Mackinder, the apostle of dominance by continental land masses, was now likely to be completely confounded by submarine-launched ballistic missiles (SLBMs) functioning as strategic deterrents. The Soviet Union, recognizing this trend, inserted submarines prominently into its naval expansion program. For a time, nuclear missiles seemed also to be about to alter surface sea battles, by replacing guns, but that development proved short-lived. Finally, following the old maxim of naval historians that "like fights like," the submarine developed an extra role as a hunter and killer of other submarines.

DEFENSE POSSIBILITIES

The chilling "balance of terror" brought on by mutual assured destruction depended on the vulnerability of both sides. If one side or the other developed an antiballistic missile (ABM) system to protect either its strategic weapons from first strike or its population from retaliation, then its vulnerability would be decreased and the balance would be upset. Both sides had researched these alternative technologies through the 1960s because they were fearful lest a breakthrough be made unilaterally by the other side. As it turned out, technological ceilings in radar and computers made ABM systems unreliable at that time, and this tended to dispose both sides to consider arms control agreements to limit nuclear missiles.

DÉTENTE AND ARMS AGREEMENTS

A measure of détente had already been achieved in some areas in a series of international agreements to limit or control aspects of nuclear-weapons development and of some other forms of warfare. In 1959 there had been a prohibition of nuclear technology in Antarctica. A year after a frightening confrontation between the Soviet Union and the United States over Russian placing of nuclear missiles in Cuba in 1962 (when Khrushchev backed down in return for a face-saving offer by Kennedy to withdraw obsolescent American missiles from Turkey, a withdrawal which had, in fact, already been planned earlier), the Soviets and the Americans set up a "hot-line" telephone communication between the Kremlin and the White House to contain crises. This system was improved in 1971. In 1963 a nuclear test ban treaty prohibited weapons testing in the atmosphere, in outer space, or under water. In 1967 twenty-one out of twenty-three South and Central American states made their continent a nuclear-free zone (Cuba and Guiana abstained). In 1968 eighty-three nations signed a nonproliferation treaty, and in 1972 thirty-one outlawed biological warfare. The culmination of this remarkable series of agreements came in 1972 when the Strategic Arms Limitation Talks produced a pair of documents known collectively as SALT I. The United States and the Soviet Union agreed to freeze their arsenals of offensive strategic missiles at the current levels. The U.S.S.R. got an advantage in numbers, but the United States had more ICBMs with multiple warheads. In a separate, and at that time to some people an even more comforting, part of the agreements, the number of defensive ABM sites in each country was limited to two, and launching pads to a hundred. Two years later the number of ABM sites was reduced to one each, for the defense of Moscow and Washington, respectively. SALT I was criticized in the United States because it did not give parity in the numbers of offensive missiles but, instead, hardened the Soviets' superiority. However, the ABM part of the treaty arrangements was regarded as a most important step toward controlling nuclear weapons. In 1973 the two superpowers agreed to continue consulting on strategic-weapons control and to discuss the renewal of SALT.

SALT I, worked out after much hard bargaining, was to have been an interim step toward the permanent limitation of strategic nuclear arms. The next step was to be completed by 1977. But it soon became clear that SALT II (and a SALT III that was also in prospect) would face great difficulties. Many in the United States thought that SALT I had yielded essential security interests. Thus, Senator Henry M. Jackson of Washington, the Senate's most vocal specialist on defense, campaigned across the country saying that the United States must never again accept anything less than parity. He ignored the fact that a requirement for nuclear parity would be difficult, perhaps impossible, because a balance between different kinds of nuclear weapons designed for different purposes could not easily be achieved. Second, the Soviets still apparently had a decided lead in conventional forces, and all attempts in discussions held in Vienna, from 1973 on, to achieve a Mutual Balanced Force Reduction agreement (MBFR), which had been overshadowed by the strategic arms talks, had proved fruitless. It was this Soviet conventional advantage that made a case for American nuclear superiority.

The strategic-weapons confrontation was still fluid. The loss of aerial surveillance when the U-2 was shot down had been quickly remedied. A year later, *Samos,* a photograph-taking satellite, was covering the Soviet Union. It had soon revealed that the missile gap had been a myth, that in the 1960s and even later the balance was significantly in American favor. But it and other intelligence information had suggested that the Soviets were making strenuous efforts to catch up. What they lacked in sophisticated technology, they made up in power and weight. So by the time of SALT I in May 1972, the Soviet Union seemed to have already become a fully fledged nuclear rival of the United States. Secretary of State Henry Kissinger therefore regarded SALT I primarily as a breathing space in which the United States had an opportunity to reaffirm or restore its former predominance in weapons development.

The root of the difficulty in making further progress toward arms control was the continuance of distrust between the superpowers. Détente lingered on fitfully until the end of the decade, but it was flawed. The Russians, as also the Chinese (and indeed many other peoples), were to see American involvement in Vietnam (which will be discussed later) as an unwarranted intrusion into the affairs of a distant Asian people. Although neither Russia nor China had openly entered the conflict, both, and especially the more remote Soviets, provided military supplies to North Vietnam and the Vietminh. In the 1950s Dulles had threatened China with a nuclear attack. In the 1960s, aided by espionage, China answered by producing and exploding its own bomb in the astonishingly short time of three years, a remarkable feat for a country that was technologically backward. Ideological disputes between Russia and China, which broke out into undeclared war on their border, while they spared the United States the danger of facing a joint Communist *démarche,* did not hide the fact that the West, with its three nuclear powers, now had two major enemies with nuclear arms. American strategic planning henceforward had to take care of the possibility of a Chinese nuclear strike.

SALT I had frozen missile numbers, but it had not limited throw-weight. In the freeze the United States had offset smaller numbers by the fact that it had more with multiple warheads, called "Multiple Independently-Targetable Reentry Vehicles" (MIRVs). After SALT I the United States continued to MIRV one-half of its missiles and also to improve their accuracy. It also continued to harden ICBM sites so that they could survive a surprise attack. Only by that means, it was believed, could the deterrent be made credible. The United States was also deploying missile-launching submarines, and by 1979 the first of the more powerful *Tridents* was operational to supplement the earlier *Polaris.* Furthermore, an improved *Ohio* class of *Tridents* was being developed and would be ready by 1982. As a means of exerting a strategic nuclear threat for deterrence, submarine launched ballistic missiles (SLBMs) have certain advantages. Although their high degree of security against being located and attacked was at first offset somewhat by some lack of accuracy caused by less certainty about the submarine's location, that disadvantage was eventually overcome by advances in navigation. The Russian SLBM program, which was being pushed vigorously to produce bigger vessels than the American ones, suffered from the disadvantages that its craft were noisier and that they also had to pass through narrow passages between

Scandinavia and Iceland to reach the Atlantic. Russian SLBM vessels were therefore more liable to detection and attack. The United States sowed detection devices on the seabed to locate them.

INTERMEDIATE-RANGE MISSILES

MIRVs, hardened missile sites, SLBMs, and a continuing edge in manned bombers might counter the threat from Soviet land-based ICBMs in the 1970s, but another danger now came to a head in Europe. Intermediate-Range Ballistic Missiles (IRBMs) were not covered by SALT. Both sides therefore continued to increase their IRBM deployment. Soviet leader Mikhail Gorbachev was to admit later that it was the Russians who began the process by stationing large numbers of their SS-20s and other SS series missiles in forward bases from which they could threaten most of the members of NATO. The increase of IRBMs in Europe raised once more a fear that in a European crisis with an exchange of IRBMs an American president might hesitate to launch his ICBMs and nuclear bombers lest that expose American cities to Russian ICBM retaliation. It was because of such a possibility that Britain and France had earlier insisted on having their own nuclear arms. In the United States some people had talked of overcoming this problem by fashioning a series of automatic responses that would trip a president's hand, a frightening thought.

The American response to Soviet escalation, strongly supported by some governments and people in Europe, was to station Pershing II intermediate-range missiles wherever they were acceptable. The United States also pushed the research and development of cruise missiles and of the stealth bomber. The cruise was a small pilotless plane like the V-1 buzz bomb pioneered by the Nazis in World War II to attack Britain; but it was very much more sophisticated. It could be launched from the ground, from an aircraft, or from a ship. It flew below the level of radar detection, followed a prearranged programmed course, and automatically adjusted to contours. The stealth bomber introduced techniques and materials to make it less visible to radar.

THE BUDGET AND THE COST
OF DEFENSE

Despite these American measures, Russian nuclear capability grew faster than that of the United States in the 1970s. One reason for this was that American defense policy and the budget were subject to severe congressional scrutiny. Russian defense policy, largely fashioned by the military, faced no such public discussion. American nuclear strategists alleged that the U.S.S.R. was spending as much as 17 percent of its gross national product (GNP) on defense, and perhaps more. The American equivalent was said to be 6 percent. Even when allowance was made for the much higher American GNP, Soviet spending on arms seemed to surpass that of the United States. But precise comparisons are not possible, partly because of Russian secrecy, but also because some American defense costs are hidden in other budgets, for instance, NASA's.

The budget crunch in the United States raised the question of how much the respective economies could bear. Economists C. J. Hitch and R. N. McKean, who studied the national balance sheet of defense spending in *The Economics of Defense in the Nuclear Age,* had said in 1961 that it was imperative that the returns from the available resources should be maximized. James R. Schlesinger, secretary of defense, had countered that there should be no "artificial" limits imposed on what was paid for security. A basic question, therefore, was whether the democratic free enterprise system could outlast Soviet autocratic communism in paying for the arms race.

THE STRATEGIC BALANCE IN JEOPARDY

Meanwhile President Richard Nixon, by dint of accepting a "two-China" policy that ignored the Taiwan government's claim to the mainland, had succeeded in making an accommodation with the communists there. But he failed to reach an agreement with the Soviets to progress toward the completion of the strategic arms limitation process. However, when he lost face over the Watergate scandal, and eventually resigned, his successor, Gerald Ford, met Secretary General Leonid Brezhnev in a summit conference in Vladivostok in 1974 and came to a tentative agreement on parity in strategic weapons and on the exclusion of forward-based strategic missile systems in Europe. This should have eased progress toward SALT II, but the numbers agreed upon were still high, around ten thousand on each side. The "cut-back" actually left armament levels higher than SALT I had permitted. At the same time Soviet activity around the world in various so-called regional disputes continued to give grounds for alarm. So, Western nuclear strategies complained that American defenses had been seriously weakened.

During the 1970s and 1980s the U.S.S.R. could deliver more strategic tonnage, but the United States still had more nuclear weapons, and its bombers and missiles were more accurate. Nevertheless, after 1975 the Soviet buildup of conventional arms sparked an American response in both the conventional and nuclear fields. By the end of the decade higher American expenditure on nuclear weapons was directed toward maintaining parity in a long-term future. As the time lag between planning new weapons and deploying them was very long, research and development reached far out into the future, creating an investment that served to check arms-control efforts. Because the Soviet ICBMs were larger than the Minuteman, it was argued that, if launched, they might irretrievably damage any possibility of an effective American land-missile response before that could get on its way. However, this argument did not take into account American aircraft and submarine launches. Systems to improve Minuteman's security, by making the missiles mobile, failed to obtain an American consensus. Meanwhile, disagreement between the United States and the U.S.S.R. about whether the American cruise missile and the Soviet Backfire bomber (which could not return from attacking the United States without being refueled in the air) should be included in the strategic talks, delayed progress in arms control negotiations.

Some nuclear strategists were of the opinion that all that détente had done was give the Soviets time to gain nuclear superiority while continuing to pursue their goal of worldwide revolution. Détente had, in fact, been discredited in 1979 by the Soviet invasion of Afghanistan to prop up an endangered communist regime. Earlier, the Soviet satellite, Cuba, armed by the Russians, had sent troops to Angola and Namibia in southern Africa; and both the U.S.S.R. and Cuba, along with the other East bloc countries, were supporting communist governments and insurgents in Nicaragua and El Salvador, places uncomfortably close to the United States. On the other hand, it had come to be widely suspected that the American CIA, acting on ideological grounds in the name of national security, and with the connivance of some members of the executive branch, was conducting operations in Central America that contravened congressional legislation. In 1984 the CIA openly admitted the mining of harbors in Nicaragua.

Jimmy Carter had won the presidency in 1976, partly because the Republicans had been tarred by Nixon's involvement in Watergate, but also because he seemed to promise a new look at the problem of ending the nuclear menace. However, Carter was careful to not reduce American defenses. In fact, he put several new weapons programs on foot. Furthermore, he ruffled the Kremlin by lecturing it on its dismal human rights record, and also on those of its Central American satellites. Then, when the Shah of Iran, an American-supported autocrat, was overthrown by a popular uprising, the Ayatollah Khomeini, a religious fanatic, returned from exile in France. Khomeini's supporters seized the American embassy in Tehran and took its staff hostage. Carter then found himself bogged down in an exhausting struggle to obtain their release. The limitations of American military power for achieving particular ends were amply demonstrated when a helicopter rescue effort came to disaster in the Arabian desert, whence it had intended to launch a surface raid into the heart of Tehran. Some critics have wondered what would have happened if the rescue party had succeeded in reaching the city and had come face to face with an unarmed fanatical crowd.

TOWARD ARMS LIMITATION

By this time mutual assured destruction had become very disturbing as a strategy for maintaining the deterrent posture. Not only was the U.S.S.R. apparently capable of wreaking untold damage in both the United States and Europe, but there was still a realization that the national policies of the superpowers could end in universal destruction. This caused some to desire to alter the rough balance in the long-standing confrontation between the nuclear strategists and the arms controllers in order to favor the latter.

The first signs of change were the renewal of proposals to resort to strategies that would limit the widespread effect of collateral damage in the event of a nuclear war. Some of these ideas had been around for a long time. Proposals for a neutron bomb for use on battlefields and invasion routes where it would kill, but not destroy property, had had a brief currency as a means of restricting damage to war fronts. Cartoonists had made much of the fact that it was just another variety of barbarity and that it was a capitalist's weapon, saving property but taking lives. "Decapitation strategy," which would aim at killing military and political leaders, had had some popular appeal but it did not carry much conviction with governments. Such ideas were, in fact, related to the long-dormant concept of "flexible response," that is, conventional defense against a limited attack, with resort to nuclear weapons only if that became necessary to avoid defeat. This provided little comfort, however, because of the belief that it would inevitably escalate through tactical nuclear warfare and would still lead to total devastation. Another proposal was that nuclear land mines should be sown along borders at invasion routes, an idea that did not get much support because of the obvious danger to civilians that it could present in peacetime.

A quite different idea received considerable popular support but did not get anywhere in official circles. The Soviets had long claimed that they would not be the first to use nuclear weapons in any future war. They could say so because they had conventional superiority. American strategy promised no first *strike* but refused to forego first *use* because of the perceived need to deter an attack on Europe by the Soviet Union's superior conventional forces.

NUCLEAR PROLIFERATION

The "no-first-*use*" proposal, like the arguments of the arms controllers, stemmed from the current popular fear that a nuclear war between the superpowers would unleash universal devastation. But the proposal could not of itself guarantee the elimination of that danger. Not merely might the commitment be broken in the event of a crisis or in the heat of battle, but the danger might come from other directions. For despite the enormous size of the superpowers' nuclear arsenals, they no longer had a monopoly of the bomb. Some people believed that, despite the Proliferation Treaty, up to 10 percent of the world's total of nuclear weapons was now in other hands. In addition to Britain and France, which were NATO allies, and to China, a communist enemy of the West, Israel, India, Pakistan, Argentina, and South Africa were known to have, or were suspected of having, or of knowing how to make, the bomb. In 1981 an Israeli air strike against Iraq, when the two countries were not actively engaged in war, destroyed a nuclear power plant that was being constructed near Baghdad. It had been launched on the grounds that the plant was for manufacturing atomic weapons.

Britain and France might be expected to behave responsibly and keep the bomb only as a deterrent, but would that hold fast if the Warsaw forces invaded Western Europe? Furthermore, China was an unknown quantity. The other small nuclear powers all had regional enemies against whom, in desperation, they might use the bomb. The prospect was fraught with dangerous possibilities. If the superpowers could not cooperate to extinguish a regional nuclear war, a solution which seemed extremely unlikely at that time because in many cases they supported different sides, they might themselves be drawn in, and then a regional conflict would become a universal one.

An even more frightening thought was that a maverick ruler such as Colonel Moamar Kaddhafi of Libya, Idi Amin of Uganda, or Saddam Hussein of Iraq might get hold of the weapon and then threaten to use it. While it might seem that no advanced technological state would allow such dangerous weapons to get into the hands of unstable or irresponsible rulers, states like Libya have in fact already obtained other modern war technology, for instance, chemical weapons, from either Eastern or Western sources. In 1982 the U.S. Air Force launched an air strike to destroy a Libyan fertilizer plant being constructed, it was believed, to provide chemical weapons. Now that the atomic genie was out of the bottle, it could not be put back. The Israeli and American raids suggested that in the future its control would require more than an arms limitation treaty between the superpowers.

SEA POWER AND THE NUCLEAR CONFRONTATION

Meanwhile a more immediate problem was the continuance of superpower confrontation. Soviet naval power, based not only on its SLBM program, in which its vessels now outnumbered those of the U.S.A., but also on other kinds of craft, continued to be a factor to be taken into the reckoning, and the Soviets had plans to build carriers. With the establishment of naval bases in places far from the Soviet mainland, for instance, in 1975 at Berbera in Somalia in the Red Sea to service ships in the Indian Ocean, they could exercise a baneful influence on politics in several parts of the world, especially in the Horn of Africa. These bases also facilitated their confrontation and rivalry with U.S. SLBMs, both to hunt them and to operate against distant centers of Western power. To counter the Soviet Red Sea base, the United States built naval dockyard facilities on the remote island of Diego Garcia in the

Indian Ocean where, many years before, it had leased rights from the British before independence was conceded to Mauritius, to which the island was politically subordinated. The U.S.S.R., in fact, lost its footing in Berbera within two years as a result of a local political upheaval. However, faced with what seemed like a Russian bid for world domination at sea, President Ronald Reagan in 1982 recommissioned the World War II battleship, *New Jersey,* (which had been used briefly to shell Vietnamese targets in 1968–69), the first of several to be reactivated in the following years.

Skepticism about the role of traditional naval vessels in modern war was allayed somewhat by the contribution made by these ships in 1982 in the Falkland Islands war. There they enabled sea power to be projected far from home, and even far from naval bases, despite some losses to air-launched missiles. A second lesson of a different dimension came when the naval powers of the Western world sent various kinds of craft to the Persian Gulf during the Iraqi–Iran War, 1980–1988. There they exercised a restraining influence, though again they suffered losses to aircraft-launched missiles. Nevertheless, Geoffrey Till, a leading British naval historian, has argued that the primary role of sea power in a future major war, apart from its function as part of the strategic deterrent, will be for commerce protection, and that new techniques and tactics will emerge. How that service could be carried out after a widespread nuclear exchange is not yet clear.

MORE ON ARMS LIMITATION

In the 1980s the superpowers again approached the problems of diminishing tensions and of limiting nuclear arms. SALT II had been negotiated by the American executive and the Soviet government, but it had failed to get ratification in the American Senate. President Gerald Ford and Secretary-General Leonid Brezhnev had agreed at Vladivostok that they would stand by its contents without ratification. But then each side began to complain that the other was violating the informal arrangement, and also violating the earlier SALT I Treaty, especially the ABM provisions. Further discussions about nuclear arms were delayed by these issues until 1982 and also by the need to await the election of a new president, and the delineation of his policies.

President Ronald Reagan was convinced that the United States was now inferior in strategic arms, intercontinental as well as intermediate-range, and also in conventional forces. He had campaigned on the theme that arms controllers had dominated previous administrations, and that if elected he would restore American strength. He had professed a war-fighting doctrine as the basis for national security. When he assumed office, he announced the furtherance of weapons production programs (some of which had, as we have seen above, been set on foot by his predecessor, Jimmy Carter). On the other hand, in Western Europe protests against nuclear arms and policies were becoming strident and were threatening the stability of NATO governments.

One of the sources of this increased agitation was a fear that a proposed deployment of American intermediate-range ballistic missiles (IRBMs) in Europe might provoke a Soviet first strike.

REAGAN AND GORBACHEV

The succession to power of a new, and younger, secretary-general in the Soviet Union, Mikhail Gorbachev, suggested an opportunity for a new attempt to resolve the nuclear arms and other problems at the summit. Gorbachev was very different from all of his predecessors in that he talked of the need for economic reform to strengthen the Soviet economy, of *perestroika* (administrative reform), and of *glasnost,* openness in government. In 1985 Reagan and Gorbachev arranged to meet at Reykjavik in Iceland for an informal face-to-face discussion.

Reagan appalled his advisers when, in a face-to-face talk with Gorbachev, he offered to work toward the elimination of all nuclear weapons within ten years; but the two came to an impasse on another issue that effectively neutralized the results of the Iceland summit. When Reagan came to power, he had been dismayed that the prevention of a nuclear conflict rested on the horrendous threat of mutual destruction. He therefore argued the need for a much better system that would guarantee absolute defense against missiles. Earlier proposals for anti-ballistic missile defense had been based on the principle of using another rocket to intercept an incoming missile and destroy it in flight, a forbidding technical problem. This difficulty was one cause of lack of belief in the possibilities of any defense. Since that time, however, improved radar and computers had increased the likelihood of interception. Strategic defense research therefore explored techniques for interception at four or five stages in the hostile missile's path. Lasers and other forms of defense directed from satellites, could aim at interception and destruction in midflight and even in the first, very vulnerable, stage of assent. Other ideas included the exploding of myriads of small pellets in the path of an approaching missile to destroy it in mid-flight in the outer atmosphere. While Reagan called his new approach to defend against the Soviet missile threat the Strategic Defense Initiative (SDI), the public, and skeptics, immediately named it "Star Wars" after a popular science-fiction movie. Despite such cynicism, Reagan flatly refused to abandon his personal attachment to this proposal for turning, as he saw it, away from the horrors of massive retaliation. So, as Gorbachev got no concession on SDI, he was unwilling to make drastic reductions in the Soviet nuclear arsenal. The Reykjavik meeting ended in a deadlock.

Skeptics argued that SDI would be unlikely to provide a hundred percent protection against a nuclear attack. Furthermore, even at its best, SDI attacked only ICBMs; it provided no protection against bombers on one-way missions or against IRBMs launched from Soviet submarines off the U.S.

coast. In these circumstances widespread destruction could still ensue. Hence, although Reagan talked of offering SDI techniques to the Soviets to give them the same degree of security, in effect, mutual assured destruction was likely to remain as a deterrent.

Despite this cooling of relations in Iceland, the two governments continued to correspond with a view to resuming talks about arms limitation, now called Strategic Arms Reduction Talks (START). One obstacle was that the Soviets were known to be constructing a facility at Knasnoyarsk in the center of their territory. They claimed that it was for communicating with satellites. But Americans suspected that it was intended to be used for defense against incoming missiles, and that therefore it was a breach of the 1972 Anti-Ballistic Missile Treaty which had been reconfirmed in the abortive SALT II agreement. The Russians responded by arguing that the Early Warning Facilities operated by the Americans in Alaska, Greenland, and Scotland, which had earlier replaced the Distant Early Warning lines stretched across North Canada when missiles had become a greater threat than manned bombers, were breaches of the ABM Treaty.

START seemed about to reach the same impasse as the SALT talks when the negotiators, Paul Nitze and Yuli Kvitinsky, took their famous "walk in the woods" and reached a substantial degree of agreement on how to work for a solution bypassing the legalistic positions maintained by their bureaucratic colleagues in Washington and Moscow. However, this new hope, too, seemed to fade away when administrators in both Washington and Moscow repudiated their proposal.

Conservative, hard-line American nuclear strategists lost some public credence when it was learned in 1986 that the CIA had permitted breaches in U.S. law in order to support Contra insurgency against the Sandinista government of Nicaragua, a government that was being supported by both Russia and Cuba. The charges came to a head when an American plane flying supplies was brought down by the rebels, and when it was learned that the United States was financing Contra operations by the sale of American arms to Iran through the medium of Israel.

THE INTERMEDIATE-RANGE MISSILES TREATY

However, pressure for progress in arms limitations to reduce the possibility of a nuclear war in Europe, stemming from moves by both sides to deploy intermediate-range ballistic missiles there, led to a breakthrough in 1982 by the negotiation of an Intermediate-Range Nuclear Forces Treaty. Reagan had proposed in 1981 that this problem should be settled on a zero–zero principle. Russian acceptance was forthcoming probably because of an awareness that the accuracy and range of the Pershing IIs being placed in Europe was such that they would be able to reach Russian ICBM bases before Soviet missiles could retaliate. An extra incentive may have been a realization that the NATO Pershings would probably be aimed at the Soviet leaders. Until they developed weapons of similar range and accuracy, which would take them the best part of a decade, the Russians would now be at a serious disadvantage in a nuclear confrontation in Europe.

The IRBM Treaty was undoubtedly a noteworthy step toward a possible solution of the nuclear dilemma. It encouraged hopes for similar moves in strategic and conventional arms. Reagan claimed that it was a product of his firm stand and arms buildup. Whatever the merits of that argument, the most important factor seems to have begun with the appointment of Mikhail Gorbachev as Communist Party Secretary-General. This was followed by a wave of reforms both in the U.S.S.R., and also in other members of the Warsaw Pact and in their relations with the West, that were to culminate in the collapse of the Iron Curtain in 1989.

However, as the 1990s dawned, when the prospects for limiting or even eliminating nuclear weapons seemed brighter than at any time since the Baruch Plan of 1946, voices of caution all over the world still expressed some concern about the stability of an international order in a new environment that would no longer be constrained by the terrifying prospect of nuclear holocaust. It was the best of times and the worst of times.

21

CONVENTIONAL WAR

The superpowers' nuclear rivalry, covered in the previous chapter, was accompanied by a recurrence of conventional conflicts that post-World War II diplomacy failed to prevent. By almost any standard, warfare in the twentieth century has been horrific. In just the first ninety years of the century, more than 100,000,000 people have died as a direct result of war. By contrast, a modest 8,300,000 died from war in the nineteenth century. There have been forty-three wars with casualties over 100,000 since 1900, compared with fourteen in the previous century. The percentage of civilian casualties has been higher in the twentieth century than in any other. The nuclear blasts of Hiroshima and Nagasaki was responsible for only a small number compared with what resulted from conventional warfare.

Yet the record of the twentieth century is not as bad as it appears at first glance, or rather it is not as bad as one might have predicted midway through it. There have been 237 wars in the twentieth century, compared to 205 in the nineteenth. A full 60 percent of the twentieth-century casualties were suffered in the two world wars; only about a quarter have been incurred since 1945. In that period, the world population has more than doubled, from about 2.1 billion in 1945 to 5.2 billion in 1990. The number of countries in the world has doubled or tripled, depending on what one counts as a country. Fifty-one nations were charter members of the United Nations in 1945; 159 were members in 1989. Many of the wars since World War II were caused by independence movements in emerging nations and by civil wars reflecting political instability in many of these new states. Not a few of the casualties were caused by ethnic and regional animosities unleashed by the breakup of empires and of autocratic regimes that had kept their subject peoples in check. Without such causes, warfare in the second half of the twentieth century would have had much less impact.

Perhaps the major reason for a leveling off in the escalation of warfare has been the comparative constraint imposed on conventional warfare. In general, conventional warfare is armed conflict between states, is fought by regular troops, is highly mechanized, and does not involve nuclear weapons. Had conventional warfare continued to grow at the rate suggested by the two world wars, casualties since 1945 would now be counted in the hundreds of millions. The terrifying prospects of nuclear war have no doubt helped to prevent a shooting war between the United States and the Soviet Union, a conflict that could have been horrendous even if fought only with the conventional arsenals of the two superpowers. Nuclear weapons have also constrained other conventional confrontations around the world; the mere existence of these weapons, for example, gave the superpowers unprecedented leverage with their client states to ensure that no regional conflict would draw the United States and the Soviet Union into confrontation, as the Arab–Israeli War threatened to do in 1973. The compelling need to avoid nuclear war trickled down the hierarchy of wars and infected even the smallest conflicts. No conventional war could be allowed to escalate into an East–West nuclear confrontation.

Thus, conventional warfare in the second half of the twentieth century is a paradox. The means of conducting it have been enhanced dramatically, largely through technological advances, but also through psychological and ideological techniques perfected by the totalitarian states in the middle decades of the century. Additionally, the opportunities for war have multiplied, with more nations and more peoples in closer contact, locked competitively in struggles for a dwindling supply of natural resources. But the incidence and severity of conventional wars have also turned out to be less than one would have predicted by historical extrapolation. The reasons for this may be illuminated by an examination of the main characteristics of conventional warfare since 1945 and by a selective review of the major wars that have occurred.

FACTORS AND TRENDS IN CONVENTIONAL WAR

The most important factor, already mentioned above, is the constraint on conventional war imposed by nuclear weapons. The second most important factor is that conventional war since 1945 has been conducted almost exclusively in the Third World, where it also took the form of low intensity conflict. Some wars, such as the conventional operations in the conflicts in Angola, Nicaragua, and Indonesia, have been proxy wars in the East–West struggle between the great powers. They have relied heavily on arms and even on troops provided by the great powers, but they were nonetheless fought outside Europe and North America. For their part, the industrialized states invested most of their preparations for conventional war in anticipation of a traditional, mechanized, high-tech war between NATO and the Warsaw Pact on the plains of central Europe. Others who made use of their weapons adapted those instruments to suit local needs.

A third factor is that technology has intensified the lethality and destructiveness of war. Precision-guided munitions, increased mechanization, electronic sensors and detectors, improved logistical capabilities, and rapid communications are just some of the developments that have multiplied the firepower and the delivery capacity of combatants. This impact of technological change is even more pronounced in air and naval warfare than it is on land, mainly because these realms have always been more dependent on technology. This pattern is rooted in World War II, the first war in which significant numbers of new weapons systems were developed and introduced during the course of the conflict. This trend seems to have accelerated since the late 1960s, when the United States bent its full technological resources to the ground war in Southeast Asia, introducing high-tech war there and what one scholar called "the electronic battlefield."

Along with high-tech war has come a fourth trend, high-cost war. War has never been cheap; in the second half of the twentieth century it has created an international arms bazaar in which only wealthy, or desperate, peoples can afford to shop. International arms sales rose from $2.5 billion in 1960 to a peak of $41.4 billion in 1984. Not only does this high premium on technology make Third World countries dependent on the industrialized states for arms and equipment, it also tends to keep them in a state of permanent subservience to get the spare parts and the servicing necessary to keep expensive weapons operating. The leverage thus available to the great powers has been used erratically, however, in part because the industrialized states have wanted to retain as many customers as possible as a means of subsidizing their own expensive arms procurement programs. The most costly weapon is the first one that is bought; after that, development costs are shared by all other copies sold. So, the more produced the better.

A fifth trend in conventional war has been an overall decline in the decisiveness of battle. This pattern began in the nineteenth century, after the Napoleonic wars, a result, it seems, of the industrialization of war. As contests between states became measures of industrial production and material strength, wars came to be decided increasingly by the ability of states to make up their losses and return to fight another day. Directing war against the civilian population and the economy, pioneered by William T. Sherman in the American Civil War, has ever since been a standard tactic in conventional conflicts. The two world wars of the twentieth century were clearly wars of industrial production, and it is not surprising that many conflicts in the second half of the twentieth century followed that same pattern. One new wrinkle, however, has been introduced. While low-tech wars have continued

THE PATRIOT MISSILE, the U.S. Army's theater air defense missile, replaced the Nike Hercules and HAWK missiles in the late 1980's. Each mobile launcher contains four ready-to-fire missiles. (U.S. Army, *1984 Weapon Systems,* U.S. Army, Washington, 1984, p. 32)

to run their courses independent of the outcome of specific battles, high-tech war has become more sensitive to single engagements. In the Arab-Israeli wars of 1967 and 1973, and in the Falklands War of 1982, the opposing sides bet their all on the most sophisticated technology—tanks, aircraft, and missiles in the first instance, ships, aircraft, and missiles in the second. These weapons are costly and, therefore, few; their ability to destroy each other is high. The result is that confrontations between them are, as one author has recently put it in a paraphrase of Thomas Hobbes, "nasty, brutish, and short." After the first salvos, the means to continue may be severely depleted. Combatants must win in the first round or seek a negotiated settlement.

The targeting of civilians, which has become a grisly hallmark of warfare in the second half of the twentieth century, derives in part from the role of industry and technology. These are often firmly rooted in civilian populations. But other factors are also at work. In low-tech wars, especially those employing the techniques of people's wars (conflicts that have, or claim to have, wide popular initiative, concern, and involvement), the enemy army is often hard to locate and destroy; in frustration or desperation soldiers often turn on the civilian population. Additionally, many of these very wars are driven by strong religious, ethnic, or ideological commitments; in such cases countries can be defeated only when their populations have been ravaged, made despondent, or are extensively devastated. Finally, new techniques of communication, mass propaganda, and transportation make it possible to isolate, indoctrinate, and relocate huge populations. The mass starvation of hundreds of thousands in the Nigerian civil war, and the extermination of comparable numbers in the civil wars in China in 1960 and Cambodia in 1979, are not events without precedent; but they are nonetheless manifestations of a new extension of military power in the second half of the twentieth century. The use of chemical and biological weapons in the 1970s and 1980s, against both soldiers and civilians, is further evidence of a new barbarity.

Set against all this is the growing influence of world opinion, perhaps the brightest hope for the direction in which conventional war seems to be moving as the twentieth century draws to a close. Improved means of communication and transportation, the emergence of a "global village," and the perception fed by ventures into space suggest that all humans share a small and ecologically fragile planet in a hostile universe. This has nursed a sense of community and concern that seem to be unprecedented. As never before, the feeling, as John Donne put it, that "every man is a piece of the continent, a part of the main," has made war in any section of the globe important to people in every other section. The nuclear sword of Damocles that hangs over the planet heightens this perception. The result has been a new sense of the community of nations, a desire of all to be a member of a common humanity, and an inclination on the part of many to shun, or sanction, those who violate accepted norms of international behavior. The rapid and widespread isolation of Iraq after its invasion of Kuwait in 1990 was the most recent evidence of the emergence of a world community. The reach of this movement is still limited. It is undermined by ideology, by competitiveness, by parochialism, and by myopia; but its strength is growing, helping to limit some of the worst excesses of conventional warfare. Iraqis may have gassed Iranians, but at least they had the shame to try to hide it.

THE EARLY WARS

As has been seen, the Cold War broke out in Europe. Driven by the paranoia of its history and of its leader, the Soviet Union at the end of World War II saw a new version of encirclement by the West. From the time of Charles XII of Sweden in 1709, through Napoleon, and most recently and disastrously with Hitler, whenever a Caesar established some control over Europe, he could not resist a temptation to invade Russia. The United States, strong and comparatively unscarred by World War II, and stationing forces all around the Soviet Union, seemed to be another in this long succession of Caesars. From bases in the Middle East, from carrier forces in the Mediterranean and North Atlantic, from its newly won posts in Japan, Korea, and especially Okinawa, and from bomber bases located just over the North Pole, the United States enjoyed a monopoly of nuclear weapons and a variety of vantage points from which to launch them at the Soviet Union. In 1945, the Soviets had nothing with which to respond except for their huge conventional military force, which remained mobilized after the war, and the territory they had occupied in the closing months of the conflict. Most of those forces, and those territories, were in Eastern Europe.

The Soviets therefore tightened their grip on Poland, Hungary, Rumania, Bulgaria, Czechoslovakia, and East Germany. They promptly outlawed noncommunist political parties in those countries, thus ensuring the establishment of puppet governments faithful to Soviet policies. Yugoslavia and Albania, liberated without Soviet assistance, became people's republics as well, aligned with, but not bound to, the Soviet Union. By 1948, if one includes the Baltic Republics of Estonia, Latvia, and Lithuania, which had been annexed in 1940, no less than eleven countries and 100,000,000 Europeans had turned communist. Western European democracies responded in kind by outlawing communist parties in their countries. An iron curtain divided Europe along the Elbe–Trieste line, which became the front line of the Cold War.

The first skirmish came along a flank of that line, in Greece. Liberated and occupied by the British at the end of World War II, Greece dissolved into civil war the following year. The royalist government, supported by the British, found itself hard pressed by communist guerrillas supplied from Bulgaria, Albania, and Yugoslavia. Using arms and supplies left over from World War II, the contestants engaged in an early version of what would soon come to be known as wars of national liberation or of low-intensity conflict. Guerrillas attacked not only at the armed forces of the state but also the infrastructure of the state. Their purpose was to destabilize the government, disrupt the economy, and spread fear and dissatisfaction among the people. They struck in small-scale raids, often inflicting civilian casualties. The government responded by trying to root them out and bring them to bay; often they too inflicted civilian casualties.

In the midst of this first shooting confrontation of the Cold War, the British had announced early in 1947 that they could no longer sustain their role in Greece. Wracked by the expense and destruction of the long war and troubled by a declining and fracturing empire, Great Britain had to retrench and consolidate. It could no longer bear the burden of world leadership that it had carried since Trafalgar.

We have seen that the United States assumed the role. Convinced that the vulnerability of European and other states to communist subversion stemmed both from military weakness and from economic distress, it formulated in 1947 a two-tiered policy designed to shore up Greece and Turkey in the first instance, and other like nations in their train. On 12 March 1947, President Truman announced the Truman Doctrine, providing economic and military assistance to any country threatened by foreign (read communist) aggression. On 5 June Secretary of State George Marshall announced the

European Recovery Program, popularly known as the Marshall Plan. With these two policies the United States assumed the role of leader of the Western world, established a policy of containment that would guide the country through four decades of confrontation with world communism, and shaped many of the conventional wars that would take place around the world over the next forty years. Wars of national liberation or attempts by the Soviet Union and other communist countries to subvert or overthrow nations by force of arms were to be countered by economic aid designed to provide material wealth and by military aid designed to bolster defense. More than once in the ensuing years, conventional wars that were only dimly related to the great East-West confrontation came to be fought with the resources, the techniques, and the agenda of the Cold War.

One such confrontation was the civil war in China, a phase of what military historian Theodore Ropp has called the Great East Asia Land War. Its roots were at least as deep as the breakup of the Chinese empire in the late nineteenth and early twentieth centuries. As European powers had jockeyed for position and influence in China, they had undermined the legitimacy of the emperor and helped to bring about his downfall. Finally, on the eve of World War I the last emperor abdicated, and a republican form of government modeled on the West assumed power. Almost from the outset, however, free government in China was challenged, externally by the Japanese and internally by local warlords and other claimants to power. By the eve of World War II, two contenders were dominant, Chiang Kai-shek, heir of the Kuomintang party that had orchestrated China's break from the imperial past, and Mao Tse-tung, leader of an indigenous communist movement predicated on the then-heretical proposition that a Marxist-Leninist revolution could take place in an agricultural country. Chiang had the upper hand when World War II broke out, and Mao was relegated to hiding in the mountains, searching for a way to defeat the Nationalists.

With the Japanese invasion of China in 1937, followed by the outbreak of general war in the Pacific in 1941, Mao and Chiang resolved to turn their attention to the new enemy, deferring their internal contest for power. Neither side, however, was true to the resolve. Partially for that reason, China contributed less than it might have to the defeat of Japan. Nonetheless, Mao developed his most influential thought in anticipation of war with Japan. By that philosophy, which is discussed in Chapter 22, he transformed warfare in the second half of the twentieth century. His contribution was primarily in the realm of low-intensity conflict, but a dimension of his theory operated in the realm of conventional war, or what he called positional warfare. By 1949, when he was in the final stages of his conflict with Chiang Kai-shek, he took up positional warfare and used it to drive the Nationalists from the mainland.

In the political aftermath of World War II, several other civil wars erupted. Many colonies read the Atlantic Charter and the Charter of the United Nations and took to heart the promises of national self-determination that they found there. The largest of these was India, where nonviolent activist Mahatma Mohandas Gandhi culminated his decades-long campaign for independence in August 1947 and then lost the ensuing civil war to religious and ethnic hatreds that were beyond the control of even his saintly ability. Felled by a reactionary Hindu for agreeing to a partition of India that allowed for a separate Muslim state of Pakistan, Gandhi left behind a pattern of domestic violence that defied the power of the British and Indian armies, killed 800,000, and forced the mass migration of millions of others. Pakistan consisted of two separate territories separated by hundreds of miles of hostile India. After independence, the first object of the formal fighting between India and Pakistan was Kashmir, a predominantly Muslim state under a Hindu prince who hoped to steer an independent course between the two rival nations. Muslim tribesmen from Pakistan invaded Kashmir in October in response to reports that fellow Muslims were being suppressed in local disturbances. When asked by the maharaja to help his resist this aggression, India drove the Pakistani invaders back within their own borders in return for a Kashmiri agreement to accede politically to India. Following mediation attempts by a United Nations commission, India and Pakistan finally agreed to a cease-fire over Kashmir in January 1949, temporarily ending one of the bloodiest and most intractable confrontations of the postwar world. But the seeds of conflict remained and produced still more poisoned fruit. Twice in the 1960s violence broke out again; war in 1971 finally freed East Pakistan to become the independent nation of Bangladesh.

Meanwhile, conventional war erupted in another corner of Southeast Asia. Vietnam, a new state formed by the Japanese during World War II out of part of French Indochina, also sought its independence when the war was over. Nationalist leader Ho Chi Minh, who had fought the Japanese throughout the war, brought Soviet training, Chinese theory, and equipment from both those countries to bear upon his own personal war of national liberation. The French, trying to reestablish their prewar colonial control throughout Southeast Asia, resisted, and the United States and other wartime allies acquiesced in that policy. The result was the first of three ten-year struggles by Ho Chi Minh's forces to drive out Western powers and win national autonomy.

Ho's military campaign, under the brilliant leadership of Vo Nguyen Giap, drove the French inexorably toward a showdown at Dien Bien Phu, an isolated military outpost in the highlands of northern Vietnam. Using the conventional techniques and equipment of European land war, the French had been unable to secure more than the major population centers. The countryside, the people, and perhaps most importantly the roads, belonged to the Vietminh. This endangered France's hold on the colony and jeopardized the economic development essential to this form of colonization. The French strengthened their position at Dien Bien Phu and challenged Giap and his partisans to dislodge them. The French were gambling that if the irregulars could be drawn into a conventional battle of positions, especially in the highlands where they supposedly could not transport their artillery and bring it to bear, superior French firepower and air power would prevail. But positional warfare was the third stage of Mao's formula; Giap and his troops were ready for the challenge. They dismantled their artillery, packed it into the mountains on their backs, reassembled it, and pounded the French position relentlessly with shells hauled by sheer manpower through the steep jungle terrain. They drove the French forces into an ever-smaller perimeter, making it easier to interdict the resupply by air that the French had counted upon. After 138 days, the French garrison, starved and out of ammunition, succumbed.

While many conventional battles in the twentieth century have proved inconclusive, this one was decisive. Just as Mao predicted in 1938, French public opinion demanded an end to the adventure in Southeast Asia. The two sides met in Geneva in 1954 and there negotiated peace. France agreed to withdraw all its forces. French Indochina was divided into four parts. Two new states, Cambodia and Laos, won their independence and autonomy. Vietnam, also now autonomous, was divided along the 17th parallel into a primarily communist north and a primarily Catholic and Westernized south. Elections were to be held within two years to determine the conditions for uniting the two artificial political entities. The United States had assisted the South Vietnamese government in reaching this peace accord, but it was not formally a party to the agreement. Its leading role in this three-part drama would come in the final stage, 1965–1975.

KOREA

The largest conventional war since World War II, and the third largest conventional war in the twentieth century, had come without warning on 25 June 1950, when North Korean troops crossed the 38th parallel in force at eleven different points. The separation of Korea along the parallel—a negotiated border between the United States, whose troops had invaded the collapsing Japanese empire in 1945, and the Soviets, who had turned their attention to the Pacific in the closing months of the war—was another artifact of the Cold War. As in Europe, the two sides agreed only with the greatest difficulty on postwar occupation and borders. Korea turned out to be one of the most intractable cases. Late in 1948, the Soviet Union announced withdrawal of its troops from North Korea, but it left behind a cache of surplus arms for the communist government in the North. Early in 1949, the United States withdrew its occupation forces, but it kept a significant contingent poised in nearby Japan, where General-of-the-Army Douglas MacArthur was overseeing occupation and reconstruction. The invasion pitted 130,000 North Korean troops, equipped with a brigade of Russian T-34 tanks and 180 Russian Yak planes of World War II vintage, against 100,000 troops of the Republic of Korea who constituted little more than a national political force, virtually devoid of tanks, artillery, or aircraft.

The war proceeded in three stages. The first was a rout. Before international assistance for South Korea could be organized, the invading North Koreans drove the defenders and their few American advisers across the southern half of the peninsula and into the Pusan perimeter, a last-ditch defensive position in the southeast corner of the country with a front of less than 150 miles. The United Nations Security Council, convening while the Soviet Union was fortuitously boycotting meetings on an unrelated issue, approved military intervention by U.N. member nations in support of the South Koreans. Subsequently a U.N. command was set up under United States leadership, with Douglas MacArthur in charge of all U.N. forces in Korea. Though many countries sent only token assistance, this nonetheless marked the first, and until 1990 the only, instance in its history when the United Nations approved the use of force in response to violations of international law.

In the first desperate months of the war, General MacArthur used the forces at hand to stem the North Korean assault. These were enough to hold the Pusan perimeter through August and to allow MacArthur a chance to seize the initiative. In a bold move acquiesced in, but not endorsed, by the United States Joint Chiefs of Staff, General MacArthur exploited his virtual monopoly of sea power in the war to land an amphibious force at the port city of Inchon, halfway up the peninsula and thirty miles from the South Korean capital at Seoul. This risky venture at a site with poor beaches and enormous tidal shifts caught the North Koreans completely by surprise. Eleven days after landing, American troops had retaken Seoul, captured 125,000 prisoners, and stranded a still greater number of North Korean troops south of the 38th parallel. The beleaguered American forces quickly broke out of the Pusan enclave and cooperated in rounding up the stranded North Koreans. The second phase of the Korean war proved as much a triumph for the U.N. forces as the first phase had been for the North Koreans. MacArthur's invasion at Inchon had been the most brilliant military stroke in a distinguished, indeed legendary, career.

But then the war changed. With the North Korean army in disarray, the United States and its U.N. allies saw an opportunity to exploit MacArthur's success by turning a defensive war into a punitive campaign north of the 38th parallel. With tacit approval of the United Nations, MacArthur's forces crossed the 38th parallel on the first of October and moved toward the Chinese border at the Yalu River, in spite of public warnings by the Chinese that they would intervene in the war. The U.N. forces now numbered over 200,000, with almost as many in supporting roles, more than enough to drive back the collapsing North Korean Army. In spite of intelligence reports that Chinese troops were crossing into Korea, MacArthur continued his pursuit, dividing his forces. On the night of 25–26 November, lead units of the U.N. offensive were hit by upwards of 180,000 Chinese troops, many of them veterans of the long and bitter civil war in China. As the surprised, demoralized, and ill-deployed U.N. troops staggered back well beyond the 38th parallel, the tide of battle had shifted once again. Another 120,000 Chinese troops attacked American forces in the eastern part of North Korea, driving them back and forcing an innovative and desperate use of air and sea forces to support their withdrawal.

In late January 1951, U.N. forces finally rallied along a line about 50 miles south of the 38th parallel. By the end of April, under the sound leadership of U.S. General Matthew B. Ridgeway, they slowly fought their way back to just north of the parallel. MacArthur, still in overall command U.N. forces as well as all U.S. forces in the Far East, demanded authority to strike Chinese air and supply bases north of the Yalu and to assist Chiang Kai-shek and his Nationalist forces in a campaign to retake the Chinese mainland. President Truman and his advisers, unwilling to risk an escalation of the war and a confrontation with the Soviet Union, steadfastly refused. The Joint Chiefs of Staff worried that Korea was just a feint to distract the United States, and they focused their attention and much of their resources on a conventional buildup in Europe. When MacArthur's impatience with these constraints was made public, President Truman relieved him and appointed in his place General Ridgeway, who had performed so brilliantly as field commander in restoring the U.N. position early in 1951. MacArthur went into retirement, a victim of his own hubris, of the Cold War, and of the subordination of conventional war to the threat of nuclear confrontation.

The Korean war dragged on for two more years, a stagnant trench war reminiscent of World War I. Tactical air power was influential but not decisive; strategic air power played only a small role. The helicopter was introduced, primarily for medical evacuation, contributing to the lowest fatality rate among casualties reaching hospital of any war up to that time. Naval power had been important at Inchon and again in the evacuation of U.N. forces trapped by the entry of the Chinese into the war. After 1951, however, sea power was used primarily for mobile airbases and for coastal patrols; it did not greatly affect the outcome of the conflict. Psychological warfare took on a modern and horrible guise in the North Korean practice of torturing prisoners of war, both physically and psychologically, to get them to renounce their country, its war aims, and its methods. Nuclear weapons were seriously considered twice. President Truman said later that he would have employed them to prevent the annihilation of a major American force, such as those who gathered with the South Koreans in defense of the Pusan perimeter. And, as seen in an earlier chapter, President Eisenhower is reported to have threatened the use of nuclear weapons in order to get the stalled peace talks moving after he took office in 1953. For whatever reason, the North Koreans finally came to terms with the U.N. negotiators in April 1953, signing an armistice the following July. After repatriation of those prisoners who agreed to it, the bloodiest engagement of the Cold War finally came to an end. Though exact figures are impossible to determine, almost 200,000 South Korean and U.N. combat personnel died in the war, and at least 1.6 million communist troops, 60 percent of them Chinese. In another grim testament to the nature of conventional war in the modern world, about 3 million South Korean and 400,000 North Korean civilians died as a direct result of the war.

OTHER CIVIL WARS

Though Korea was a war between two parts of a divided country, a kind of civil war, it was also a proxy confrontation between the super powers, East and West. The involvement of the United States on one side, and the Soviets and Chinese on the other, ensured that the combat would be conducted with ample supplies of modern arms and equipment. Still, those material resources were not much different from what had been available in World War II. Both sides flew jet aircraft, and there were some improvements in other technologies, such as helicopters, communications equipment, and medical care. But anyone who had fought in World War II would have recognized the battlefield in Korea.

That similarity was less clear in numerous other civil wars conducted around the world in the 1950s and 1960s. Usually the government side in a conflict had access to modern military technology, but very often the rebels had to rely on relatively primitive arms and equipment. Occasionally, however, those positions were reversed, and rebels, supported by one or the other of the superpowers, found themselves better equipped than a government that they were trying to overthrow. In either case, such wars were usually conducted without weapons parity, making them difficult to classify. They were most often fought as conventional wars on the part of the established power, but as a kind of people's war or war of national liberation by the rebels. This disparity in equipment and technique helped shape the warfare and determine its outcome.

In Colombia, for example, political division over control of the postwar government devolved into violence in 1948, touched off by the assassination of liberal leader Jorge Gaitán. Fighting continued sporadically for the next fourteen years, abating finally only after the national constitution was amended to provide for joint rule by liberals and conservatives. Seizure of power by General Gustavo Rojas Pinilla in 1953, and succession of a military junta in 1957, were insufficient to restore order and guarantee internal security. Before the conflict was done, an estimated 300,000 Colombians, mostly civilians, had lost their lives in "La Violencia." This was typical of many modern conflicts in that it was hardly traditional warfare at all, but rather a rash of domestic violence made more deadly by modern technology and by the intensity of the political struggle. Such violence is analyzed in more detail in Chapter 22 on "Low-Intensity Conflict."

China engaged in a different kind of bloodletting in 1950–51. Even while losing the first of its million casualties in the Korean War, Mao's government was conducting an internal purge that resulted in the extermination of as many as a million of its own citizens. In many ways, this campaign was an extension of the civil war that had just ended. Over a five-year period, the communists established the apparatus of a totalitarian state, launched a campaign to shift the economy from an agricultural to an industrial base, and eliminated its internal enemies. Many of these last were landlords of the old regime; they were liquidated and their land redistributed to former peasants. The program was successful to the extent that it secured the revolution and helped to eliminate famine for the first time in the recorded history of China. But the price paid was virtually a conventional war of extermination by one class of Chinese against another class of Chinese, a horror of modern war reminiscent of the totalitarian purges of Hitler and Stalin. As Hitler had discovered with the Jews, the sheer mechanics of murder on such a scale taxes the killing capacity of any modern state.

Algeria presented a more traditional example of a war driven by external strife and escalated by the participation of a highly industrialized state. The results were hardly less bloody. Though run as a virtual colony since 1830, Algeria had been since 1848 an actual part of the French state. When Muslim nationalists, who made up a large majority of the population, demanded independence in 1949, they challenged French honor and national integrity and the interests of French settlers even more than had been the case in Southeast Asia. On 1 November 1954 the dispute broke into open war, as guerrillas of the Front Libération Nationale (FLN) attacked French targets. Operating from bases in neighboring Tunisia and Morocco, the rebels forced the French to introduce approximately 400,000 troops into Algeria to maintain order. Terrorism by the rebels was met by harsh counterterrorist actions by the French, which undermined public support for national policy at home and created a political crisis that finally swamped events on the battlefield. Though the French army was successful in containing FLN activity and in closing the borders with Tunisia and Morocco by the use of sophisticated modern equipment—electrified fences, barbed wire, electronic surveillance, helicopters, and radar—it could not entirely eliminate the rebel threat, nor could it prevent Algerian terrorist attacks in France itself. Charles de Gualle was elected president of France in 1958, largely in the expectation that he would take a hard line in suppressing the rebellion. When he proposed a political settlement including autonomy for Algeria, the die was cast. Algeria achieved independence in 1962, at the cost of a bitter and divisive civil war that had taken over 300,000 lives and had brought the French army to the brink of open rebellion against its own government. As Mao had predicted in 1938, conventional military force had proven effective against partisan forces; the rebels won politically, not militarily.

Other African nations followed similarly violent paths to independence. The Belgian Congo (renamed Zaire in 1972) won independence in 1960 but quickly dissolved into civil war which required the intervention of United Nations troops. As many as 100,000 people died before some order was imposed on the country in 1965. Another 100,000 died in neighboring Rwanda, in spite of that nation's efforts to steer a more peaceful course toward independence from Belgium; not even U.N. assistance could prevent tribal rivalries from breaking into open conflict. Sudanese independence brought on a civil war that claimed 300,000 lives between 1963 and 1972, witnessed large-scale intervention by the Soviet Union and Egypt, and sent thousands of refugees fleeing into neighboring Zaire, Ethiopia, Uganda, and the Central African Republic.

Civil wars also shook the Middle East in the 1950s and 1960s. Yemen experienced a costly civil war between republicans and leftists in the 1960s. Egypt sent as many as 60,000 troops to fight for the republicans; King Faisal of Saudi Arabia provided arms and equipment for the royalists. A negotiated settlement finally ended the bloody conflict, which had cost 100,000 lives. The revolt of Kurdish tribes in northern Iraq in the 1960s ended in a measure of autonomy for the Kurds, but only after the loss of another 100,000 lives. Other wars, more numerous but often less costly in lives, marked the retreat of empire in the postwar world and revealed the instability that frequently came with autonomy and independence.

Two of these wars in the 1950s and 1960s are particularly worthy of note here. The Nigerian civil war of 1967–70 has the odious distinction of being the bloodiest of the African wars of the twentieth century, and also the most cruel. Nigeria, the most populous black African nation and, with its oil-based economy, one of the wealthiest, achieved its independence from Great Britain peacefully in 1960. However, regional and tribal conflicts marred the early years of independence, even after a republic was declared in 1963. Finally, the Ibo tribe of southeastern Nigeria declared themselves the independent state of Biafra in 1967. After initial military successes, the outnumbered Ibo were driven back and isolated, causing mass hardship and even starvation. But the Ibo would not give up and the military government of Nigeria refused to recognize their demands. In spite of, or perhaps because of, military aid from Great Britain and the Soviet Union for the Nigerians, and from France for the Biafrans, the war dragged on for two-and-a-half years and took an estimated two million lives, mostly Biafran civilians who perished of disease and starvation. Though the Nigerians seem not to have purchased an intentional policy of genocide, the result of the war was nonetheless much the same as if they had; it was one more example of the extremes to which regional and ethnic rivalries could push conventional war in the modern world.

VIETNAM

A second stage in the Vietnam conflict began almost before the ink was dry on the Geneva accord of 1954. The United States was already deeply involved in Vietnam, providing 70 percent or more of the French military budget there at the time of Dien Bien Phu. Unwilling to use nuclear weapons to save the French, as some members of a divided Joint Chiefs of Staff recommended, and unable to secure British support for a joint rescue mission, the United States had reluctantly acceded to the Geneva agreement, but it never formally signed it. Even before the French surrender at Dien Bien Phu in 1954, President Eisenhower was publicly expounding a "domino theory" of communist expansion in Asia; he likened the countries of Southeast Asia to a row of dominoes; if one fell to the communists, the others were sure to follow. This would propel the United States and Vietnam into the next phase of this quagmire. Refusing to see the unfolding events in Vietnam as a war of independence or a civil war for the control of an ancient and reborn nation, Eisenhower and his advisers saw it as another attempt at communist expansion, a continuation of the adventure in Korea and the communist insurgency in Malay, one more test for the U.S. policy of containment. President Eisenhower was already looking for ways to avoid the national elections set for 1956 in the Geneva accords, for he believed that Ho Chi Minh would win as much as 80 percent of the vote nationwide and carry the unified country into the communist bloc.

The United States increased its financial and military aid to South Vietnam, intervened in the establishment and running of the successive governments there, and sought excuses to avoid the coming elections. In 1956, the government of South Vietnam refused to hold elections on the plausible, if not entirely candid, grounds that the North had violated the agreement by leaving some Vietminh forces to operate in the South. With the electoral route to national unification blocked, Ho Chi Minh resumed the civil war. The Viet Cong, an indigenous guerrilla arm of the North Vietnamese army, began a campaign of guerrilla and mobile warfare in the South, which was designed to undermine the government, to win the support of the people through conversion or coercion, and to gain military control of the countryside. As the campaign increased through the late 1950s and into the 1960s, the United States tried unsuccessfully to shore up the corrupt and inefficient government of South Vietnam and to provide material and training assistance for the South Vietnamese army.

By the time President Kennedy took office in 1961, the situation in Vietnam had deteriorated alarmingly. Conditions in neighboring Laos were worse. Kennedy, determined to "bear any burden," as he said in his inaugural address, took steps to enhance America's competitiveness in the Cold War. He introduced a policy of "flexible response," replacing the massive retaliation and the overreliance on nuclear weapons of the Eisenhower administration with an ability to fight two-and-a-half wars at the same time—a strategic confrontation with the Soviet Union, a full-scale conventional war, and a brush-fire war. To help the United States compete in the latter, he created the Special Forces within the United States Army and also significantly enhanced American capacity and training for unconventional war. At the same time he began sending more "advisers" to Vietnam, increasing their numbers from a few hundred when he took office to about 10,000 when he was assassinated in autumn of 1963. His successor, President Lyndon Johnson, saw no alternative but to continue this escalation of American involvement.

On 4 August 1964, President Johnson announced that, for the second time in two days American naval vessels in international waters in the Gulf of Tonkin had been attacked by North Vietnamese torpedo boats. Acting on his own authority, the president ordered attacks on oil and naval installations in North Vietnam. The next day Congress passed the "Tonkin Gulf Resolution," empowering the president "to take all necessary measures to repel any armed attack against the forces of the United States and to prevent further aggression," and further "to take all necessary steps, including the use of armed force, to assist any member or protocol state of the Southeast Asia Collective Defense Treaty requesting assistance in defense of its freedom." The second stage of the Vietnam War was over; the third stage was about to begin.

With broad popular support, and on the advice of virtually all of his military and civilian advisers, President Johnson, who had assumed office in November 1963 when Kennedy was assassinated, committed ground troops to Vietnam in the spring of 1965, initiating a policy of "gradual escalation" in military pressure that was intended to force Ho Chi Minh to desist from his attempt at a military overthrow of the South Vietnamese government. Unwilling to exert the full might of the United States on a Third World country of only 19 million people, Johnson elected instead to punish the North Vietnamese and the Vietcong for military aggression against the government of the South, while at the same time holding out a carrot of reconciliation and financial aid should the parties come to terms. He and the Pentagon had at their disposal a new and largely untried arsenal of weapons and tactics developed since the Kennedy inauguration to fight just this kind of war.

What the United States policy miscalculated was the depth of the commitment of Ho Chi Minh to independence, and also the inability of America's ground warfare techniques to solve the puzzle of people's war. The North Vietnamese and the Viet Cong had significant political advantages which they exploited brilliantly to undermine America's military strategy. They used sanctuaries in Cambodia and Laos to protect their bases of operations and lines of supply down the so-called Ho Chi Minh Trail into South Vietnam. They used their political and ideological ties to China and the Soviet Union to obtain the supplies and munitions that they needed to cover their losses. And they used the political divisions in the United States, and in the free world, about U.S. policy in Vietnam as a means to impose constraints upon American use of force.

The United States tried several different approaches. It built up its troop strength, from 16,000 in 1964 to a peak of 543,482 in 1969. It tried bombing strategic targets in the North, both to punish that country in general, and to interdict its production and transshipment of war material to the south. It attempted various programs of civic action with the South Vietnamese population, in an attempt to undermine the base of support so essential to guerrilla activity; it even went so far as to experiment with forced relocation of populations into strategic hamlets, a technique that had been tried successfully by the British in Malaya but unsuccessfully by the French in Indochina. The Americans attempted "search and destroy" techniques, built around sophisticated surveillance equipment and the unprecedented mobility provided by helicopters, to trap enemy units and to force them into positional warfare. In perhaps the most desperate attempt to return to conventional warfare, the Americans even tried to lure the North Vietnamese into a positional showdown modeled on Dien Bien Phu. This one, staged at Khe Sanh, a small mountain outpost just below the demilitarized zone along the 17th parallel, failed to bring on the decisive military engagement that the Americans sought.

For their part, the North Vietnamese twice misjudged the state of the war and moved prematurely into positional warfare, once during the Tet Offensive of 1968 and again during the spring offensive of 1972. Both campaigns failed after initial successes, victims of superior American firepower. In both instances the North Vietnamese and the Viet Cong retired in disarray with severe losses that took many months to replace. U.S. military commanders reported confidently, and with some accuracy, that these were great battlefield successes for the United States and its South Vietnamese allies, and that they marked turning points in the war. What the military analysts always failed to appreciate, however, was that this was a political war, and it would be settled politically, not militarily. The Tet Offensive, though it cost the Vietcong dearly, won the war. American public opinion, having been led to perceive a "light at the end of the tunnel," turned against the administration and the Pentagon. President Johnson announced in March 1968 that he would not seek reelection the coming November. Instead he would devote the rest of his presidency to the search for a peaceful solution to the conflict. The spiral of ever-increasing troop deployments to Vietnam was halted; and American troop strength there began to decline the following year when President Richard Nixon announced a new policy of Vietnamization, that is, of leaving the South Vietnamese to conduct the operations. America thus backed out of a war that it had not lost, but could not win.

In an effort to salvage what he called "peace with honor," President Nixon and his national security adviser, Henry Kissenger, played the same carrot-and-stick game that the Johnson administration had invented. Even while reducing American troop levels in Vietnam, President Nixon increased supplies to the South Vietnamese government, secretly invaded Cambodia to root out Viet Cong bases there, negotiated actively with Communist China and the Soviet Union to try to isolate North Vietnam politically, and finally resumed the bombing of North Vietnam that President Johnson had suspended in an attempt to lure the enemy to the bargaining table. The war dragged on through President Nixon's first term, until the fall of 1972, when Kissinger, now secretary of state, announced that as a result of his secret negotiations with Le Duc Tho, "peace is at hand." In the event, it still took several more months and another round of bombing in the north before a settlement acceptable to the United States and to both North and South Vietnam was finally reached. This provided for what one cynic called "a decent interval" before the North Vietnamese resumed their campaign against the South. When this renewal materialized, it took the form of positional warfare. On 5 March 1975, the communists launched a conventional offensive in the central highlands; the South Vietnamese army panicked and fled before it. On 30 April the South Vietnamese government surrendered, ending what many have come to call the modern Thirty Years War.

Vietnam makes clear the difficulties in talking about conventional war in the modern world. For the North Vietnamese and the Viet Cong this was from start to finish a people's war, a combination of low-intensity conflict and of conventional or positional war. For the United States, during its period of greatest involvement, from (1965 to 1973), it was an attempt to respond by conventional means and weapons to the challenge of unconventional war. The Green Berets (Special Forces) and "flexible response" notwithstanding, it was still primarily for the United States a conventional war, fought by and large with conventional weapons developed for mechanized warfare on the plains of Europe. The tanks and planes and trucks and rifles and radios were designed for other wars on other terrain and in other climates. Some weapons, like the helicopter gunship, which evolved out of this struggle, were particularly well adapted to unconventional warfare in a jungle environment. Too many, however, proved ill suited to the demands made upon them.

This was also a war of high tech against low tech; in this case low tech won. The confident predictions of the mid-1960s about an electronic battlefield, in which the marvels of modern science would overcome the will and wiles of the guerrilla fighter, proved overly optimistic. The Viet Cong were tough, aggressive, clever, and relentless fighters, well trained, well disciplined, and battle hardened. They used the night, the jungle, the people, and their sanctuaries in Cambodia and Laos to mask their operations and to choose the time and place of battle. The initiative, which they held throughout the war, allowed them to attack when they had local advantage and to flee when the Americans brought their superior firepower to bear. Most of all, they had a strategy, inherited from Mao, and adapted to the conditions in Vietnam by Giap. The Americans never developed a plan to adapt their military superiority to secure the political objectives that they sought.

The United States nevertheless did achieve a number of battlefield successes, and they did introduce significant new or modified technologies into the modern arsenal. The so-called "smart bombs" introduced in the late 1960s were laser-guided munitions that locked on their targets and carried their weapons to them with pinpoint accuracy. They proved especially effective in air attacks on North Vietnam. There the enemy had emplaced the heaviest anti-aircraft resources ever seen in warfare and had placed important military targets close to civilian facilities, targets which they knew the Americans would want to avoid. The helicopter, already mentioned above, transformed ground warfare. It provided unprecedented mobility and, by making possible a true "air cavalry" and as a helicopter gunship, it became a weapons platform with the capability of hanging suspended above the battlefield. While electronic sensors never solved the problem of finding and fixing an elusive enemy, they did experience some significant successes,

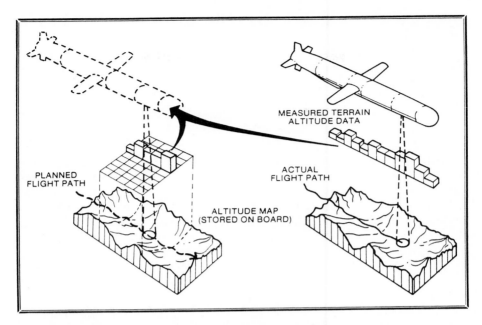

THE CRUISE MISSILE is a programmed, self-guiding missile that can be fired on land, in the air, or at sea. This drawing illustrates its navigation system, i.e., following contours to its target, which allows it to fly below enemy radar. (Kenneth P. Werrell, *The Evolution of the Cruise Missile,* U.S. Air Force, Washington, 1985, p. 138)

such as the infrared scopes that allowed riflemen to see their targets in the dark. The B-52, backbone of the air force's strategic bombardment arsenal, was employed to drop conventional armaments, both on strategic targets in North Vietnam and on the battlefield and infiltration routes to the South. In the latter role it was a big hammer for a small nail, but it was terrifying and effective nonetheless.

The high technology weaponry introduced in Vietnam was not enough to change the course of that war, but it did mark a significant new direction in the evolution of conventional warfare. The research and development skills of the industrialized states, hitherto concentrated on strategic weapons, now turned increasingly to the development, employment, and sale of high-tech conventional arms, thus changing the face of war throughout the world and opening an international arms bazaar of unprecedented size and complexity.

When the shooting stopped, 3 million people were dead as a direct result of this Thirty Years War, 2 million of them in the last third of the war, when the United States was most deeply involved. More than 50,000 of those dead were Americans, another 5,000 were Australians, South Koreans, New Zealanders, Philippinos, and Thais who had joined the United States in what was seen by them as a campaign of containment against communist aggression. The rest were Vietnamese, northern and southern, communist and noncommunist. Vietnam had paid an enormous price, almost 7 percent of its total population, to achieve Ho's dream of independence.

As if the cost were not enough, they soon invested more by exporting their revolution to neighboring Cambodia and Laos, an apparent confirmation of the domino theory about which President Eisenhower had warned in 1954. Vietnam invaded Cambodia in 1977 and again in 1979, finally setting up a puppet regime to replace Pol Pot and the hated Khmer Rouge, a rebel group which had bathed the country in blood after coming to power in 1975. The North Vietnamese assisted the communist Pathet Lao to establish and maintain control in Laos. Conflicts with China and Thailand also marred the early years of the Socialist Republic of Vietnam, finally forcing the new and impoverished country to retreat from its external commitments and cut back on its armed forces, which totalled 1.23 million men in 1984, the fourth largest army in the world after China, the Soviet Union, and the United States.

THE ARAB–ISRAELI WARS

In stark contrast to the Vietnam war, the four Arab–Israeli wars were uniformly high tech on both sides, making them the model testing ground of the evolving conventional technology and a bloody war game of the East–West confrontation. The United States was the primary guarantor of Israel and the source of most of its arms. The Soviet Union backed the shifting alliance of Arab states that sought to dislodge the Israelis, or at least to establish an independent homeland for the Palestinians, who had been displaced when Great Britain's withdrawal led to the creation of the Israeli state after World War II.

The Arab–Israeli wars were not primarily wars of religion or of ethnic animosity, though elements of both are evident in the relations between the two peoples. At heart the dispute has always been territorial. The World Zionist Congress, established in 1897, had long sought a permanent homeland for the Jews, a displaced people that had suffered persecution for nearly two thousand years in virtually every country of Europe. In the Balfour Declaration of 1917, the British government endorsed "the establishment in Palestine of a national home for the Jewish people." This commitment grew more urgent and also more poignant in the wake of the Holocaust in World War II, and in the light of a flood of Jewish refugees from Eastern Europe. After the war, however, when Great Britain had to retreat from its earlier role of world leadership, it could offer the Jews no more than residential rights in a Palestinian state that would soon be granted its independence. The United Nations then took over and proposed to divide Palestine into Jewish and Arab states. But it could not prevent the outbreak of violence. On 14 May 1948, the day the British mandate ended, the Jewish state of Israel was proclaimed. Immediately, Lebanon, Syria, Jordan, Egypt, and Iraq attacked, only to be repulsed by their new and unwanted neighbor. Early in 1949, armistices were finally signed that provided for Palestinian enclaves, but nothing was resolved.

The second phase of the Arab–Israeli conflict was precipitated by the Suez Crisis of 1956. When Egyptian President Gamal Abdel Nasser nationalized the Suez Canal in July 1956, Britain, France, and Israel planned secretly to recover control of that important international waterway and to discipline a troublesome Arab state. By prior arrangement, the Israelis would attack through the Sinai Desert and threaten the Canal. The British and French would then intervene as "peacemakers," to secure and internationalize the Canal. It took the British and French four months to assemble the 90,000 troops and the extensive naval and air forces needed for the operation. The Israeli force of 45,000 men, 180 tanks, 130 airplanes, and 150 guns was ready at the appointed hour on 29 October to face a roughly comparable Egyptian force in the Sinai Peninsula. In spite of tough resistance by the Egyptians, the Israelis finally secured their hold on the Peninsula, inflicting about 1,500 deaths on the Egyptian army for 181 of their own lives lost. They captured 6,000 prisoners and almost $50 million worth of military equipment. The British and French were equally successful. From distant bases such as Malta, a multifaceted airborne invasion was followed by an amphibious assault on Port Said, the coastal city at the Mediterranean entrance to the canal. The allies then drove toward the southern end of the canal. But the operations were plagued by problems and met with stiff resistance. Nonetheless, they did succeed in gaining military control of the Suez Canal with far less loss of life on either side than was experienced in the Sinai. The attackers were entirely unprepared, however, for the worldwide political protest that arose over the operation. The British and French were forced by international and domestic pressure to withdraw entirely; the Israelis gave up all their conquests in the Sinai save the Gaza Strip, a Palestinian enclave, and the area of 'Aqaba. Here was yet more evidence that international public opinion was becoming a strong force on the battlefield.

The next chapter in the continuing Arab–Israeli conflict was dramatic and decisive. More than any other operation, it secured a reputation for the Israelis as the acknowledged masters of conventional warfare in the second half of the twentieth century. A further deterioration in the already-strained relations between Israel and her Arab neighbors had begun in late 1966 and continued through the spring of 1967. Increased shipments of Soviet arms, mobilization of military forces in several Arab states, hostile moves by Egypt in the Sinai, and increasingly combative rhetoric convinced the Israelis that an attack by another Arab alliance was imminent. Outnumbered 2–1 in personnel, 2½–1 in planes, and 3–1 in tanks, the Israelis launched preemptive air strikes, first into Egypt and then into Jordan and Syria. Their targets were aircraft. By the end of the first day, 5 June 1967, two-thirds of the Egyptian air force had been destroyed, most of it on the ground. Those planes that did get into the air, including Soviet-provided MIG fighters, proved to be no match for Israel's French and American planes, which were piloted by better-trained Israeli airmen. When Syria and

Jordan entered the combat, their combined total of 89 planes was wiped out entirely, along with 17 Iraqi planes that ventured within flying distance of Israel. During the entire six days of the war, Israel claimed to have destroyed 418 enemy planes at the cost of 27 of her own.

With air cover secured, Israel moved to the next phase of its planned operation, armored assaults launched into the Sinai, Jordan, and Syria. Though meeting firm resistance on all fronts, the Israelis nonetheless prevailed. By using close air support, *Blitzkrieg* tactics, and brilliant planning against superior numbers and equipment, they achieved simultaneous victories on all three fronts. When a cease-fire was finally arranged on 10 June 1967, six days after the Israeli invasion, Israel was in control of the Sinai Peninsula, the Jordan Valley, and the Golan Heights, all of the strategic objectives that it had sought. It had suffered 689 deaths while inflicting more than 14,000 on its combined Arab enemies. It lost 100 tanks to 954 for the Arabs. Only two Israeli planes had been lost in aerial combat, as against fifty of the enemy's. By any criteria this was the most rapid, lopsided, and decisive conventional war since the German army swept through France and the Low Countries in 1940. It suggested that high-tech conventional war, when both sides staked all on sophisticated, expensive weaponry, could be both quick and decisive. After six days the Arabs simply did not have the resources, let alone the will, to resist Israel's onslaught.

In the Six-Day War the Israelis developed some admiration for the Jordanians, who had defended the Jordan Valley with real heroism against overwhelming odds. For the Syrians and Egyptians, the Israelis had little more than disdain, an opinion that would come beck to haunt them. In 1972, frustrated by the failure of diplomacy to roll back Israel's territorial gains of 1967, President Anwar Sadat of Egypt determined to roll the military dice once more in an attempt to restore Arab pride and self-confidence. He also hoped to return the Middle-East problem to international prominence. He and Syrian President Hafiz Assad, with whom he had set up a loose United Arab Republic, planned a surprise attack on the Israelis to take place in 1973. They launched coordinated attacks in the Sinai and the Golan Heights to catch the Israelis off guard and to regain some of their lost territory. They intended to dig in for a war of attrition, relying on superior numbers and world public opinion to secure their gains.

The surprise attack was launched on 6 October 1973, the Jewish Yom Kippur holiday from which the war took its name. The Egyptians successfully crossed the Suez Canal with the help of Russian bridging equipment and seized Israeli defensive positions on the East Bank. Israeli air power was less effective than in the 1967 war because the Egyptian offensive was covered by effective Soviet ground-to-air missiles. Meanwhile, Syria launched an equally rapid and intensive armored assault on the Golan Heights, severely pressing the outnumbered Israeli forces there but failing to achieve the massive breakthrough it had sought. In both theaters, fierce resistance by front-line defensive forces gave the Israelis time to recover from their initial reverses and to bring up reinforcements. Superior armored tactics and equipment then prevailed in both the Golan Heights and the Sinai. Soon the Israelis were employing accurate intelligence, electronic countermeasures, American smart bombs, and squadron-sized attacks to destroy the enemy's anti-aircraft missiles and open up avenues of counterattack. This culminated in a crossing of the Suez Canal and the capture of new Egyptian territory. By the time a cease-fire finally took hold on 24 October, Israel had made further territorial gains in both Egypt and the Golan Heights.

The Israelis had triumphed once more, but the costs were enormous on both sides. On the Golan Heights alone the Arabs lost 1,400 tanks and the Israelis 250. The Israelis admitted losing 115 aircraft in the war, but U.S. estimates ran as high as 200. The expenditure of ammunition on both sides was prodigious, and expensive weapons like surface-to-air missiles and smart bombs were used up at a rate that taxed the resupply capabilities and inclinations of both the United States and the Soviet Union. Once more high-tech war had demonstrated that it cannot last long. And once more, political objectives had proved to be more important than battlefield objectives, for the world remembered the early Arab successes more than the final Israeli victory.

Israel unleashed its potent military force yet once again in 1982. It invaded Lebanon in a punitive raid designed to root out Palestinian terrorists operating there and to destroy their sanctuaries. Withdrawal was swift, but the Israelis retained control of a demilitarized zone at the southern end of Lebanon, a buffer against terrorist raids in the future. The Israelis also used their potent forces in a conventional attack against an Iraqi nuclear plant, which the Israelis claimed was designed to produce nuclear weapons. The world condemned this unilateral attack within the territory of another country, but no international sanction could restore the Iraqi plant. Subsequent behavior by Iraqi President Saddam Hussein in Iran, Kurdistan, and Kuwait was to lend weight to the Israeli argument that aerial bombardment of the plant was an act of prudent self-defense. Whatever the political and diplomatic wisdom of Israeli policies, none doubted that this small country of 4.5 million people packed more conventional war–fighting ability for its size than any other nation since World War II. The Israelis' effectiveness was based on what was virtually a state of permanent mobilization, on a high rate of per capita military spending, on ample supplies of sophisticated Western arms, on brilliant strategic planning, on heroic combat by its battle-hardened troops, on the development of innovative armored tactics, and on an unflinching will to use force to ensure national survival in a hostile environment. Many other countries, especially small ones, have modeled their military on Israel's, but none have achieved her level of effectiveness.

OTHER CONVENTIONAL WARS

Elsewhere in the world, especially in Africa and Southeast Asia, the civil wars and wars of national liberation that characterized the 1950s and 1960s

THE ABRAMS TANK was one result of the armor race between the United States and the Soviet Union in the 1970s and 80s. The model shown here, the M1E1, features the 120-mm gun that distinguished the tanks produced after 1985. (U.S. Army, *1984 Weapons Systems,* U.S. Army, Washington, 1984, p. 2)

continued through the following two decades—but with a difference. Increasingly after the Americans had developed and used high-technology weapons in Vietnam, the countries of the Third World turned to the burgeoning international arms bazaar to support their military objectives. As the volume of international arms sales grew, the percentage of those purchases made by Third World countries soared, from 44 percent of the total in 1960 to 71 percent in 1985. Where NATO forces spent only 1.3 percent of their combined military budgets on international arms purchases, and the Warsaw Pact countries spent 1.2 percent, the countries of the Middle East spent 27 percent, and those of Africa spent 38 percent. Arms from abroad were more than ever before a desiderata of war in the Third World.

The reasons for this trend are many. Industrialized countries like the United States developed such weapons in the first instance for their own use, not only for conventional wars among themselves, but also for police actions and brush-fire wars that they anticipated in the less developed world. To subsidize their own development costs, they were more than willing to sell these weapons, especially the obsolescent versions, to anyone who would buy. Second, many of the wars in the Third World were proxy wars in the greater East–West struggle; the superpowers often armed their client states with the latest weapons as a way of advancing their own cause in the world and as a way of testing them. Third, the high-tech weaponry, especially the planes and missiles, became status symbols for petty dictators and insecure governments that wanted to impress their neighbors or to cow their own populations; it

AFGHAN GUERRILLA DEMONSTRATES THE USE OF THE U.S. STINGER MISSILE, which
helped the rebels drive the Soviet Union from Afghanistan. Thus, the revolution in electronics
has helped place in the hands of individuals compact weapons systems that can engage the
huge conventional forces fielded by the industrialized states. (*Soviet Military Power: An
Assessment of the Threat, 1988,* Department of Defense, Washington, 1988, p. 25)

served purposes of internal control as well as external defense or aggression.
Finally, these weapons really were force multipliers on the battlefield, as Israel had demonstrated so clearly by its use of planes and tanks in its series of
encounters with numerically superior Arab forces.

For all these reasons, arms spread throughout the world in the 1970s and
1980s at an unprecedented rate and made the warfare that ensued more
bloody and costly. In the early 1970s, for example, war flared up again between India and Pakistan. The Bengalis of East Pakistan, separated from
the Pakistani central government by geography, language, religion, culture,
and ethnicity, declared their independence in 1971. India, sympathetic to
the East Pakistanis, was inevitably drawn into the civil war, since its territory separated the two combatants. The peak fighting in what became a new
India–Pakistan war took place in November and December of 1971. Both
sides suffered heavy casualties on a battlefield supplied with weapons by
the Soviet Union, China, the United States, Saudi Arabia, Jordan, Libya, and
Iran. The Indian air force attacked every major city in West Pakistan; its
navy devastated Karachi harbor, and its army experienced similar victories
on land. A cease-fire declared on 17 December was followed by the resigna-
tion of West Pakistani President Yahya Khan and the transformation of East
Pakistan into the new independent state of Bangladesh. One-and-a-half million people are estimated to have died in this short and deadly encounter, a

grisly toll caused by the effectiveness of the new weapons and the brutality of the fighting between these bitter enemies.

Even that death count was exceeded, however, by the grim events that took place in Cambodia later in the decade. Caught up in the Vietnam War and radicalized by the American incursion of 1970, Cambodia fell into a dizzying succession of civil conflicts and external interventions that left the country in turmoil and its people in devastation. The U.S.-backed government of Lon Nol was overthrown by a communist insurrection in 1975. Over the next five years the Cambodian people experienced forced collectivization, mass murder at the hands of their own government, military intervention by Vietnam, mass migration, and a political instability that was still not settled at the end of the 1980s. The price in human suffering was unimaginable; the cost in lives ran as high as 2 million. Not all of these were victims of conventional war by any means, but they nonetheless attest to the power of the state to kill people—its own and others—at a sickening pace in the modern world.

Africa witnessed more than its share of the same kind of carnage in the 1970s and the 1980s. In Burundi, Tutsi and Hutu tribesmen fought for political control of the new republic, killing as many as 100,000 in 1972 alone. In Ethiopia, the military regime that overthrew emperor Haile Selassie, the longest-sitting national monarch at the time of his deposition, found itself faced with a secessionist movement in Eritrea, a territorial dispute with Somalia in Ogaden, and a drought that devastated an already-reduced population. To compound her economic and political woes, Ethiopia suffered more than half a million war deaths between 1974 and 1986. In nearby Uganda, military strongman Idi Amin seized control and conducted a reign of terror against his own people that resulted in 300,000 deaths. Tanzanian troops and Ugandan exiles finally forced Amin from power in 1979, but the country still could not escape a period of civil disorders and guerrilla fighting that cost another 100,000 deaths in the first half of the 1980s. Similar disorders accompanied the early years of independence of many other Third World countries, especially in Africa, and though the death tolls were not always as high, they were nonetheless sobering. Cruelty, barbarism, ferocity, and even genocide—including against one's own people—were not unprecedented in war, but the new weapons of the late twentieth century seemed to have put into the hands of guerrillas, dictators, partisans, rebels, and freedom fighters an unprecedented power to inflict death and destruction.

Nor was the phenomenon limited to the Third World. The Soviet Union found itself drawn into its own Vietnam in 1979, when the imminent collapse of its puppet government in the crucial buffer state, Afghanistan, made it risk intervention. Aside from the political costs of a U.S. boycott of Soviet trade and the 1980 Olympics, and the enmity of Muslims around the world, the Soviet Union also found that the military cost was high. Afghan tribesmen, the Mujahideen rebels, used American-supplied arms, especially deadly hand-held stinger surface-to-air missiles, to overcome the Soviet advantages in arms and numbers and to conduct a war of attrition similar to that experienced by the Americans in Vietnam. Soviet President Mikhail Gorbachev finally negotiated a Soviet withdrawal, in part because of his new policies of *glasnost'* (openness) and *perestroika* (restructuring) and in part because the Soviets were paying too high a price internationally and domestically for the war. But curiously, the same arms, tactics, and nationalist zeal that allowed the rebels to drive out the Soviets proved inadequate to topple the Soviet-backed government left behind, leaving a stalemated civil war and yet another country torn by conflicts that cannot be terminated.

Great Britain also felt the power that the new weaponry has placed in the hands of Third World powers. In April 1982, the military government of Argentina attacked the British-held Falkland Islands in the South Atlantic off the Argentinian coast. Long desirous of acquiring the islands that had been claimed for Britain in 1765 and reclaimed in 1833, the Argentinians risked war, apparently on a calculation that the British would be hard pressed to respond militarily across thousands of miles of Atlantic Ocean. The British did respond, however, and sent a major amphibious force to retake the islands. Weight of numbers and material finally prevailed, even across stretched lines of supply, but not before some of the high-tech weaponry that Argentina had bought on the international market had exacted a painfully high price. A French Exocet air-to-surface missile struck the British destroyer HMS *Sheffield* on 4 May, before the ship's sophisticated electronics equipment could give adequate warning for defensive measures. Though the missile did not explode, its fuel ignited and set off a fire that forced abandonment of the ship with the loss of twenty lives. Modern British weapons and equipment exacted an even higher toll from the Argentinians before the short and violent confrontation ended on 14 June. At least 746 Argentinians died (compared with 255 Britons), and Argentina lost the cruiser *Belgrano,* several smaller ships, and perhaps 100 aircraft. The failure of this gamble toppled Argentina's oppressive military government, chastening others who might seek to roll the iron dice in the modern world.

If one conflict can be said to capture the essence of conventional war in the second half of the twentieth century, it is the Iran-Iraq War. Launched on 2 September 1980 by Saddam Hussein of Iraq, a Sunni Muslim Arab country, against the revolutionary government of the Ayatollah Khomeini of Iran and his Shiite Muslims, it pitted superior equipment against superior numbers. In the course of the eight-year war, the Iranians fielded an army of about 2.75 million men, about 6 percent of its population; Iraq fielded a comparable percentage of its population (6.3) but never surpassed a million troops. Iraq's numerical advantage was in tanks (4,000 to 1,040), armored fighting vehicles (3,000 to 750), and combat aircraft (632 to 70). The contrast favored neither side, though Iran finally wore down the Iraqis in a fierce, bloody, but inconclusive ground war. Iraq responded with measured escalation of the conflict, using tear gas, mustard gas, and even nerve gas when faced with human wave assaults by Iran's fanatical fundamentalist Shiite revolutionaries. It conducted terror bombings of Iran's cities and finally attacked the lifeline of the Iranian war effort: petroleum exports. To each escalation, Iran responded in kind, turning the conflict into a bitter stalemate that ruined both countries economically and at times threatened to spread throughout the entire Persian Gulf region.

The course of the war revealed the indecisiveness of low-tech warfare in the modern world and the effectiveness of high-tech weaponry. In 1980, in pursuit of what he viewed as legitimate political, territorial, and economic objectives, Hussein had invaded oil-rich Iranian territory on the Persian Gulf coast and had then sued quickly for a settlement. Iran responded by declaring a Muslim holy war against its neighbor and by recapturing much of its lost territory in the course of 1980 and 1981. By the middle of 1982, Iranian forces were making their first incursions into Iraq, prompting the first use of gas warfare by the Iraqis. The following year, still besieged on the ground, the Iraqis used their superior air power to initiate first the "war of the cities," with an attack on the civilian population of Tehran, and then the "tanker war," with assaults on Iranian oil ships. There were finally five "wars of the cities," conducted at first by Iraqi airplanes, but escalating at last to the use of missiles by both sides, against Baghdad, Tehran, and other civilian population centers in both countries. The assault on petroleum expanded from Iraqi attacks on Iranian tankers to assaults on Iran's critical Kharg Island and Sirri Island facilities in 1985 and 1986, respectively, and finally to attacks on tankers belonging to third parties.

The last escalation finally brought the United States and some other countries to send naval forces in 1987 to protect neutral ships in the Gulf. This raised anew the specter of superpower confrontation. In fact the Soviet Union, the United States, China, France, and other countries had all been supplying arms to the combatants, some to both sides. Some countries considered this a proxy war in the East-West confrontation. Others simply wanted the revenue generated by selling their weapons. Some wanted to maintain a balance of force in the Middle East, hoping neither side would win the conflict. In the end, neither side did win. They accepted instead a cease-fire under the terms of United Nations Security Council Resolution 598. Out of economic and moral exhaustion, the countries returned essentially to the 1980 status quo ante, having expended more than a million lives and countless billions of their national treasuries. Sadam Hussein emerged from the conflict with the largest army in the Middle East and a huge war debt. Both of these played a part in his decision to invade oil-rich Kuwait in 1990. War, it seems, even disastrous and inconclusive war, begets still more war.

22

LOW-INTENSITY CONFLICT
AND PEACEKEEPING

The last chapter of this history of warfare and its relationships with Western civilization deals with two distinct topics, low-intensity conflict and peacekeeping. During the Cold War both of these activities came to have major worldwide significance for the first time. They also have other things in common. Both are now ways of applying force, threat of force, or a show of force to manipulate, adjust, or control political dispositions. But neither is an aspect of conventional or nuclear warfare. Yet both may be associated with, may be affected by, or may affect those very different forms of strategy and warfare. These two topics will be treated separately in this chapter.

DEFINITIONS AND OCCURRENCE

Current confusion about the nature of low-intensity conflict is demonstrated by the fact that official definitions differ (see table). But it is generally accepted that in low-intensity conflict (sometimes called "unconventional war" despite a possible ambiguity, since nuclear war is also "unconventional"), force, or threat of force, is applied without the sophisticated technology of nuclear strategy, and without modern conventional war's elaborate military administration, command structures, and mechanization. Two authorities on low-intensity conflict have described it as a new form of warfare. Theodore Shackley called his book on unconventional warfare *The Third*

TWO OFFICIAL DEFINITIONS OF "LOW INTENSITY CONFLICT" AND ONE UNOFFICIAL DEFINITION

The White House*	Office of the Chairman, U.S. Joint Chiefs of Staff[†]	Trevor Dupuy, Military Historian[‡]
While high intensity conflict has been successfully deterred in most regions of primary strategic interest to the United States, low intensity conflicts continue to pose a variety of threats to the achievement of important U.S. objectives. As described in last year's report, low intensity conflict typically manifests itself as political-military confrontation below the level of conventional war, frequently involving protracted struggles of competing principles and ideologies, and ranging from subversion to the direct use of military force. These conflicts, generally in the Third World, can have both regional and global implications for our national security interests. . . .	*Low intensity Conflict:* Political-military confrontation between contending states or groups below conventional war and above the routine, peaceful competition among states. It frequently involves protracted struggles of competing principles and ideologies. Low intensity conflict ranges from subversion to the use of armed force. It is waged by a combination of means employing political, economic, informational, and military instruments. Low intensity conflicts are often localized, generally in the Third World, but contain regional and global implications. Also called LIC.	Low-intensity conflict involves sporadic or limited hostilities between military forces which may, or may not, be conventionally organized, in which there may be prolonged lulls between combat encounters, or in which the weapons systems of one or both sides are limited in number, or in type, or in sophistication.

* *National Security Strategy of the United States* (Washington, D.C.: G.P.O., The White House, 1988), p. 34.
† Joint Chiefs of Staff, *Department of Defense Dictionary of Military and Associated Terms* (Washington, D.C., G.P.O., 1, December 1989), p. 212.
‡ Trevor Dupuy, Curt Johnson, and Grace P. Hayes, *Dictionary of Military Terms: A Guide to the Language of Warfare and Military Institutions* (New York: H. H. Wilson, 1986), p. 38.

Option. Professor Sam Sarkesian entitled his work on the same topic *The New Battlefield.*

Low-intensity conflict occupies the lower end of a spectrum of warfare that is arranged on the basis of the degree of destructive potential, weapons-sophistication, amount of organized manpower, extent of state financing, intensity of effort, and impact. Low-intensity conflict employs smaller forces, less formal organization, and fewer resources than do the other two components of the spectrum, namely, nuclear strategy and conventional warfare. Perhaps the most obvious characteristic of low-intensity warfare is the nature of the weapons it employs, which contrasts sharply with nuclear technology, and also with the heavy military equipment of modern conventional war. Low-intensity weapons are typically bombs, handguns, and hand-held rockets. Its tactics are sudden forays from concealed bases and quick withdrawals, assassinations, kidnappings, and armed robberies committed to obtain funds. These are the methods of both guerrillas and terrorists.

Though guerrilla insurgency and terrorism differ in some respects, these two forms of violent confrontation have enough in common to suggest that they are both aspects of the same phenomenon, low-intensity conflict. In insurgencies the operators are more numerous. They are usually located in a known base or bases from which they launch persistent, though sometimes sporadic, campaigns aimed at achieving a political objective. Terrorist groups, on the other hand, are smaller, sometimes only a few persons or even a single individual. Their bases are likely to be secret. Their operations are primarily separate incidents intended to terrify, or to disrupt, with less expectation of immediate results in the form of the overthrow of the enemy.

Insurgents and terrorists may both obtain financial support from fellow citizens who can be persuaded or coerced into giving, from foreign sympathizers (who may not know that violence is intended), or from foreign governments. Thus Libya and Syria financed Palestinian terrorists against Israel, the U.S.S.R. bankrolled Syrian activities and also guerrilla movements in Central America and elsewhere, and the American CIA financed Contra rebels against the Sandinistas in Nicaragua, and also the right-wing forces that overthrew and assassinated President Salvador Allende in Chile. Low-intensity conflict may occur independently or in conjunction with conventional operations, like the partisan resistance in occupied territories during World War II.

F. M. Mickolus and two collaborators in his book *International Terrorism in the 1980s* calculated the number of incidents of terrorism between 1968, when they say terrorism began a dramatic increase, and 1988. Excluding acts of a criminal nature with no known political objective, and not counting peaceable protest demonstrations, they estimated 3,856 key domestic occurrences. Among 24 distinguishable types of incident, explosive bombings total 1,364; assassinations, 468; threats, 401; kidnappings, 335; armed attacks with mortars, etc., 340; and aerial hijackings, 61. Sixty-one percent were in the Middle East, with lesser numbers in West Europe, Latin America, and Africa. Mickolus and his colleagues identified forty-four terrorist groups. They said that guerrillas are especially active in Africa, where they have operated against the governments of nineteen countries, and that they have been effective in Angola, Mozambique, and Zimbabwe. The African National Congress (ANC), with its headquarters in Zambia, has operated against South Africa from Mozambique, as the Pan-African Congress has done from Tanzania. Mickolus reported that terrorism increased in the 1980s and that in that decade it became more likely than in the previous one to culminate in bloodshed.

Two of the definitions in the table defining low-intensity conflict (p. 360) stressed that it tends to be "sporadic" and "protracted." On the other hand, low-intensity conflicts may differ from conventional wars in that they may often last much longer. They differ from conventional wars also in other important respects. Their operators are usually, though not always, civilians or paramilitary personnel, rather than members of a regular military establishment. But regular military personnel can be used to oppose them, preferably after special training. Finally, and perhaps most important of all, low-intensity conflict lacks whatever legitimacy international law bestows on conventional military operations. This may be so even when it is undertaken in association with a conventional campaign.

Although we placed low-intensity conflict at the lower end of a spectrum of conflict intensity, this must not be taken to mean that it is always warfare "writ small." For, although it usually involves fewer operators than conventional warfare, and although it costs much less, its effect can be proportionately very great, out of all relation to numbers and cost. In a small state, the loss of life and the material damage that it can cause can be devastating. Low-intensity conflict's psychological effect can also hurt larger states; for while conventional war tends to draw people together in defense of the fatherland, in low-intensity conflict many in the attacked state may sympathize and even opt for concessions. It may thus be an important corollary of more conventional conflicts or of the Cold War nuclear confrontation.

TERRORIST ORGANIZATION

There are several reasons that this kind of warfare has been defined in a number of different ways. In the first place, we often know all too little about the organizations that practice it. Second, there is no consensus on how particular operators should be described. There is a much-quoted cliché, "One man's terrorist is another's freedom fighter." Third, the terrorist groups that are involved are very informal, as also are many guerrilla forces. Some of these consist of tiny cells operating quite independently of other cells. Thus the two cells of the Front de Libération de Quebec (FLQ), which in 1971 kidnapped Jasper Cross, a British trade commissioner in Montreal, and Pierre Laporte, a Quebec government cabinet minister (who was later murdered), did not know of each other's existence. Their relation to a higher command, if any, is still uncertain. Finally, some terrorist and insurgent groups are secret military wings of known political parties or organizations, but often relatively independent. For instance, the Palestine Liberation Organization (PLO), an exile political group headed since 1968 by Yasir Arafat, with its headquarters first in Lebanon and later in Tunisia, appears to have several clandestine terrorist subgroups that it controls only intermittently, if at all, and which it disclaims at will. Similarly, the Irish Republican Army (IRA) of today, which must not be confused with the Eire regular army, has a military terrorist wing known as the Provisionals (Provos).

Insurgent military guerrilla groups, although at first quite unmilitary in organization, may adopt a hierarchical structure of command nominally modeled on traditional military establishments. In some cases, when their cause moves toward conventional operations, and perhaps victory, they may become, or be absorbed into traditional military forces. Members of the earlier IRA, which fought against the British Army and the locally raised paramilitary force, the Black and Tans, to win independence for Eire in 1922, were taken into the new army of the Eire Republic. The Black-uniformed *Schützstaffel* (SS or storm troopers), originally Hitler's personal bodyguard, used in 1934 to suppress the *Sturmabteilung* (SA or Brown-shirts, street fighters who had won him political power), became shock divisions in the Second World War, fighting alongside the *Wehrmacht* (German Army) on both the Eastern and Western fronts. Two secret Zionist underground forces, *Irgun Tz'vai L'Umi* and *Haganah,* which were set up in the early 1920s to protect Zionist settlements from Palestinian Arab attacks, were absorbed by the Israeli Defense Force when the new state was declared in May 1948.

Low-intensity conflict organizations and operations, especially those we can class as "terrorist," are usually either covert, meaning that the identity of the agencies involved is concealed, or even clandestine, which means that the origin of the perpetrators is entirely hidden. Rural guerrillas base themselves in remote, often mountainous, areas or outside the country. Terrorists and urban guerrillas seek anonymity in crowed cities. As the immediate purpose of low-intensity conflict is often largely psychological, namely to gain a political advantage by the spread of concern and fear, so incidents carried out by clandestine organizations sometimes lead to multiple claims of responsibility by a variety of groups or individuals. Secrecy makes low-intensity cooperations difficult to counter.

Although insurgencies, and to a lesser extent terrorism, have been frequent in the course of history, their identification as a distinct form of warfare dates only from events in 1934–35 when the Chinese Communist leader, Mao Tse-tung, being unable to maintain a conventional campaign in his attempt to overthrow the Chinese Nationalists led by Chiang Kai-shek, withdrew his forces by the "Long March" to a remote part of North-West China. Mao's lectures, "On Protracted War," delivered in 1938, posited that an underdeveloped country like China could not hope to defeat an industrialized nation like Japan using only the techniques of traditional war. In its place Mao proposed "people's war," a three-stage, nonconventional war that exploited China's natural advantages of land and population to overcome Japan's edge in weapons and material.

MAO, PIONEER OF LOW-INTENSITY CONFLICT

Guerrilla war, the first but by no means the most important stage of Mao's people's war, entailed partisans who infiltrated enemy lines and operated behind them without bases to sabotage enemy facilities and to inflict casualties and damage on targets of opportunity. When the enemy sent out patrols to locate and destroy the partisans and their bases in the countryside, the partisans would engage in a second stage of conflict, "mobile warfare." They would attack enemy columns in hit-and-run strikes and then, before reinforcements could be brought up, would melt once more into the countryside. The result of these two stages, Mao believed, would be demoralization of the enemy, attrition of his forces, and public reaction against him in both the world community and among his own people. All these would set up the final stage of people's war, in which a depleted and demoralized enemy could be engaged by strengthened and emboldened partisans who would now resort to positional war. This phase of the conflict would become conventional war of a kind indistinguishable from conflicts between industrialized powers.

Drawing his insights from Clausewitz via Lenin, Mao understood that war was a continuation of politics by other means. He expected to win protracted war in the court of public opinion. His soldiers would trade space for time (a bargain in the case of China) and cultivate the minds and hearts of the populace. They would move among the people as "fish in the sea," winning them to the cause by propaganda and good works, and using the people in turn as a source of intelligence, material support, and passive resistance.

Mao and his theory, penned in the mountains of Kwantung province in 1939, would have been little noted had it not been for the course of events in China following World War II. He and Chiang renewed their civil war after the United States rid them of the Japanese invaders. In spite of massive U.S. aid to Chiang, Mao's forces steadily gained ground, pushing out from their mountain retreats, and finally in 1949 pushing the Nationalists from the continent. Chiang and his army fled to Taiwan and the neighboring islands and set up a government in exile. Mao established a stormy, but nonetheless permanent, regime on the mainland and set about exporting his brand of communist revolution and his brand of warfare. Though he had not actually followed his own prescription for protracted war in seizing control of China, he nonetheless offered to Third World nations a model for a war of national liberation in an agricultural and underdeveloped country. He thus set himself up as a potential competitor to China's sister in socialism, the Soviet Union. Mao's philosophy became the inspiration and model for many future low-intensity operations elsewhere, and especially those conducted in rural areas. He may properly be described as "the father of low-intensity conflict." As a result it is often assumed—quite incorrectly, as will become clear in the course of this chapter—that all low-intensity conflict is Marxist or communist.

While Mao was moving toward victory after the defeat of the Japanese and his break with the Nationalist leader, Chiang (with whom he had loosely cooperated during the war), extremist elements, around the world, in what came to be called movements of national liberation, began to adopt his tactics. The Atlantic Charter issued by Roosevelt and Churchill at their meeting in Placentia Bay, Newfoundland, in 1941, had promised peoples freedom from foreign dominations. We have seen that colonial peoples everywhere assumed that this should apply to them as well as to the victims of Nazi and Japanese conquests. This sometimes led to low-intensity conflict.

INDIA AND INDONESIA

The liberation of India in 1947 is usually hailed as a triumph for Ghandi's program of nonviolent protest. But in his political campaign before World War II Gandhi had verged on the use of low-intensity conflict, for instance, when, in the "March to the Sea" to defy the collection of taxes on salt manufacture, he had used sheer numbers to overcome a restrained military opposition. After the war the continuance of Gandhi's program of noncooperation led to a serious decline in law and order, and to the renewal of "communal" (that is inter-religious) friction to the point where India became ungovernable and the British conceded independence. What was perhaps even more important for the long-term future was that the partition of the subcontinent between a Hindu India and a Muslim Pakistan opened the way for a continuation of confrontation, including that over the possession of Kashmir, and that that confrontation was marked by low-intensity raids by tribesmen, as well as sporadic conventional campaigns.

Low-intensity conflicts thus surfaced prolifically in what came to be known after the Second World War as the Third World. Colonies that had been overrun by the Japanese were often unwilling to return to their earlier bondage after being liberated. Thus in 1945 the Indonesians proclaimed their independence and, with arms seized when the Japanese left, resisted British forces sent to pave the way for their return to Dutch rule. In this case, resistance quickly moved from low-intensity patterns to conventional. War-weary imperialists lacked the will to reassert imperial rule by force, especially as the United States was professedly anti-colonial. Indonesia won an agreement for the creation of a "union" with Holland in 1949. In 1956 that agreement was abrogated. Indonesians now celebrate their independence as a conventional military victory over the British and Dutch, but it had begun as a low-intensity conflict, and the Indonesians had triumphed because supplies of arms were available for them, and even more because they did not face the full weight of heavy components of British conventional arsenals and forces.

THE WORLDWIDE
ANTI-COLONIAL REVOLUTION

A most significant result of the attainment of independence by India and Indonesia was that they inspired other liberation movements throughout the overseas empires of the Western powers. Lacking the huge populations, and often also the degree of historic sophistication, of those two great new states, other colonial liberation efforts elsewhere were usually more prolonged. Frustrated by the slowness of political agitation to achieve the desired results, some impatient peoples resorted to violence, especially those that professed leftist tendencies and probably obtained support from communist powers. Where there were large populations of white settlers, the process of liberation was often more prolonged and more violent, being marked by raids, farm burnings, and murders. At times rebellion developed from low-intensity conflict into conventional operations such as, for instance, the war against the French in Indochina in 1949–54, and in Algeria, 1954–62. These were covered in an earlier chapter.

It is clear, then, that a primary cause of the growth of low-intensity conflict after Mao had shown the way, and after India and Indonesia had become independent, was the continuance of Western colonialism, which appeared to contradict the principles of democratic self-determination and self-government for which World War II had ostensibly been fought, and the subsequent violent reaction of colonial peoples to it. However, it would be wrong to assume that it was low-intensity conflict alone that subsequently won freedom for a large part of the world's peoples. Important simultaneous contributing factors were the sympathy of liberal-minded people in the West, and the contemporary existence of the Cold War which fueled ideas of liberation, and in some cases, also provided financial and technical aid for dissidents and weakened the will of the colonial powers.

POST-LIBERATION LOW-
INTENSITY CONFLICTS

Liberation did not bring an end to low-intensity conflict in the Third World, which had now come to include many newly emerged states. In fact, civil disorder often became more frequent than it had been under imperial rule, or even during the struggle for independence. One reason for this was that when they carved out their empires, the imperialists had often ignored tribal, linguistic, and cultural boundaries. The new states followed the old colonial divisions, often grouping together in a single state peoples who were ancient enemies. Once imperial control was removed, old hatreds bred new conflicts, and these would often be low-intensity conflicts.

The Belgian Congo, for example, which had not been prepared for self-rule by education and training, immediately began to disintegrate when the Belgians withdrew in 1960. It relapsed into tribal anarchy and disorder and drew in foreign mercenaries. We shall see later in this chapter that this outbreak of chaotic low-intensity conflict in the Congo called for peace-keeping action by the United Nations.

Tribalism had been a primary cause of the Mau Mau rebellion in Kenya toward the end of British rule, a movement largely restricted to the Kiu-Kiu people. In Zimbabwe, the former British subdominion of Southern Rhodesia, the intense political rivalry between Prime Minister Mugabe's National Union party, which represented Shona interests, and Joshua Nkomo's Nde-bele people, two African parties which had cooperated in the fight for inde-pendence, was in effect a renewal of the wars between the Mashona and Matabele at the end of the nineteenth century, before the British took con-trol. It now led to incidents of farm burning, that is, to low-intensity conflict. Although accusations of the continuance of tribalism in the emergent states in Africa are, of course, often inspired by opponents who seek to disparage, it is nevertheless clear that the inexperience of African peoples in coping with the working of modern national democracy was one cause of the growth of low-intensity conflict after the Second World War.

Another example of this is Uganda, a prosperous, predominantly black, model colony during British colonial days. When Idi Amin, a former British African NCO from a remote tribe, the KAKWA, seized power in 1971 and ruled brutally, partisan raids followed. In 1979, when those raids were aided by regular troops from neighboring Tanzania, Amin was overthrown and ex-iled. That did not stop the continuance of outrages, now apparently the work of the "liberators" who overthrew him and then engaged in sporadic violence against the population. Ghana, the first British dependency in tropical Africa that had become free (in 1959), and Kenya, both of which were at first regarded as very successful transitions to indigenous democracy, also eventu-ally saw declining law and order leading to some terrorism and insurgency.

SOUTHERN AFRICA

Low-intensity conflict came to be even more prevalent in southern Africa than it was in the tropical part of the continent. This was because of the intransigence of the white settlers in Rhodesia, and of the Portuguese in Mozambique and Angola, and also because of the Republic of South Africa's attempt to retain control of its League of Nations mandate for the former German South West Africa (now Namibia) despite a United Nations ruling that that should be surrendered. Low-intensity conflict in the Republic also developed in response to the policy of *Apartheid,* which denied nonwhite peo-ples political, social, and economic equality with whites. When the African National Congress, the African party fighting *Apartheid,* was outlawed in South Africa for refusing to disavow violent tactics, it established its head-quarters in Zambia and from there masterminded violent opposition, not only in South Africa but also in the former Portuguese colony of Angola and in the South African mandate now called Namibia. Guerrilla operations in

Southern Africa, and violence in the Afrikaaner Republic provoked by state-directed brutality and discrimination, but often also between rival black groups or tribes, spread widely during the 1970s and 1980s. South African security forces responded by operations in Namibia aimed at suppressing the South West African People's organization (SWAPO) and by raids into Angola to attack UNITA (the National Union for the Total Independence of Angola). Low-intensity conflict drew support for the African cause from Cuba, which sent troops to bolster UNITA. Cuba was supported in this by the U.S.S.R.

THE MIDDLE EAST

Terrorism has been even more common in the Middle East. Mickolus reported in 1989 that almost two-thirds of the known incidents since 1968 had occurred there. The Arabs in Palestine had been in revolt before the Second World War, and Zionist settlements had been furnished with arms by the British for their protection against Arab attacks. Although low-intensity conflict thus began in the struggle for Palestine, other factors prolonged it. Palestinian Jews fought with the British Army against the greater enemy, Hitler. They thus acquired military experience. When the British attempted to restrict the immigration of the survivors of the Holocaust to Palestine in an attempt to placate Palestinian Arabs, the Zionists transformed their defensive organizations, *Irgun* and *Haganah,* into underground resistance groups. Then, when attempts at partition had again failed, and the British were moving toward abandoning the mandate, the neighboring Arab states boasted that they would drive the Jews into the sea. Instead, a new Israeli Army composed of veterans from British wartime forces and former members of *Haganah* and *Irgun,* quickly defeated the Arab invaders in conventional campaigns.

This motivated many Palestinian Arabs to flee the country. These Palestinian refugees, often living in United Nations relief camps in Lebanon, in the Gaza Strip, which was administered by Egypt, and on the West Bank of the Jordan, which was ruled by the Emir of Jordan (the former British mandate of Transjordan), are now a fruitful source of recruits for Palestinian terrorist organizations that seek to win back their homeland by violent measures. When Yasir Arafat became PLO chairman in 1968, support for terrorism became its primary business. Although some of its subgroups are Marxist, the PLO is a nationalist enterprise. It operates throughout the world, striking at such targets as the Israeli athletes at the 1972 Olympic Games and allegedly the Pan-American jet that was destroyed over Lockerbie, Scotland. But many of its activities are attacks in Israel, frequently conducted from bases in neighboring Lebanon. The Israeli response has been to bomb, and even invade, Lebanon to strike at those bases. Finally, the PLO has now promoted a new form of insurgent activity in the Gaza Strip and on the West Bank (which were occupied by Israel after the 1977 war). In the

1980s, in the Intifada, Palestinian children were encouraged to throw stones at armed Israeli patrols and so to incite them to reply with rifle fire.

Low-intensity conflict is also epidemic elsewhere in the Middle East where the whole Muslim world is torn by feuds based on racial and dynastic rivalries, ideology, religious division, and the historic clash between city dwellers and desert peoples. The United States had supported the Shah of Iran, ruler of a Middle East Islamic non-Arab state as a barrier against the spread of Soviet influence into the oilfields of the Persian Gulf area. When the Shah was exiled by a revolution that brought the fanatical Ayatollah Ruhollak Khomeini to power, the distant United States became known in Iran as the "Great Satan." The occupation of the American Embassy in Tehran with the taking of its employees hostage from 4 November 1979 to 21 January 1981, was an extreme example of low-intensity conflict. The seizure of other Americans and Europeans in Lebanon by Iranian-affiliated terrorist gangs is another example of Mid-East terrorist low-intensity conflict.

A most deplorable example of low-intensity conflict in the area is that which plagues Lebanon, a state artificially created as a French mandate after World War I, and its capital, Beirut, once the jewel of the Eastern Mediterranean. The basic cause there is the existence of internal schisms. A traditional Christian ruling minority in the city is faced by the aspirations of the Muslim majority for a dominant role in government. Pressures from Syria, which has ancient claims to Lebanon, and from Israel, which aims to suppress PLO terrorism, have emasculated Lebanese sovereignty. Sporadic endemic fighting between the Muslims and Christians, with the traditionally semi-independent Druse mountain peoples as a third force, and fighting even between rival Christian groups, has destroyed a beautiful city and has demolished its once-great business prosperity, a tragic example of the impact of low-intensity conflict on normal ways of life and also of its frequent tendency to become something close to conventional warfare.

VARIOUS CAUSES OF CONFLICT

As insurgents often claim to be striving for the relief of social evils, low-intensity conflict is often assumed to be a left-wing revolutionary, or Marxist, phenomenon. Of course some low-intensity conflicts do have left-wing origins. Fidel Castro's rising against Fulgencio Batista in Cuba in 1959, the Sandinista insurgency that overthrew the Anastasio Somoza government in Nicaragua in 1979, and the Farabundo Marti de Liberación National (FMLN) that supports efforts to oust General Alfredo Cristiani in El Salvador, are all Marxist, as were disturbances in Thailand and Malaysia. Those in Oman are Marxist-Leninist, and the Guerrilla Army of the Poor in Guatemala is socialist. Such revolutionary movements often fought what Mao called people's wars.

On the other hand, the Contras, who pursue low-intensity operations against President Daniel Ortega's Sandinistas in Nicaragua, are clearly not leftist. Many other low-intensity conflicts are also not "revolutionary" in the colloquial leftist sense of the word. The Irish Republican Army (IRA) of today, one of the most persistent of terrorist organizations, recruits from the working classes but scorns intellectuals and leftist dogma. Britain's intervention, protecting the right of the Protestant majority in the North of Ireland to self-determination, has given the IRA cause a reputation as anti-imperialist or anti-colonial. Economic circumstances in that depressed area, where employment opportunities may be monopolized by "religious" groups, serve as an irritant, but the primary cause of the trouble in Northern Ireland is neither socioeconomic nor religious. It is nationalist. The IRA's opponents, hard-shell Protestants, include "Ulster Freedom fighters" who, like the IRA, also resort to low-intensity conflict tactics. They too are primarily nationalist, though in a more limited geographical sense, roughly for control of the former province of Ulster.

Other insurgent and terrorist campaigns can similarly be classified under headings other than Marxist or revolutionary—for instance, as secessionist, reformist, restorationist, reactionary, or even conservationist. The Bay of Pigs invasion of Fidel Castro's Cuba in 1961 has been called "restorationist." The Eritreans in Ethiopia, the Basques in Spain, and Quebec's FLQ are either nationalist or secessionist, depending on one's point of view. The Kurds' endemic struggle in Iran and Iraq, organized by the Patriotic Union of Kurdistan (PUK) to secure a homeland for a people long divided among those two countries and Turkey, is also either secessionist or nationalist. Thus revolutionary Marxism is far from being the sole, or even the most common, cause of the extraordinary rise of low-intensity conflict since World War II.

Low-intensity conflicts are sometimes fought in distant places or times to revenge ancient or traditional grievances. For instance, Armenians carried out terrorist attacks against Turkish diplomats in Canada in the 1980s to avenge massacres perpetrated in 1915. Canadian residents of Sikh origin are believed to have destroyed Air India Flight 182, which was blown up off southern Ireland on 23 June 1985; and they are said to have also on the same day planted a bomb on Canadian Pacific Flight 101, which was to have connected with an Air India flight in Tokyo. That bomb killed a Japanese baggage handler. The Sikhs were campaigning against the alleged repression of Sikhs in the Punjab and they were protesting against Indian armed intrusion into their sacred temples in search of Sikh terrorists.

EXTERNAL AGENTS

Low-intensity conflict may be seen on the one hand as an outlet for tensions, but it can also be viewed as a dangerous condition that might bring on a superpower conflict. It can be an aspect of the Cold War.

Foreseeing danger from that in the Middle East, the United States made a defensive treaty with Saudi Arabia to promote stability and protect access to its oil. It also financed the state of Israel as a Western outpost. (Important elements of the American electorate, of course, had a strong religious and kinship interest in supporting Zionist settlement there.) The United States also made a bilateral treaty with Iran and supported the Shah of Iran by supplying arms. That treaty lasted until, as we have seen, he was deposed by the Ayatollah Khomeini.

On its part, from about 1957–58 the Soviet Union supplied arms to Syria, Israel's most implacable enemy, and it made little or no effort to restrain Syria from interfering in the perpetual civil wars in Lebanon, or from attacking Israel. The Soviets also aided Egypt from about 1955, and in 1970 they made an alliance (which was short-lived) to train the Egyptian Army. Finally, as we saw in an earlier chapter, they moved to get a permanent naval base on the Red Sea in 1960 but lost it a few years later as a result of a local government upset.

On the other hand, the Soviets kept out of the United State's Iranian hostage crisis. They were also restrained in their conduct and utterances when the United States sent warships to patrol the oil routes through the Persian Gulf because tankers were being attacked during the Iran-Iraq War. However, when the communist government of Afghanistan was threatened by fanatically fundamentalist Muslim guerrillas (the Mujahideen, or Holy Warriors), the U.S.S.R. sent its army in on 27 December 1979. Pakistani and American support for the rebels prolonged the war until the Soviets withdrew in 1988. But that did not lead to the collapse of the government in Kabul. The guerrilla campaigns continued.

The Soviet Union had acted vigorously in Afghanistan because the low-intensity conflict there seemed too close to home. The situation was reversed in Central America and the Caribbean, where Third World states, many of which had been independent for over a century, were troubled by incipient revolution. Castro's overthrow of Batista in Cuba on Christmas Day 1956 gave the Soviet Union a proxy to stir up revolts in the vicinity of the United States. One of Castro's chief aides, an Argentinian doctor, Che Guevara, was captured and shot when campaigning in Bolivia in 1967. Castro also supported Marxist uprisings in Nicaragua (the Sandinistas) and in El Salvador, and also the Marxist Guerrilla Army of the Poor in Guatemala. The United States reacted strongly to those communist intrusions into the Western Hemisphere that it regarded as inimical to its security. President Kennedy challenged Secretary General Khrushchev in the 1962 Cuban missile crisis. When Chile elected a Marxist government in 1970, the American CIA allegedly aided a revolt in which President Allende was killed. From 1980 to 1989 it bolstered a shaky democratic government in Peru to resist the Marxist *Sendero Luminoso* (Shining Path) guerrillas.

THE CIA

At times it seemed as if much of Central and South America might become communist neighbors of the United States. Hence, when Cuba subverted Grenada in October 1983 and dissidents killed its prime minister, Maurice

Bishop, a socialist, the United States invaded the island. It also propped up the government of Guatemala against a long-continuing terrorist campaign. When Daniel Ortega's Sandinista government drove out Anastasio Somoza, an American protégé, in 1979 and then won an election in 1984, the United States reacted by supporting his opponent, alleging that the election was flawed. It also laid mines to obstruct the supply of Soviet planes and arms to Nicaragua by sea. The International Court of Justice judged this to be a breach of international law.

Daniel Ortega was accused of supporting rebellion in El Salvador and Guatemala, and President Reagan was determined to halt any spread of communism in the Americas. The CIA therefore subsidized the Contras, insurgents operating against Nicaragua from neighboring Honduras (without that country's consent). (It was alleged that planes hired by the CIA to send supplies to the Contras carried drugs from drug dealers to markets in the United States.) When Congress banned further government aid, Reagan's known attitude led some elements in the White House to make secret plans to raise money for the Contras by private subscription.

In December 1987 an American pilot, James Denby, was shot down over Nicaragua. Documents taken from him unveiled a bizarre story of White House involvement in arms trafficking. It was organized by National Security Advisers Robert McFarlane and Rear-Admiral John Poindexter and by National Security Council Staff member, Marine Lieutenant-Colonel Oliver North, through the agency of a retired Air Force general, Richard Secord, and a Middle East businessman, Albert Hakim. Despite public statements which the president had made that the United States would never made deals to secure the release of Americans held hostage in Lebanon, Poindexter and North had funneled arms through Israel to Iran, sponsor of Middle East terrorism, with a vague hope that that would facilitate the release of the hostages held by Iranian protégés. Payments made by Israel for the arms were deposited in secret Swiss bank accounts, supposedly to be used to buy arms for the Contras, who by law could no longer be supplied from U.S. government stocks. American efforts to counter hostile low-intensity conflict in the Middle East, and to support it in Central America, had thus led to the adoption of practices that matched those of their opponents in their immorality and illegality. However, the defeat of Ortega and the Sandinistas in February 1990, in the first free election in Nicaragua's history, seemed to support Reagan's assertion that the government lacked popular support. On the other hand, Ortega's willingness to accept the mandate of the people and turn the government over to his opponent belied Reagan's characterization of him as a totalitarian dictator.

TECHNOLOGY

Before further exploring ways by which low-intensity conflict may develop and can be met, we should first attempt to analyze some of the factors that caused this vast surge of a new level of warfare in the second half of the twentieth century. It was undoubtedly due in part to technological development, innovations introduced during the Second World War and after. That war accelerated the invention and production of arms of all kinds, including many that are suitable for the kind of operations undertaken by insurgents and terrorists. Resistance groups in occupied countries were supplied with many such weapons, and they learned guerrilla techniques. In some areas caches of weapons stored for resistance fighters were still available for insurgents after the war. Superpower tensions and other potential sources of conflict maintained a perpetual incentive for further development, and also inspired preparation for insurgency and terrorism.

Among the technical advances during and after World War II that furthered low-intensity conflict were explosives with vastly greater destructive capability, plastic explosives, plastic construction that made weapons harder to detect, miniaturization, simplification that made weapons easier to transport and operate, and radio- and pressure-detonators. Many of these led directly to suitcase and parcel bombs, land mines, and car bombs, all weapons ideal for guerrillas and terrorists. Other weapons that were especially suitable for low-intensity operations included automatic handguns, which were much more sophisticated than the primitive wartime Sten. Later came the hand-held rocket launcher. The Soviet SAM-20, a simplification of earlier SAM (surface-to-air missile) weapons, is ideal for terrorists since it can be used from somewhere in the vicinity of an airport, to shoot down an aircraft that is approaching to land. The equivalent American rocket launcher had proved even more effective, for instance, in Vietnam. Dissidents, often with the aid of larger countries, organized schools of instruction to teach weapons construction, maintenance, and use, and also low-intensity tactics.

Improved communications was almost as relevant to the growth of low-intensity conflict as were new weapons. Air Commodore Frank Whittle's jet aircraft, which first flew in 1941, and the jet engine used by the Nazis toward the end of the war, opened up a new chapter in the history of flight. A vast postwar increase of aerial intercontinental travel served to make terrorism a global operation. Groups in different countries that had little in common made contact with one another and shared training and experiences. At the same time the expansion of air travel provided terrorists with tempting targets in the form of airliners to hijack or destroy. The development of radio and television was also an assist for insurgent and terrorist operators, because they gave their operations instant publicity, which often was the operators' main objective. When outer-space satellites began to distribute television news pictures almost instantaneously to a worldwide audience, the publicity impact of a terrorist outrage was immeasurably increased. The news media, feeding on sensation, had thus become unintentional abettors of the terrorists.

SOCIAL AND ECONOMIC BASES

What we have said already about Mao's innovations and about the movements of national liberation suggests, however, that new technology was not
the primary explanation of the vast increase of low-intensity conflict that
came after the Second World War. It seems that we must seek a fundamental explanation in contemporary social, political, and economic changes in
addition to the anti-colonial revolution. These occurred on a global scale. It
was noted in an earlier chapter in this book that World War II reached a
"climax in a new degree of totality" in its objectives, methods, and impact.
In much of Europe and parts of Africa and Asia it brought loss of life,
material damage, and physical and social dislocation to an unprecedented
extent. Countries that were theaters of war suffered most; but with the long
reach of air power, and the long tentacles of European imperialism, the
war's material, moral, and political impact had spread infinitely wider than
those of other wars in earlier times. Belligerents had had to alter their normal mode of life, drafting their peoples to other tasks, many of which included resorting to, or becoming aware of, violence. Colonial troops, enlisted
or drafted by contending powers, traveled far from their homes and saw
material living standards, and also material destruction, much beyond anything that they had previously imagined. They carried back home ideas that
helped to fertilize the demands for national liberation, and also military
skills that could be used in low-intensity conflict.

The war also prepared the way for social changes that led to a rejection
of many traditional values and of authority, first in the West, and later
throughout the world. The second half of the century saw a vast increase in
several kinds of criminal activity in many parts of the world. There was a
departure from the standards of propriety and authority that had been established, apparently briefly and temporarily, during what we call "the Victorian age" in the second half of the nineteenth century. Noticeably, these
conditions, now sometimes attributed to drug abuse, began to occur before
the increase of narcotics became a significant problem. The loosening of
authority and of standards in advanced Western democracies and their dependencies was subsequent to, perhaps a consequence of, the war, of the
anti-colonial revolution of social and ethnic unrest, and of the immigration
of populations from the Third World into previously fairly static Western
societies.

The principles of the Western tradition, based on acceptance and
observance of the law, had been weakened by economic and social changes
that had decreased the moral influence of parents and the home, and by the
spread of concepts of individual equality and individual rights. Unrest was
a logical conclusion in a Western society that preached individualism but
practiced social discrimination. Not all countries were affected. Communist
totalitarian states for a long time ruthlessly imposed their version of law
and order. Mao's insurgency imposed strict modes of conduct on a society
that had had a long tradition of petty disorder. But low-intensity conflict

may possibly have been a natural consequence of a decline of traditional standards in those areas of the West and its dependencies when impatient and idealistic individuals sought to obtain advantages that society seemed still to deny. Eventually, that general trend also penetrated the Iron Curtain and affected the East.

LOW-INTENSITY CONFLICT
IN THE WEST

This explanation of the rise of low-intensity conflict may be supported by the fact that it appeared in some forms in the West where social and political changes were quite as significant as in the Third World, and where political and social violence seemed at times to be increasing. Some specific low-intensity conflict incidents in the West were, of course, spillovers from Third World violence, for instance, the assassinations of Middle East politicians in Paris and elsewhere, the bombing of a nightclub frequented by American servicemen in Berlin, the murder of a Turkish diplomat in Ottawa by Armenians, and the downing of the Pan-American plane in Scotland, and of the Air India plane off Ireland, and other incidents that have been mentioned earlier. Guerrillas and terrorists in the Third World thus attempted to strike at their enemies, including Western supporters, in the West itself; and among the migrants to the West there were individuals and groups that carried with them a desire to continue to pursue the causes that were agitating their fellows in their native countries.

But some examples of low-intensity conflict in the West had indigenous roots. There were ancient grievances, for instance, the desire of the Basques for freedom from Spain, which acquired a new impetus with the downfall of totalitarianism after the death of the Caudillo, Francisco Franco, and then led to clashes with the authorities and to terrorist incidents. More extensive was the earlier-mentioned campaign of the IRA to unite Ireland and expel the British forces sent there to maintain law and order when clashes between the IRA and Ulster Protestants led to a state of virtual warfare in Londonderry and Belfast. Finally, race riots proliferated in British cities in the 1970s after the huge postwar and postindependence migration of Caribbeans, Asians, and Africans seeking economic advantages not yet available in their homes. However, despite the influence of ideas about racial and other forms of equality that were endemic in the West, these tensions in Britain did not, apparently, lead to as extensive secret organizations for fomenting rebellion (except for the IRA) as occurred in Third World low-intensity conflict.

On the continent of Europe, clashes deriving from the overthrow of traditional privilege and authority, and especially from the quest for obtaining rights for youth, often at first took the form of worker-student alliances. On 1 May 1968, noticeably in the year when incidents of terrorism around the world began to proliferate, a huge rally in Paris protested reforms in education that would limit its availability. When a large crowd seemed to threaten

A SUSPECTED VIET CONG SOLDIER IS CAPTURED BY SOUTH VIETNAMESE REGULARS during the Viet Nam war. Thus, low-intensity conflict often pits irregular forces against conventional troops. (Jeffrey J. Clarke, *Advice and Support: The Final Years, 1965–1973*, United States Army, Washington, 1988, p. 180)

the Chamber of Deputies, which was considering the proposed innovations, it was met by force. One student was killed. The student-worker alliance was broken. Some students of a more extreme temperament then turned to clandestine organizations. From that time on, small "Red Brigades" flourished in Europe, and especially in Germany and Italy, many of them with leftist affiliations. They carried out bombings and assassinations, sometimes without expressed objectives but that were often vaguely related to general political or social change.

APPROACH TO CONFLICT IN THE UNITED STATES

In the United States, civil rights movements that sought to eliminate racial discrimination, also youth campaigns and efforts to end gender inequalities, developed in the 1960s and 1970s; they included similar manifestations of a revolt against traditional social structures such as those which marked the second half of the twentieth century elsewhere. Americans, however, were especially embittered by the Vietnam War, first because conscription for it fell more heavily on Blacks and later, in 1967, when the elimination of undergraduate exemption from the draft spread disaffection also among white youths. It should be noted in passing, however, that the Vietnam War also stimulated student protest in other countries, for instance in Canada and various European countries, that were not affected by the draft. The war itself was thus an irritant both in the United States and elsewhere.

The first American organizations whose activities inclined toward low-intensity conflict included the Students for a Democratic Society (SDS), set up by Paul Booth, Tom Hayden, Al Haber, and Stokely Carmichael in 1962, and the Black Panthers, formed by Huey Newton and Bobby Seale in 1966. In the summers of 1965, 1966, and 1968, overcrowded conditions in the Black ghettos in Los Angeles, Detroit, and other cities led to outbreaks of arson and looting that seemed to point to racial war. Meanwhile the Reverend Martin Luther King, Jr., an Atlanta preacher following on Ghandi's example, had been conducting a nationwide protest against racial discrimination for which he was awarded the Nobel Peace Prize in 1964. King's assassination on 4 April 1968, the motive for which has not yet been fully explained, led immediately to more riots, torchings, and curfews in American cities. Buildings within sight of the White House and the Capitol were set on fire.

When Malcolm X (Malcolm Little) began to preach violence blatantly in the early 1960s, and when the professedly extremist Weathermen were organized in 1968, some movements in the United States seemed to be moving toward revolution. Then, in May 1970 the Ohio National Guard, called out to oversee a peaceful protest demonstration at Kent State University, as a result of inadequate training for dealing with such situations, shot four students to death. In 1974, a hitherto unknown organization that preached revolution, the Symbionese Liberation Army (SLA), kidnapped a granddaughter of William Randolph Hearst, the newspaper magnate, and then recruited her to participate in a bank robbery which was recorded by the bank's security cameras and broadcast on national television. Many wondered whether the SLA was a localized affair or the surfacing of a nationwide conspiracy. These incidents brought a general realization that it was time to consider whether the United States was heading for outbreaks of low-intensity conflict such as had shaken much of the rest of the world, and if so how they should be met.

Some time elapsed before the nation was able to breathe more easily. The end of the war in Vietnam, following American withdrawal in 1973, coupled with the ending of conscription, removed some sources of unrest. The gradual elimination of Jim Crow laws, and other forms of racial discrimination, also contributed. By the late 1970s the possibility of increasing low-intensity conflict in the United States had faded away. Like most other countries in the Western world, the United States had been able to contain it.

REMEDIAL ACTION

Nevertheless, low-intensity conflict elsewhere still remains as a worldwide problem for which a more permanent answer remains to be found. Clearly an elimination of justifiable basic grievances is the ideal solution. This calls for effective national political and international diplomatic action. But in some cases the differences that produce conflict seem irremediable, for instance, the rival aspirations of Ulster hard-line Protestants and those of pro-Irish nationalists; and, for instance also, the differences between those Israelis who want all the historic territories of biblical times and the Palestinian exiles in a new diaspora (dispersion or exile) who want their historic homeland.

A primary requirement for dealing with low-intensity conflict is, of course, good intelligence. But experience has shown that conventional military force can have only limited success in following up with measures to deal with low-intensity conflict situations. Israeli air and ground attacks on PLO bases in Lebanon, and even a major invasion of that country, have not eliminated the PLO menace. Similarly, a 1986 American air strike against a suspected chemical warfare plant in Libya, which was apparently also designed to achieve what has been called "decapitation," that is, the killing of the head of government, Moamar Kaddhafi, failed in that latter objective, but nevertheless seemed to cool his enthusiasm for sponsoring international insurgencies. An Israeli commando-type air strike in 1976 did free hostages held in an airliner in the remote Entebbe, Uganda, airport. But the deployment of the British Army in Ulster, and the patrols of the Israeli Army in the Occupied Territories, while they may have held those situations in check, have also brought on more violence; and they have not put an end to the terrorism and the insurgency. Similarly, the deployment of units of the Indian Army in Sri Lanka to aid that country to control dissident Tamils has been accompanied by further outbreaks of violence. The United States's approach to the low-intensity conflict problem has been to train Special Forces, successors of Commando units first introduced in World War II, but actually also heirs of a long historic tradition of light infantry forces. It remains to be seen how far they can succeed in this very different kind of conflict.

International measures to deal with some of the problems of low-intensity conflict include stepped-up security at airports everywhere, but incidents still occur through lapses in control. Another international countermeasure has been diplomatic negotiation to secure international cooperation against sky-jacking. On 26 January 1973, the International Civil Aviation Organization (ICAO), based in Montreal, announced that, as thirty states had now ratified conventions for the suppression of hijacking, those regulations would go into force. Further ratifications including that of Cuba, Mecca of many hijackers, followed later. Incidents still occur but seem to be declining in frequency. It may be that in this as in many other areas low-intensity conflict can be contained. But it still remains a problem to the solution of which the military must contribute.

PEACEKEEPING

Peacekeeping is the use of neutral forces to prevent the outbreak of or resumption of war.

At their meeting at Placentia Bay, Newfoundland, in 1941, President Franklin D. Roosevelt and Prime Minister Winston Churchill had talked of the need for "a wider and permanent system of general security" after World War II. Article 42 of the United Nations Charter, drafted in 1945, authorized the organization to take "such action by air, sea, or land forces as may be necessary to maintain or restore international peace and security"; and Article 47 provided for a military staff committee to advise the Security Council and the secretary-general on such matters. The Charter gave primary responsibility for initiating military action for collective security to the Security Council; but its resolutions on substantive matters needed the unanimity of the five great powers, Britain, China, France, the U.S.S.R., and the U.S.A. But in the circumstances of the Cold War, the great powers' vetoes made effective collective action to enforce peace by those means very unlikely. Until 1990, the sole occasion on which the United Nations endeavored to oppose aggression by this procedure was in 1950, when North Korean forces invaded South Korea. It was possible to act on that occasion because the Soviet Union was boycotting the United Nations and so could not cast its veto. The Korean War experience thus provided no hopeful precedent for a system of collective security to keep the peace.

UNITED NATIONS COMMISSIONS

However, in 1947 two developments had pointed to another way in which military action can help the United Nations to limit potential causes of war. In that year, on the recommendation of the Security Council, the U.N. Secretary-General, Trygve Lie, appointed a consular commission to be provided by those members of the Security Council which had consulates in the vicinity to report to him on problems arising from the transfer of sovereignty to Indonesia from the Dutch. He also formed a special committee on the Balkans to deal with similar problems relating to Greece and its neighbors. The Security Council members who furnished personnel for both of those bodies employed members of their armed forces to assist in matters where military expertise was needed. In each case, however, those service personnel worked as members of their national delegations. They were not directly under the Secretary-General.

The following year, 1948, the Security Council established the United Nations Supervision Organization (UNTSO) to assist its mediator and a truce commission to supervise arrangements in Palestine after the creation of the State of Israel within its borders. In this case military officers from the countries who formed the commission served under the Secretary-General. UNTSO has since carried out a number of related tasks, including supervision of a cease-fire after the Arab-Israeli War in June 1967. Because these and other armistices have not yet led to a peace treaty, UNTSO still exists. By 1985 eighteen different countries had provided officers for it, including the United States from the beginning in 1948, and the Soviet Union from 1973.

UNTSO became the prototype for more United Nations Observation Groups. The United Nations Military Observer Group in India and Pakistan (UNMOGIP) was set up in the same year, 1948, to supervise a cease-fire between India and Pakistan, which were contesting possession of the State of Jammu and Kashmir. It, too, is still extant because that issue has also not yet been settled. A second India-Pakistan Observer Mission (UNIPOM) operated for six months in 1965–66 after the two countries had gone to war again. UNIPOM operated on borders where UNMOGIP did not function.

UNITED NATIONS OBSERVERS

Meanwhile, in 1958 the United Nations Observation Group in Lebanon (UNOGIL) had been created to ensure that there was no illegal infiltration of arms or personnel across the Lebanese borders. It thus supplemented UNTSO. In 1963 the United Nations Yemen Observation Mission (UNYOM) was appointed to observe and certify the implementation of a disengagement between Saudi Arabia and the United Arab Republic. It functioned for fourteen months. A mission of the representative of the secretary-general in the Dominican Republic (DOMREP) worked for seventeen months from May 1965 to report on breaches of a cease-fire between two de facto authorities in that country.

U.N. observer forces consist of officers who are unarmed. They have ranged in size from two officers in DOMREP, to three hundred in UNTSO. Observer work is often tedious and lonely, and there have been fatalities through accidents, or in action. One of the most publicized instances was the kidnapping of U.S. Marine Corps Lieutenant Colonel William Higgins, who was in charge of a U.N. truce supervisory team of officers. He was kidnapped by terrorists on 17 February 1988, and on 12 December 1988 he was reported to have been "executed."

Other international observer groups have been established outside the United Nations framework. In 1963 the Geneva Conference that negotiated the cessation of hostilities in Indochina when the French withdrew set up international control commissions for the supervision of the terms of the agreement in each of the three successor countries, Vietnam, Cambodia, and Laos. The commissions were to arbitrate conflicting claims by communists located to the north and the regimes left by the French in the south. Canada, India, and Poland provided diplomatic personnel of ambassador rank, and military officers, to constitute separate commissions for each of the three countries. The membership selection was an attempt to balance opposing sides in the Cold War, with India holding the balance. But the Poles "did not observe" incidents that the Canadians wanted to report, and the Indians sat on the fence. An uneasy peace was established in Indochina but, as seen earlier, when North Vietnam began to infiltrate into South Vietnam from about 1966 in support of Vietnam guerrillas, the United States was drawn in. A second truce supervisory commission (for Vietnam only) had to be constituted in 1973 with Canada, Poland, Indonesia, and Hungary as members. With a 50 percent East Bloc representation, it was ineffective. Canada withdrew after six months. Cold War hostilities made this particular peace observation work futile.

UNITED NATIONS EMERGENCY FORCE I

In 1956, the Suez crisis called for a response more drastic than observers could provide. Britain and France had invaded Egypt in collusion with Israel to seize control of the Suez Canal, which Egyptian President Nasser had nationalized. (See pp. 350–351.) When the invaders agreed to abandon their effort because world opinion, including that of the United States, had turned sharply against them, an organized military force was needed to supervise the withdrawal. It was also necessary to place a buffer force between the Israelis and the Egyptian army. We have seen that Nikita Khrushchev, whose Soviet forces had almost simultaneously invaded Hungary to suppress a rising against the communist government there, was loudly proclaiming his intention to resort to nuclear rockets if the Suez crisis was not quickly resolved. However, as Britain and France had a veto in the Security Council, which had priority in directing the settlement of disputes, discussion of action there would undoubtedly have been prolonged. Fear of a Third World war spurred the General Assembly to act in that instance. In 1950, during the Korean War, when the return of the U.S.S.R. to the U.N. Security Council had appeared to threaten the continuance of U.N. operations, the General Assembly on 3 November 1950 had passed what was called the "Uniting for Peace Resolution." This stated that, in the event of the Security Council's being unable to act because of a voting deadlock, the General Assembly, by a two-thirds majority, could act to preserve peace. On 1 November 1956, Dag Hamarskjold, U.N. secretary-general, summoned an unprecedented emergency general session of the

Assembly which voted on 3 November to set up a peacekeeping force for the Middle East, the United Nations Emergency Force (UNEF I). Several countries volunteered troops. Units of various kinds, hurriedly equipped with the new blue helmet that henceforward was to be the distinguishing uniform of U.N. forces, were flown to Egypt in American aircraft.

The establishment of UNEF I set several precedents that were expected to govern future U.N. action and to a large extent have so done. In the first place, peacekeeping forces would only be dispatched if the host country approved them. When Nasser objected that Canadian troops might be mistaken for British, and also to the selection of the "Royal" Canadian Rifles because the name of that regiment sounded "British," a different Canadian regiment was substituted. When General E. L. M. Burns of Canada, who was selected to command UNEF I, was told that he looked like a British Army Officer, he wired London for an RAF headdress without the badge. Second, the UNEF force was instructed not to bear arms except for self-defense. Third, it was not to use force to carry out its mission; and, fourth, it was not to interfere in the administration of the country. In 1967, when Nasser ordered UNEF I to leave, it did so in forty-eight hours, thus confirming the principle that the host country must approve.

THE CONGO

Meanwhile, in 1960, the United Nations responded to an appeal for help to establish law and order in the new Congo Republic, from which Belgium had withdrawn precipitously, and where there was widespread bloodletting. The Security Council acted promptly, and the first U.N. contingents, Tunisians and Ghanese, arrived four days after the request was received. Before the United Nations Operation in the Congo (UNOC) ended in 1964, thirty nations had contributed forces to it. Contingents from some countries had been vetoed by the Congo government or by other parties there.

Unfortunately, UNOC became involved in power struggles between Prime Minister Patrice Lumumba, President Joseph Kasavubu, Chief of Staff General Seko Mobotu, and Moise Tshombe. The latter, with the help of foreign mercenaries and Belgian troops, tried to detach Katanga, the richest province. Some U.N. units were attacked and suffered casualties. On at least one occasion they were authorized to take the offensive. In the course of four years the ONUC peacekeeping forces suffered 234 fatalities, including Secretary-General Dag Hamarskjold, who was killed in an air accident that has never been explained. UNOC had thus shown up many deficiencies and weaknesses in the U.N. peacekeeping system, but it was an international military operation of a kind never before undertaken. By 1964 it had given the Congo Republic, which Belgium had inadequately prepared for independence, a chance for future stability.

NEW GUINEA

The United Nations operation in West New Guinea merits special attention because it is the only instance in which the world organization took responsibility for the administration of a territory. The dispute occurred because the Netherlands claimed that West Irian, which was ethnically different from Indonesia, should remain under its jurisdiction after Indonesia itself became independent. Indonesia denied the claim on the grounds that the New Guinea territory had always been administered from Djakarta. In 1962, U Thant, secretary-general of the United Nations, appointed Ellsworth Bunker, a former American ambassador to India, to negotiate a settlement. Bunker arranged for the Dutch to hand West Irian over to the United Nations for at least a year, but not for more than two, and for the territory then to be transferred to Indonesia, which was to arrange for an election to decide its future. The United Nations General Assembly authorized the Secretary-General to set up a United Nations temporary executive authority (UNTEA) and a United Nations special force (UNSF) to assist it. Because there had been naval and military activity in the area, U Thant, at that time still acting Secretary-General, without awaiting the authorization of the General Assembly or the Security Council, sent military observers. Pakistan provided 1,500 infantry for the observers' protection, and the United States and Canada provided air transportation. Within the allotted time the West Irians voted virtually unanimously to join Indonesia.

CYPRUS; AND THE MIDDLE EAST AGAIN

In 1964, when Turkey intervened in the dispute between Cypriots of Turkish and Greek descent, the Security Council again authorized the dispatch of a U.N. force. Because the British still had bases on the island from which it could provide technical assistance, the recently adopted principle that the great powers should not serve on U.N. peacekeeping forces was waived to permit the inclusion of a British contingent. Eight other countries have also contributed forces to UNCIFYP, and some countries have given other assistance without cost to the United Nations. UNCIFYP is still operating to control the border between the Turkish and Greek populations on the island.

   In 1973, when another cease fire between Egypt and Israel had to be policed, the Security Council set up a second United Nations Emergency Force (UNEF II). This showed that peacekeeping forces could be very quickly established by the United Nations if the two superpowers were in agreement. In 1974 and 1975, UNEF II was authorized to supervise the deployment of Egyptian and Israeli forces and to control a buffer zone between them.

Thirteen countries supplied contingents. Its maximum strength was 6,973 in February 1974. It withdrew in July 1979.

In 1974 the Security Council set up an observer group again, the United Nations Disengagement Observer Force (UNDOF), to supervise a cease-fire between Israel and Syria and the redeployment of Syrian and Israeli forces, with the usual policed buffer zone. Seven countries supplied contingents. Its authorized strength was 14,800. Finally, in 1978 the United Nations Interim Force in Lebanon (UNIFIL) was given the task of confirming the withdrawal of Israeli forces from Southern Lebanon, which they had invaded to attack PLO bases. Fourteen countries provided personnel of various kinds for it to make up an authorized strength of 7,000.

NON-U.N. PEACEKEEPING

By this time, however, there were alternative peacekeeping forces not under U.N. auspices. After the Yom Kippur War in 1975, Israel and Egypt agreed to electronic early warning stations located in the Sinai passes to prevent another surprise attack. In this case they asked for the expertise of the United States. A United States Support Mission commissioned civilian firms and technicians to erect the stations. In 1979 the work of inspecting military sites on the border between Israel and Egypt was allocated to a United States field mission. That mission operated until 1982 when the task became the responsibility of a multinational force of observers (MNFO). On Israel's northern border, a similar task was entrusted to a multinational force in Lebanon (MNF) which included U.S. Marines and French and Italian troops. These multinational forces were not under United Nations control because the Soviet Union was now opposed to a U.N. presence in the Sinai, and because Israel feared that a U.N. force on its northern border might allow the Soviet presence there to bolster Syria, its bitter enemy and a Soviet protégé. These were the largest non-U.N. peacekeeping forces up to that time. United Nations peacekeeping thus seemed to be on the decline. In 1978, however, the Security Council authorized the establishment of a United Nations transition group (UNTAG) to arrange for the evacuation of South Africans from the former Portuguese colony, Angola. But the deployment of that force was delayed because of the presence of Cuban troops and because of the dislike of the United States for their presence there.

THE POTENTIAL OF PEACEKEEPING

The troops who participated in peacekeeping forces or observer groups, whether under U.N. or other auspices, always retained their national identity, as was inevitable in a world made up of sovereign states. They differed from each other in their mode of operations and their interests, yet their common military expertise usually enabled them to work together harmoniously for

CANADIAN SOLDIER ON U.N. PEACEKEEPING PATROL. (Directorate of History, Department of National Defence, Canada)

their common purpose. This was so even though nations differed sharply in the way in which they responded to requests for troop or officer contingents. Some sent their regular forces, even in certain cases earmaking particular units in advance for possible future U.N. requests, as was the case in Canada. Others, especially the Scandinavians, always sent reservists who had been called upon specifically to volunteer for U.N. duty. The work of the reservists did not suffer unduly when compared with that of regulars; indeed, as individual reservists often tended to repeat their U.N. service if opportunity offered, they became experience peacekeepers. However, perhaps the most skilled of all the national contingents for this work were the Irish. Ireland had a combination of Western skills and an anti-imperialist alignment that made Irish troops desirable in the eyes of Third World peoples. Furthermore, Eire adopted the policy of asking for volunteers from its regular army for its U.N. peacekeeping assignments.

Current U.N. peacekeeping is a far cry from what the United Nations Charter proposed for the use of military forces to keep the peace. It has been plagued by Cold War rivalries, and it has not solved intractable problems. This latter failure has given some of the U.N. operations an appearance of permanence, for instance, that in Cyprus. Although costs are supposed to be borne on the U.N. budget proportionate to the rates established for general contributions, their incidence has fallen more heavily on some countries than on others. U.N. peacekeeping is a pale substitute for the international standing army that some United Nations idealists dreamed of in the early days of the Charter. It has, however, done some admirable work in easing situations that might have led to wider conventional conflagrations, and possibly even to nuclear war. The fact that conventional wars have been fewer than might have been expected when one considers the number of international hatreds and rivalries that still persist must perhaps be attributed to other circumstances than to the work of the U.N. peacekeepers and observers. Nevertheless, the employment of national military force for international peacekeeping is an innovation that may have great significance in the future.

EPILOGUE

On 7 October 1989, television showed the world an unexpected, almost unbelievable, sight—thousands of West Berliners seated on the wall that had divided their city for twenty-eight years. While East German Security Police watched quietly, they began to tear it down. That wall, where many East Germans attempting to escape had died, had been the most visible evidence of the Iron Curtain which, a few hundred miles to the west, separated Soviet-dominated Eastern Europe from the free nations of the West. Its breaching was the most dramatic episode in a widespread series of relatively peaceful revolutions that swept the communist world in the last years of the 1980s, overthrowing the communist autocracies of Europe and undermining the Warsaw Pact. The Cold War was coming to an end. No military, political, or academic specialist had expected that to happen so soon or in this way.

It is not a function of history to forecast the future. Predictions are especially risky in a time and circumstance where social scientists and leaders responsible for policy have so recently been confounded. It is, however, appropriate that, in a conclusion to this study of the interrelationships of warfare and Western society, we take cognizance of the long-term trends that have culminated in these developments, for they must be taken into account in future military planning, organization, and function.

This book is based on the premise that, to be meaningful, military history must not be studied, as too often in the past it has been, in a vacuum, but must be set in the context of political, economic, social, and technological change. Some scholars have called this approach "the new military history." This so-called new military history has been marked in recent years by the emergence of war and peace studies that have appealed to audiences beyond the military professionals and buffs who, almost alone, were interested in traditional military history. War and peace studies have also been able to attract more financial support for research. The new military history has come to stay. But this does not diminish the continued need for the study of military organization, strategy, tactics, and leadership. On the contrary, the new military history must take these special military fields fully into its scope.

We have seen that from the fifteenth to the nineteenth century of the Christian era, the peoples of Western Europe fashioned a unique society out of the feudal relationships of the Middle Ages. That society had no known precedent in human history. It comprised sovereign nation states, mainly monarchies, all of which to a greater or lesser degree had developed economies based on capitalist free enterprise, and all of which, more or less, subscribed to a rule of law. In some of them the monarchs were bound by constitutional restraints that gave their subjects some voice in government. Some republics, notably the United States, a North American outpost of Western society, and France after its revolution, similarly put power within the grasp of their citizens. This European state system was unique because of its material progress, because no single state was able to dominate the whole and build an all-inclusive empire and because of the extent of its individual liberties. By the end of the nineteenth century some of the nations of Western Europe had extended their rule, or their influence, into virtually every part of the globe. Peoples everywhere began to emulate, or aspire to, those aspects of Western society that seemed to bring prosperity, strength, and individual satisfaction. Although these aspirants retained many of their traditional cultural patterns, they had, in effect, become Westernized in certain ways. The society of Western Europe has thus become potentially unique in another respect. It has become global.

Wars played an important part in both the building of Western society and in its expansion. Not only did wars between competing states decide the course of its internal and external development, but warfare brought technological, political, and social changes that contributed to its achievements. So warfare, although in some respects often seen only as wasteful and destructive, was also a constructive agent. However, over the centuries war's capacity for destruction has increased. Some philosophers and others therefore came to regard it as an exclusively destructive force, but they have noted that at times the capacity for destruction has been restrained. Thus the eighteenth century has been called a period of "limited war." Jack Levy, in his book *War in the Modern Great Power System, 1495–1975,* has corrected those political scientists who measured the growing intensity of wars by statistics based on all wars. He pointed out that, as the great powers have been involved in warfare to a greater extent than the small powers, the sum total of all wars in not conclusive. Levy questioned the description of the eighteenth century as a period of limited warfare by noting that the great powers were even less involved in war in the nineteenth century, and that great-power wars in the twentieth century have been few and short. He thus appeared to challenge the accepted concept that war is growing in frequency and intensity.

Men in Arms uses "limitation" in the Clausewitzian sense. Clausewitz called the eighteenth century warfare "limited" because its perpetrators carefully limited their objectives, partly in order to avoid endangering their own thrones. The impact of war was limited in the nineteenth century by a different factor, namely Britain's command of the sea, which discouraged military adventuring on land.

In the twentieth century, although great-power wars have been few and relatively short, their impact has been great. The First World War spread destruction in many parts of the globe; and on European fronts it caused vastly more casualties than any previous conflict. Its effect was so severe that it led to the establishment of an international agency, the League of Nations, primarily for the purpose of imposing on sovereign states restraints they would not voluntarily undertake themselves. The Second World War spread its casualties widely among civilian, as well as military, victims. It, too, was followed by the creation of an international agency to attempt to keep the peace, the United Nations. Once more, however, nations were reluctant to surrender the prerogative to use armed force, and the United Nations failed in its primary objective, although it was able to provide other useful services that were necessary in an increasingly interdependent world. Until 1990 its direct contribution to maintaining the peace was limited to the provision of international peacekeeping forces where the parties have agreed to accept them.

The Cold War came about when the Soviet Union, suspicious of Western intentions and determined to spread its influence, insisted on retaining control of Eastern European countries liberated from the Nazis. The West interpreted the move as a betrayal and a threat. Under American leadership, some of the states of Western Europe and North American resorted to a defensive alliance, NATO, to which the Soviets responded with the Warsaw Pact.

In the past a confrontation of defensive alliances usually resulted in war. One reason that this did not happen in this case was that there was a new factor in warfare, nuclear weapons, which had brought the possibility of mutual, even universal, destruction. War's intensity had thus moved toward totality. Nuclear deterrence now seems to have done what the philosophers' wisdom and international organization could not achieve, namely, restrain nations armed to the teeth.

However, the cost of maintaining both nuclear and conventional armaments was high. It therefore came to be a question, if war did not come by accident or the action of deranged individuals, which side would crack first under the economic strain, the free enterprise system of the United States, or the totalitarian communist regime of the U.S.S.R. Labor strikes in Polish shipyards and Soviet coal mines that the state could not suppress were the first signs of a decision. When Mikhail Gorbachev introduced policies of *glasnost'* and *perestroika,* and then refused to bolster the regimes of satellite states threatened by popular movements, the question appeared to be answered. The East European communist governments tried to restore their bankrupt economies by large doses of political democracy and market economics. Soviet "republics" talked of more freedom to operate independently or even of sucession. Economics seemed as important as armed might in the arena of international competition. It seemed as if the cycles of intensity and restraint in warfare that had lasted for several centuries might now have ended. This left open the question whether future Western planning should stress continued strong defense or economic rehabilitation.

But the restraint imposed by fear of a nuclear holocaust in the second half of the twentieth century had not put an end to all wars. It had, in fact, only affected the superpowers, and even in their case only in their relations with each other. Both the United States and the Soviet Union had been involved directly in bloody wars, the United States in Korea and Vietnam and the U.S.S.R. in Afghanistan. They had also supported other military clashes by their satellites, client states, and allies. They had thus had need for military forces apart from the requirements of their Cold War confrontation. Smaller powers, including former great powers, had not been greatly inhibited by fear of the bomb, and their wars were numerous and costly in lives and money. Despite the restraining influence of the United Nations, and even of the fear that any small war might become the total third world war that no one wanted, resort to arms remained a part of the human condition. Nuclear weapons had not brought an end to war, only a constraint on the pursuit of war.

An end of the Cold War seems unlikely to end the mutual suspicion between the superpowers. Many in both camps will not lightly cast off long-held distrust of the other's intentions. Many in the East still cherish the ideals presented by the concept of a classless world, and the struggle in the U.S.S.R. between hard-line conservative communists and reformers who wanted to go farther than Gorbachev is indicative of a problem that might in the end bring a return to a Stalinist-like autocracy or lead to military intervention. In view of the possibilities, there has been a reluctance in the West to disarm precipitately, even though significant progress in still being made toward international agreements on arms limitation at both the nuclear and conventional level.

Nevertheless, with the diminution of the perception of a Soviet threat, it is clear that there will be pressure to abolish, reduce, or radically change NATO. President George Bush hinted in this direction in June 1990, when he made reference to the long-ignored Article 2 of the treaty, which said that NATO should foster mutual economic relations. Bush recalled that aspect of the treaty in an effort to ease Soviet fears of a reunited Germany. It is also clear that in the long run the possibility of eliminating the danger of a renewal of Soviet intransigence may depend on the rescue of the U.S.S.R. from bankruptcy. But will that be possible when it retains elements of the centrally planned economy and lacks the free markets that have flourished in the West? Workers in the Soviet Union might be finding it difficult to get consumer goods, but they are still relatively free from the blight of unemployment. The transition from a planned economy to a free market economy will undoubtedly be painful. That pain might bring disenchantment and backsliding. The reduction of NATO forces and redeployment of Americans from Europe to home bases in North America would weaken the possibility of the alliance continuing to act as a restraining influence. Last, powerful military and business interests will argue against a reduction of military preparedness. International agreements may now be easier to negotiate, but they will still need to be guaranteed by an inspection and control that will be composed of military experts. That might be difficult to arrange in a world of sovereign states.

With the continuance of unrestricted national sovereignties in every part of the world, there will still be national rivalries and the retention of ancient grievances. Few states are likely to disarm completely. The futility of past attempts to control wars by international agreements and organization suggests that war will continue to be an option for the settlement of disputes. Peacekeeping by the United Nations or other agencies may serve to diminish the impact of military intervention, but is unlikely to abolish it. The superpowers, hitherto almost completely excluded from peacekeeping operations, will continue to be suspected of imperialist ambitions if, or when, they attempt to operate abroad.

On the other hand, with lesser powers inclined to pursue their ambitions or free themselves from fears by resort to arms, the superpowers will be impelled to retain superior strength. This is especially the case because of the persistence of apparently insoluble differences, as for instance between Arab and Israeli in the Middle East, with all their possibilities for friction and for enlargement. Finally, a relatively new feature of this source of international conflict is the emergence of activity by individuals and groups as the new phenomenon of terror on the international scene. Terrorism may become a major problem for armed forces, which formerly considered such action a problem for the police.

The three kinds of warfare described in the closing chapters of this book have been influenced very differently by recent events. Nuclear war seems less likely than at any time since 1950. This is especially true of the great nuclear exchange between the superpowers, which was expected then and dreaded ever since. The possibility of a nuclear exchange between other nations is perhaps higher, but nonetheless still small. Not even proliferation seems to have raised the risk appreciably, though it does increase the chance for accident or the irrational act of a madman or terrorist.

Conventional war also appears unlikely between the superpowers, or even between major industrialized states. Most have come to recognize that between comparable states with substantial arsenals of today's weapons, it has simply become too indecisive, too costly, and too destructive. In the Third World, however, there seem to be many states, like Argentina and

Iraq in the 1980s, still willing to take the risk. As the Iran–Iraq conflict showed, such wars with the new weapons are horrible and useless on an unprecedented scale.

Low-intensity conflict shows no signs of abating. Indeed, the end of the Cold War seems to have sparked a host of ethnic, regional, religious, and cultural conflicts that have been simmering for decades beneath constraints imposed by empire or by the Cold War itself. Additionally, as the disparity in wealth between the world's haves and have-nots grows, the temptation for the poor and displaced, be they individuals, groups or countries, to strike out against the rich and the powerful can only increase. The new technology of war places in terrorists hands weapons far more powerful than ever before. This suggests that they must be given a chance to compete fairly and peacefully in economic and political matters. It will be a dangerous world indeed if they continue to view war as their only recourse.

So, despite the end of the Cold War, if it has indeed ended, military organizations are likely to continue, although in somewhat different forms and for somewhat different purposes. This will undoubtedly impose great strains and will call for a new form of dedication. Those who choose to follow a military career, the political groups who decide policy, and the people in every country who ultimately make final dispositions will need to understand the continued requirement for military organization and the ways in which it must be adapted without losing its efficiency. We may be standing on the threshold of a new stage in the relations between armed force and the development of civilized society.

BIBLIOGRAPHY

GENERAL WORKS
AND THOSE COVERING MORE
THAN ONE PERIOD

Addington, Larry, *The Patterns of War since the Eighteenth Century.* Bloomington, Ind.: Indiana University Press, 1984.

———, *The Patterns of War through the Eighteenth Century.* Bloomington, Ind.: Indiana University Press, 1990.

Aimone, Alan C. (comp.), *Military History Bibliography and Guide,* 5th ed. West Point, N.Y.: United States Military Academy, 1987.

Anderson, Romola, *Oared Fighting Ships: From Classical Times to the Coming of Steam.* London: Percival Marshall, 1962.

Anderson, Romola, and R. C. Anderson, *The Sailing Ship.* New York: Norton, 1963.

Andreski, Stanislav, *Military Organization and Society,* 2d ed. London: Routledge & K. Paul, 1968.

Aron, Raymond, *War and Industrial Society.* New York: Oxford University Press, 1958.

———, *On War.* Transl. by Terence Kilmartin. New York: Norton, 1968.

Asprey, Robert, *War in the Shadows: The Guerrilla in History.* Garden City, N.Y.: Doubleday, 1975.

Banks, Arthur, *Atlas of Ancient and Medieval Warfare,* rev. ed. New York: Hippocrene Books, 1982.

Barnet, Richard J., *Roots of War.* Baltimore: Penguin, 1973 [1972].

Barnett, Correlli, *Britain and Her Army: 1509–1970: A Military, Political, and Social Survey.* New York: Morrow, 1970.

Beebe, G. W., and M. E. De Bakey, *Battle Casualties: Incidence, Mortality and Logistic Considerations.* Springfield, Ill.: Thomas, 1952.

Best, Geoffrey, *Humanity in Warfare.* New York: Columbia University Press, 1980.

Bohannan, P. (ed.), *Law and Warfare: Studies in the Anthropology of Conflict.* New York: Natural History Press, 1967.

Bond, Brian, *War and Society in Europe, 1870–1970.* New York: Oxford University Press, 1986.

Borowski, Harry R. (ed.), *The Harmon Memorial Lectures in Military History, 1959–1987.* Washington, D.C.: Office of Air Force History, 1988.

Bramson, Leon, and George W. Goethals (eds.), *War: Studies from Psychology, Sociology, Anthropology,* rev. ed. New York: Basic Books, 1964.

Breadnell, Charles M., *An Encyclopedic Dictionary of Science and War.* Ann Arbor, Mich.: Gryphon Books, 1971 [1943].

Brice, Martin H., *Stronghold: A History of Military Architecture.* New York: Schocken Books, 1985.

Brodie, Bernard, *A Guide to Naval Strategy,* 4th ed. Westport, Ct.: Greenwood Press, 1977 [1958].

——, *Sea Power in the Machine Age.* Princeton, N.J.: Princeton University Press, 1941.

Brodie, Bernard, and Fawn Brodie, *From Cross-bow to H-bomb,* 2d ed. Bloomington, Ind.: Indiana University Press, 1973.

Carman, W. Y., *A History of Firearms: From Earliest Times to 1914.* London: Routledge & Kegan Paul, 1963.

Carthy, J. D., and F. J. Eibling (eds.), *The Natural History of Aggression.* New York: Academic Press, 1964.

Clausewitz, Carl von, *On War.* Ed. by Michael Howard and Peter Paret. Princeton, N. J.: Princeton University Press, 1986.

Coblentz, Stanley A., *From Arrow to H-bomb: The Psychological History of War.* Cranbury, N.J.: A. S. Barnes, 1966.

Cohen, R., and E. Service (eds.), *Origins of the State.* Philadelphia: Institute for the Study of the Human Issues, 1978.

Colby, Elbridge. *Masters of Mobile Warfare.* Princeton, N.J.: Princeton University Press, 1943.

Corbett, Julian S., *Some Principles of Maritime Strategy.* London: Brassey's Defence Publishers, 1988 [1911].

Craig, Gordon, *The Politics of the Prussian Army, 1640–1945.* New York: Oxford University Press, 1964.

Cunliffe, Marcus, *Soldiers and Civilians: The Martial Spirit in America, 1775–1865.* Boston: Little, Brown, 1968.

Davie, Maurice, *The Evolution of War: A Study of Its Role in Early Societies.* Port Washington, N.Y.: Kennikat Press, 1985 [1929].

Delbrück, Hans, *History of the Art of War Within the Framework of Political History.* 4 vols. Westport, Ct.: Greenwood Press, 1975–85.

Denison, George T., *A History of Cavalry from the Earliest Times, with Lessons for the Future.* Westport, Ct.: Greenwood Press, 1977 [1866].

Divale, William T., *Warfare in Primitive Societies: A Bibliography.* Santa Barbara, Calif.: ABC-Clio, 1973.

Doorn, Jacobus A. A. van, *The Soldier and Social Change: Comparative Studies in the History and Sociology of the Military.* Beverley Hills, Calif.: Sage Publications, 1974.

Douhet, Giulio, *The Command of the Air.* Trans. by Dino Ferrari. New York: Coward-McCann, 1942 [1921, 1927].

Duffy, Christopher, *Fire and Stone: The Science of Fortress Warfare, 1660–1860.* Newton Abbot, Eng.: David & Charles, 1975.

——, *Russia's Military Way to the West: Oorigins and Nature of Russian Military Power.* London: Routledge & Kegan Paul, 1981.

Dupuy, R. Ernest, and Trevor N. Dupuy, *The Encyclopedia of Military History from 3500 B.C. to the Present,* 2d ed. New York: Harper and Row, 1986.

Dupuy, Trevor N., *Numbers, Predictions and War.* Indianapolis, Ind.: Bobbs Merrill, 1979.

Earle, Edward Mead (ed.), *Makers of Modern Strategy: Military Thought from Machiavelli to Hitler.* Collab. Gordon A. Craig and Felix Gilbert. Princeton, N.J.: Princeton University Press, 1971 [1943].

Emeny, Brooks, *The Strategy of Raw Materials: A Study of America in Peace and War.* Wilmington, Del.: International Academic Publishers, 1979 [1937].

Falls, Cyril, *The Art of War: From the Age of Napoleon to the Present Day.* London: Oxford University Press, 1961.

Feld, Maury D., *The Structure of Violence: Armed Forces as Social Institutions.* Beverley Hills, Calif.: Sage Publications, 1977.

Ferguson, R. B. (ed.), *Warfare, Culture and Environment.* Orlando, Fla.: Academic Press, 1984.

Finer, S. E., *The Man on Horseback: The Role of the Military in Politics.* New York: Praeger, 1962.

Fried, M., M. Harris, and R. Murphy (eds.), *The Anthropology of Armed Conflict and Aggression.* Garden City, N.Y.: Natural History Press, 1968.

Fuller, J. F. C., *Armament and History: A Study of the Influence of Armament on History from the Dawn of Classical Warfare to the Second World War.* New York: Scribner's Sons, 1945.

———, *The Conduct of War, 1789–1961: A Study of the Impact of the French, Industrial, and Russian Revolutions on War and Its Conduct.* New Brunswick, N.J.: Rutgers University Press, 1961.

———, *A Military History of the Western World.* 3 vols. New York: Funk & Wagnalls, 1954–1956.

Gaier, Claude, *Four Centuries of Liege Gunmaking.* Transl. by F. J. Norris. London: Sotheby Parke Bernet, 1977.

Garber, Max B., *A Modern Military Dictionary: Ten Thousand Technical and Slang Terms of Military Usage.* Detroit: Gale Research Co., 1975 [1942].

Goerlitz, Walter, *History of the German General Staff, 1657–1945.* Transl. by Brian Battershaw. Boulder, Colo.: Westview Press, 1985 [1953].

Gray, J. Glenn, *The Warriors: Reflections on Men in Battle.* New York: Harper and Row, 1973 [1959].

Griffith, Paddy, *Forward Into Battle: Fighting Tactics from Waterloo to Vietnam.* Chichester, Eng.: Sussex, 1981.

Haas, J. (ed.), *The Anthropology of War.* Cambridge, Eng.: Cambridge University Press, 1990.

Hagan, Kenneth J., *This People's Navy: The Making of American Sea Power.* New York: Free Press, 1990.

Hagan, Kenneth J., and William R. Roberts (eds.), *Against All Enemies: Interpretations of American Military History from Colonial Times to the Present.* Contributions in Military Studies, Number 51. Westport, Ct.: Greenwood Press, 1986.

Hall, John A., *Powers and Liberties: The Causes and Consequences of the Rise of the West.* Berkeley, Calif.: University of California Press, 1986.

Hamley, Edward Bruce, *The Operations of War Explained and Illustrated by General Sir Edward Bruce Hamley.* London: Blackwood, 1923.

Hammond, Paul Y., *Organizing for Defense: The American Military Establishment in the Twentieth Century.* Princeton, N.J.: Princeton University Press, 1961.

Handel, Michael, *War Termination: A Critical Survey.* Jerusalem: Hebrew University of Jerusalem, 1978.

Hawthorne, Daniel, *For Want of a Nail: The Influence of Logistics on War.* New York: Whittesey House, 1948.

Haycock, Ronald, and Keith Neilson (eds.), *Men, Machines, and War.* Waterloo, Ont.: Wilfrid Laurier University Press, 1988.

Henderson, G. F. R., *The Science of War: A Collection of Essays and Lectures, 1892–1903.* Ed. by Neill Malcolm. London: Longmans, Green, 1927.

Hittle, James, D., *The Military Staff: Its History and Development.* Westport, Ct.: Greenwood Press, 1975 [1961].

Hogg, Ian V., *Fortress: A History of Military Defense.* New York: St. Martin's, 1975.

———, *The Illustrated History of Ammunition.* Secaucus, N.J.: Quarto, 1985.

Hogg, Oliver F. G., *Clubs to Cannon: Warfare and Weapons Before the Introduction of Gunpowder.* London: Duckworth, 1968.

Holmes, Richard, *Acts of War: The Behavior of Men in Battle.* New York: Free Press, 1986 [1985].

———, *Soldiers: A History of Men in Battle.* London: Hamilton, 1985.

———, *The World Atlas of Warfare: Military Innovations That Changed the Course of History.* New York: Viking Studio Books, 1988.

Hovgaard, William, *Modern History of Warships.* London: Conway Press, 1971 [1921].

Howard, Michael, *The Causes of War and Other Essays.* Cambridge, Mass.: Harvard University Press, 1983.

———, *War in European History.* New York: Oxford University Press, 1976.

——— (ed.), *Soldiers and Governments: Nine Studies in Civil-military Relations.* Westport, Ct.: Greenwood Press, 1978 [1957].

——— (ed.), *The Theory and Practice of War.* Bloomington, Ind.: Indiana University Press, 1965.

Hughes, B. P., *Firepower: Weapons Effectiveness on the Battlefield, 1630–1850.* London: Arms and Armour Press, 1974.

Hughes, Quentin, *Military Architecture.* London: Hugh Evelyn, 1975.

Huntington, Samuel P., *The Soldier and the State: The Theory and Politics of Civil-military Relations.* Cambridge, Mass.: Belknap Press of Harvard University Press, 1957.

Ikle, Fred, *Every War Must End.* New York: Columbia University Press, 1971.

International Bibliography of Military History. Berne, Switzerland, 1974–   .

Jessup, John E., Jr., and Robert W. Coakley, *A Guide to the Study and Use of Military History.* Washington: U. S. Army Center of Military History, 1982.

Jones, Archer, *The Art of War in the Western World.* Urbana, Ill.: University of Illinois Press, 1987.

Kahn, David, *The Codebreakers: The Story of Secret Writing.* New York: Macmillan, 1967.

Karsten, Peter, *Military Threats: A Systematic Historical Analysis of the Determinants of Success.* Westport, Ct.: Greenwood Press, 1984.

Keegan, John, *The Face of Battle: A Study of Agincourt, Waterloo, and the Somme.* New York: Viking, 1976.

——, *The Mask of Command.* New York: Penguin, 1987.

Keep, J. L. H., *Soldiers of the Tsar: Army and Society in Russia, 1462–1874.* Oxford, Eng.: Clarendon Press, 1985.

Kennedy, Paul M., *The Rise and Fall of British Naval Mastery.* New York: Scribner, 1976.

——, *The Rise and Fall of the Great Powers: Economic Change and Military Conflict from 1500 to 2000.* New York: Random House, 1987.

Keylor, William R., *The Twentieth-century World: An International History.* New York: Oxford University Press, 1984.

Kiernan, Frank A., Jr., and John K. Fairbanks (eds.), *Chinese Ways in Warfare.* Cambridge, Mass.: Harvard University Press, 1974.

Kinnell, Susan K. (ed.), *Military History of the United States: An Annotated Bibliography.* Santa Barbara, Calif.: ABC-CLIO, 1986.

Knorr, Klaus, *The War Potential of Nations.* Princeton, N.J.: Princeton University Press, 1956.

Kohn, Richard H., *Eagle and Sword: The Beginnings of the Military Establishment in America.* New York: Free Press, 1975.

Lambert, Andrew, *Battleships in Transition: The Creation of the Steam Battle Fleet, 1815–1860.* Annapolis, Md.: United States Naval Institute Press, 1984.

Laquer, Walter, *Guerrilla: A Historical and Critical Study.* Boston: Little, Brown, 1976.

——, *Terrorism.* Boston: Little, Brown, 1977.

——, *A World of Secrets: The Uses and Limits of Intelligence.* New York: Twentieth Century Fund, 1985.

Levy, Jack S., *War in the Modern Great Power System, 1495–1975.* Lexington, Ky.: University of Kentucky Press, 1983.

Lewis, A. R., and T. J. Runyan, *European Naval and Maritime History, 300–1500.* Bloomington, Ind.: Indiana University Press, 1985.

Lewis, Michael, *The History of the British Navy.* Baltimore: Penguin, 1962 [1957].

——, *The Navy of Britain: A Historical Portrait.* London: Allen and Unwin, 1948.

Liddell Hart, Basil H., *Strategy: The Indirect Approach.* 4th ed., rev. and enlarged. London: Faber, 1967.

Lloyd, Christopher, *Atlas of Maritime History.* New York: Arco, 1975.

Lloyd, Ernest M., *A Review of the History of Infantry.* The West Point Military Library. Westport, Ct.: Greenwood Press, 1976 [1908].

Macksey, Kenneth, *Technology in War*. New York: Prentice Hall, 1986.

McNeill, William H., *The Pursuit of Power: Technology, Armed Force and Society Since AD 1000*. Chicago: University of Chicago Press, 1982.

McPherson, James, *Battle Cry of Freedom: The Civil War Era*. New York: Oxford University Press, 1988.

Mahan, Alfred Thayer, *The Influence of Sea Power upon History*. Boston: Little, Brown, 1890.

Mandelbaum, Michael, *The Fate of Nations: The Search for National Security in the Nineteenth and Twentieth Centuries*. Cambridge, Eng.: Cambridge University Press, 1988.

Marsden, E. W., *Greek and Roman Artillery: Historical Development*. New York: Oxford University Press, 1969.

Masland, J. W., and Lawrence I. Radway, *Soldiers and Scholars: Military Education and National Policy*. Princeton, N.J.: Princeton University Press, 1057.

Maurice, F. B., *British Strategy: A Study of the Application of the Principles of War*. London: Constable, 1938.

Mendelsohn, Everett, Merritt Roe Smith, and Peter Weingart (eds.), *Science, Technology and the Military*. Sociology of the Sciences, A Yearbook, Vol. XII/1-2. Dordrecht: Kluwer Academic Publishers, 1988.

Millett, Allan, *Semper Fidelis: The History of the United States Marine Corps*. New York: Macmillan, 1980.

Millett, Allan R., and Peter Maslowski, *For the Common Defense: A Military History of the United States of America*. New York: Free Press, 1984.

Millis, Walter, *Arms and Men: A Study in American Military History*. New Brunswick, N.J.: Rutgers University Press, 1981 [1957].

Montgomery, Bernard L., *A History of Warfare*. London: Collins, 1968.

Montross, Lynn, *War Through the Ages*, 3d ed. New York: Harper, 1960.

Morrison, John, *Long Ships and Round Ships, Warfare and Trade in the Mediterranean, 3,000 B. C.–500 A. D.* London: HMSO, 1980.

Mumford, Lewis, *Technics and Civilization*. New York: Harcourt Brace, 1934.

Nef, John U., *War and Human Progress: An Essay on the Rise of Industrial Civilization*. Cambridge, Mass.: Harvard University Press, 1950.

O'Connell, Robert L., *Of Arms and Men: A History of War, Weapons, and Aggression*. New York: Oxford University Press, 1989.

O'Sullivan, Patrick, and Jesse W. Metler, *The Geography of War*. New York: St. Martin's, 1983.

Otterbein, Keith, *The Evolution of War: A Cross-cultural Study*. New Haven, Ct.: H.R.A.F. Press, 1970.

Padfield, Peter, *The Battleship Era*. London, 1972.

Parks, Oscar, *British Battleships, "Warrior" 1860 to "Vanguard" 1950: A History of Design, Construction and Armament*. London: Seeley Service, 1966.

Peltier, Louis, and G. Etzel Pearcy, *Military Geography*. Princeton, N.J.: Van Nostrand, 1966.

Pemsel, Helmut, *A History of War at Sea: An Atlas and Chronology of Conflict at Sea from 480 B.C. to the Present.* Annapolis, Md.: U. S. Naval Institute Press, 1977.

Perret, Geoffrey, *A Country Made by War: From the Revolution to Vietnam — the Story of America's Rise to Power.* New York: Random House, 1989.

Phillips, Thomas R., *Roots of Strategy: A Collection of Military Classics.* Harrisburg, Pa.: Stackpole Books, 1985 [1940].

Potter, E. B. (ed.), *Sea Power: A Naval History,* 2d ed. Annapolis, Md.: Naval Institute Press, 1981.

Pryor, John H., *Geography, Technology, and War: Studies in the Maritime History of the Mediterranean, 649–1571.* Past and Present Publications. New York: Cambridge University Press, 1988.

Ralston, David B. (ed.), *Soldiers and States: Civil-military Relations in Modern Europe.* Studies in History and Politics. Englewood, N.J.: D. C. Heath, 1966.

Redlich, Fritz, *The German Military Enterpriser and His Work Force: A Study in European Economic and Social History.* Wiesbaden: Franz Steiner, 1964–1965.

Reynolds, Clark G., *Command of the Sea: The History and Strategy of Maritime Empires.* New York: William Morrow, 1974.

Richardson, Lewis Fry, *Statistics of Deadly Quarrels.* Pittsburgh, Pa.: Boxwood Press, 1960.

Richmond, Herbert, *Sea Power in the Modern World.* London: G. Bell, 1934.

Ritter, Gerhard, *The Sword and the Scepter: The Problem of Militarism in Germany.* Transl. by Heinz Norden. 4 vols. Coral Gables, Fla.: University of Miami Press, 1969–1973.

Robertson, Frederick, *The Evolution of Naval Ordnance.* London: Constable, 1921.

Ropp, Theodore, *War in the Modern World,* rev. ed. Westport, Ct.: Greenwood Press, 1981 [1959].

Rusbridger, James, *The Intelligence Game: The Illusions and Delusions of International Espionage.* London: Bodley Head, 1985.

Sandler, Stanley, *The Emergence of the Modern Capital Ship.* Newark, Del.: University of Delaware Press, 1979.

Shaw, R. Paul, *The Genetic Seeds of Warfare: Evolution, Nationalism, and Patriotism.* Boston: Unwin, Hyman, 1989.

Sherry, Michael S., *The Rise of American Air Power: The Creation of Armageddon.* New Haven, Ct.: Yale University Press, 1987.

Shotwell, James T., *War as an Instrument of National Policy, and Its Renunciation in the Pact of Paris.* Intro. by Gerald E. Markowitz. New York: Garland Publishing, 1974.

Small, Melvin, and J. David Singer, *Resort to Arms: International and Civil Wars, 1816–1980,* 2d ed. Beverley Hills, Calif.: Sage Publications, 1982.

Smith, Perry M., *The Air Force Plans for Peace, 1934–1945.* Baltimore: Johns Hopkins University Press, 1970.

Snyder, Glenn, *Conflict Among Nations: Bargaining, Decision Making, and System Structure in International Crises.* Princeton: Princeton University Press, 1977.

Spaulding, Oliver, Jr., Hoffman Nickerson, and John W. Wright, *Warfare: A Study of Military Methods from the Earliest Times.* New York: Arno Publishing, 1972 [1925].

Sprout, Harold, and Margaret Sprout, *The Rise of American Naval Power, 1776–1918.* Princeton: Princeton University Press, 1939.

Stanley, George F. G., *Canada's Soldiers: The Military History of an Unmilitary People.* Toronto: Macmillan of Canada, 1974.

Stockholm International Peace Research Institute. *The Law of War and Dubious Weapons.* Stockholm: Almqvist & Wiksell International, 1976.

Strachan, Hew, *European Armies and the Conduct of War.* London: Allen Unwin, 1983.

Strachey, Alix, *The Unconscious Motives of War: A Psychoanalytical Contribution.* New York: Hillary, 1957.

Sun Tzu, *The Art of War.* Transl. by Samuel B. Griffiths. New York: Oxford University Press, 1963.

Terraine, John, *The Smoke and the Fire: Myths and Anti-myths of War, 1861–1945.* London, 1980.

Toy, Sidney, *A History of Fortification from 3000 B.C. to A.D. 1700.* London: Heinemann, 1955.

Toynbee, A. J., Jr., *War and Civilization.* New York: Oxford University Press, 1950.

Tsangadas, Bryon C. P., *The Fortifications and Defense of Constantinople.* East European Monographs, Boulder. New York: Columbia University Press, 1980.

Turner, G. B. (ed.), *A History of Military Affairs in Western Society Since the Eighteenth Century.* New York: Harcourt Brace, 1953.

Vagts, Alfred, *A History of Militarism: Civilian and Military,* rev. ed. Westport, Ct.: Greenwood Press, 1981 [1959].

van Creveld, Martin, *Supplying War.* Cambridge, Eng., Cambridge University Press, 1977.

———, *Technology and War: From 2000 B.C. to the Present.* New York: Free Press, 1989.

Waltz, Kenneth, *Man, the State, and War.* New York: Columbia University Press, 1959.

Weigley, Russell F., *The American Way of War: A History of the United States Military Strategy and Policy,* enlarged ed. Bloomington, Ind.: Indiana University Press, 1984.

———, *A History of the United States Army.* New York: Crowell, Collier & Macmillan, 1967.

Weigley, Russell F. (ed.), *New Dimensions in Military History.* San Rafael, Calif.: Presidio Press, 1975.

Williams, T. Harry, *The History of American Wars from 1745 to 1918.* Baton Rouge: Louisiana State University Press, 1985 [1981].

Willoughby, Charles A., *Maneuver in War*. Washington: U. S. Marine Corps, 1986 [1939].

Winter, J. M. (ed.), *War and Economic Development*. Cambridge, Eng.: Cambridge University Press, 1975.

Wintringham, Tom, and J. N. Blashford-Snell, *Weapons and Tactics*. Harmondsworth, Eng.: Penguin, 1973 [1943].

Wright, Quincy, *A Study of War,* 2d ed. Chicago: University of Chicago Press, 1965 [1942].

Zook, David H., and Robin Higham, *A Short History of Warfare,* 2d ed. New York: Twayne Publishers, 1967 [1965].

THE CLASSICAL PERIOD
AND BYZANTIUM

Adcock, F. E., *The Greek and Macedonian Art of War*. Berkeley, Calif.: University of California Press, 1957.

————, *The Roman Art of War under the Republic,* rev. ed. New York: Barnes & Noble, 1963.

Amit, M., *Athens and the Sea: A Study in Athenian Sea Power*. Brussels: Collection Latomus, no. 75, 1965.

Anderson, J. K., *Military Theory and Practice in the Ages of Xenophon*. Berkeley, Calif.: University of California Press, 1970.

Baynes, N. H., "The Emperor Heraclius and the Military Theme System," *English Historical Review,* LXVII. 1952.

Bell, M. J. V., "Tactical Reform in the Roman Republican Army," *Historia,* XIV. 1965.

Bishop, M. C. (ed.), *The Production and Distribution of Roman Military Equipment*. Oxford: B.A.R., 1985.

Brunt, P. A., "The Army and the Land in the Roman Revolution," *Journal of Roman Studies,* 52. 1962.

Burn, A. R., *Persia and the Greeks: The Defense of the West, 546–478 B.C.* New York: St. Martin's. 1962.

Bury, J. B., *A History of Greece,* 3d ed. London: Macmillan, 1959.

Caesar, *War Commentaries: De Bello Gallico and De Bello Civili*. New York: Dutton, 1953.

Cambridge Medieval History, IV, Parts I and II, *The Byzantine Empire,* new ed. 2 vols. Cambridge, Eng.: Cambridge University Press, 1966–1967.

Campbell, J. B., *The Emperor and the Roman Army, 31 BC–AD 235*. New York: Oxford University Press, 1984.

Carter, John Marshall, *Arms and the Man: Studies in Roman and Medieval Warfare and Society*. Manhattan, Kan.: MA/AH Publishers, 1983.

Casson, L., *The Ancient Mariners: Sea Farers and Sea Fighters of the Mediterranean in Ancient Times*. New York: Funk & Wagnalls, 1967.

————, *Ships and Seamanship in the Ancient World*. Princeton: Princeton University Press, 1986 [1971].

Caven, Brian, *Dionysius I: Warlord of Sicily.* New Haven, Ct.: Yale University Press, 1989.

Clark, F. W., *The Influence of Sea Power in the History of the Roman Republic.* Menasha, Wis.: Banta, 1915.

Connolly, Peter, *Greece and Rome at War.* Englewood Cliffs, N.J.: Prentice Hall, 1981.

Dilke, O., *Greek and Roman Maps.* Ithaca, N. Y.: Cornell University Press. 1985.

Dodge, T. A., *Caesar: A History of the Art of War Among the Romans down to the End of the Roman Empire.* 2 vols. New York: Houghton Mifflin, 1892.

————, *Hannibal: A History of the Art of War Among the Carthaginians and Romans down to the Battle of Pydna, 168 B.C.* 2 vols. New York: Houghton Mifflin, 1891.

Eadie, J. W., "The Development of Roman Mailed Cavalry," *Journal of Roman Studies,* 57, 1967.

Engels, Donald W., *Alexander the Great and the Logistics of the Macedonian Army.* Berkeley, Calif.: University of California Press, 1978.

Ferrill, Arther, *The Fall of the Roman Empire: The Military Explanation.* London: Thames and Hudson, 1986.

————, *The Origins of War from the Stone Age to Alexander the Great.* New York: Thames and Hudson, 1985.

Finley, M. I., *Ancient History: Evidence and Models.* New York: Viking, 1985.

Foss, Clive, and David Winfield, *Byzantine Fortifications: An Introduction.* Unisa 1986, number 22. Pretoria, S.A.: University of South Africa Press, 1986.

Fox, Robin Lane, *Alexander the Great.* London: Allen Lane, 1973.

Fuller, J. F. C., *The Generalship of Alexander the Great.* New York: Funk & Wagnalls, 1968.

————, *Julius Caesar: Man, Soldier, and Tyrant.* New York: Funk & Wagnalls, 1969.

Garlan, Yvon, *War in the Ancient World: A Social History.* Transl. by Janet Lloyd. New York: Norton, 1975.

Garland, Robert, *Thi Piraeus from the Fifth to the First Century B. C.* Ithaca, N.Y.: Cornell University Press, 1987.

Glover, R. G., "The Elephant in Ancient War," *Classical Journal,* XXXIX. Feb. 1944.

Graves, Robert, *Count Belisarius.* London: Cassell, 1938.

Green, Peter, *Alexander the Great.* New York: Praeger, 1970.

Griffith, G. T., *Mercenaries of the Hellenistic World.* Chicago: Argonaut, 1935.

Grundy, G. B., *The Great Persian War.* New York: AMS Press, 1969.

————, *Thucydides and the History of his Age.* 2 vols. Oxford: Blackwell, 1948.

Hackett, Sir John (ed.), *Warfare in the Ancient World.* New York: Facts on File, 1990.

Hanson, Victor Davis, *The Western Way of War: Infantry Battle in Classical Greece.* New York: Knopf, 1989.

Harris, W. V., *War and Imperialism in Republican Rome, 327–70 B. C.* Oxford, Eng.: Clarendon Press, 1979.

Henderson, B. W., *The Great War between Athens and Sparta.* London: Macmillan, 1927.

Herodotus, *The History.* 2 vols. New York: Dutton (Everyman's Library), 1936–1937.

Hignett, C., *Xerxes' Invasion of Greece.* Oxford: Oxford University Press, 1963.

How, W. W., "Arms, Strategy and Tactics in the Persian Wars," *Journal of Hellenic Studies.* 1923.

Humble, Richard, *Warfare in the Ancient World.* London: Cassell, 1980.

Jenkins, Romily, *Byzantium: The Imperial Centuries, AD 610–1071.* New York: Random House, 1967.

Johnson, Stephen, *Late Roman Fortifications.* Totowa, N.J.: Barnes & Noble, 1983.

Jordan, B., *The Athenian Navy in the Classical Period.* Berkeley, Calif.: University of California Press, 1975.

Jouguet, Pierre, *Macedonian Imperialism.* New York: Knopf, 1928.

Keppie, L., *The Making of the Roman Army.* Totowa, N.J.: Barnes & Noble, 1984.

Kromayer, J., *Antike Schlachtfelder.* Berlin: Weidmann, 1907–1931.

Laffont, Robert, *The Ancient Art of Warfare: Antiquity, Middle Ages, Renaissance, 1300 B.C.–1650 A.D.* Greenwich, Ct.: New York Graphics Society, 1966.

Landels, J. G., *Engineering in the Ancient World.* Berkeley, Calif.: University of California Press, 1978.

Launey, Marcel, *Recherches sur les Armées Hellénistiques.* 2 vols. Paris: Boccard, 1949.

Lawrence, A. W., *Greek Aims in Fortification.* Oxford, Eng.: Clarendon Press, 1979.

Liddell Hart, B. H., *A Greater than Napoleon, Scipio Africanus.* London: Blackwood, 1930.

Luttwak, Edward N., *The Grand Strategy of the Roman Empire: From the First Century A.D. to the Third.* Baltimore: Johns Hopkins University Press, 1984 [1976].

MacMullen, R., *Enemies of the Roman Order: Treason, Unrest and Alienation in the Empire.* Cambridge, Mass.: Harvard University Press, 1966.

———, *Soldier and Civilian in the Later Roman Empire.* Cambridge, Mass.: Harvard University Press, 1963.

Marsden, E. W., *The Campaign of Gaugamela: Alexander's Most Decisive Battle.* Chicago: Argonaut, 1964.

———, *Greek and Roman Artillery.* Oxford: Clarendon, 1969.

Mellersh, H. E. L., *Roman Soldier,* New York: Taplinger, 1964.

Morrison, John S., and J. F. Coates, *The Athenian Trireme.* Cambridge, Eng.: Cambridge University Press, 1986.

Morrison, John S., and R. T. Williams, *Green Oared Ships, 900–322 B.C.* London: Cambridge University Press, 1968.

Ormerod, H. A., *Piracy in the Ancient World.* Chicago: Argonaut, 1924.

Ostrogorsky, G., *History of the Byzantine State,* rev. ed. New Brunswick, N.J.: Rutgers University Press, 1969.

Parker, H. M. D., *The Roman Legions,* 2d ed. New York: Barnes and Noble, 1958.

Partington, J. R., *A History of Greek Fire and Gunpowder.* Cambridge: Heffner, 1960.

Polybius, *The Histories of Polybius.* Transl. by S. Shuckburgh and F. Huitisch. Bloomington, Ind.: Indiana University Press, 1962.

Pritchett, William Kendrick, *The Greek State at War.* 4 vols. Berkeley, Calif.: University of California Press, 1971–1985.

Rodgers, W. L., *Greek and Roman Naval Warfare.* Annapolis, Md.: United States Naval Institute, 1937.

Rouge, J., *Ships and Fleets of the Ancient Mediterranean.* Middletown, Ct.: Wesleyan University Press, 1981.

Runciman, Steven, *The Fall of Constantinople, 1453.* Cambridge, Eng.: Cambridge University Press, 1965.

Sandars, N. K., *The Sea Peoples,* 2d ed. London: Thames & Hudson, 1985.

Sanders, Lionel Jehuda, *Dionysius I of Syracuse and Greek Tyranny.* London: Croon Helm, 1987.

Scullard, H. H., *Scipio Africanus in the Second Punic War.* Cambridge: Cambridge University Press, 1930.

Smith, Richard Edwin, *Service in the Post-Marian Army.* Manchester: Manchester University Press, 1958.

Snodgrass, A. M., *Arms and Armour of the Greeks.* Ithaca, N.Y.: Cornell University Press, 1967.

Starr, Chester G., *The Emergence of Rome as Ruler of the Western World.* Westport, Ct.: Greenwood Press, 1982 [1950].

———, *The Influence of Sea Power on Ancient History.* New York: Oxford University Press, 1989.

———, *The Roman Imperial Navy,* 2d ed. New York: Barnes & Noble, 1960.

Tarn, W. W., *Alexander the Great,* 2 vols. Boston: Beacon Press, 1956.

———, *Hellenistic Military and Naval Developments.* Cambridge: Cambridge University Press, 1930.

Thiel, J. H., *A History of Roman Sea Power in Republican Times.* Amsterdam: North-Holland Publishing Co., 1954.

Thucydides, *Complete Writings.* New York: Random House (Modern Library), 1934.

Toynbee, Arnold J., *Constantine Porphyrogenitus and His World.* New York: Oxford University Press, 1973.

————, *Hannibal's Legacy: The Hannibalic War's Effect on Roman Life.* 2 vols. London: Oxford University Press, 1965.

Treadgold, Warren, *The Byzantine Revival, 780–842.* Stanford, Calif.: Stanford University Press, 1988.

Tsangadas, Bryon C. P., *The Fortifications and Defense of Constantinople.* East European Monographs, Boulder. New York: Columbia University Press, 1980.

Ure, P. N., *Justinian and His Age.* Harmondsworth, Eng.: Penguin, 1951.

Vegetius Renatus, Flavius, *The Military Institutions of the Romans.* Westport, Ct.: Greenwood Press, 1985.

Walbank, Frank William, *The Awful Revolution: The Decline of the Roman Empire in the West.* Liverpool, Eng.: Liverpool University Press, 1969.

Wallinga, W. T., *The Boarding-bridge of the Romans.* Groningen: J. B. Wolters, 1956.

Warry, John, *Warfare in the Classical World: An Illustrated Encyclopedia of Weapons, Warriors, and Warfare in the Ancient Civilizations of Greece and Rome.* London: Salamander Books, 1980.

Watson, G. R., *The Roman Soldier.* Ithaca, N.Y.: Cornell University Press, 1969.

Webster, Graham, *The Roman Imperial Army.* London: Black, 1969.

Wheeler, Everett L., *Strategem and the Vocabulary of Military Trickery.* Mnemosyne, Biblioteca Classica Batava, number 108. New York: E. J. Brill, 1988.

Winter, Frank, *Greek Fortification.* Toronto: University of Toronto Press, 1971.

Xenophon, *The Persian Expedition.* Harmondsworth, Eng.: Penguin, 1949.

Yadin, Yigael, *The Art of Warfare in Bibliocal Lands in the Light of Archaeological Study.* New York: McGraw-Hill, 1963.

THE MIDDLE AGES

Allmand, C. T. (ed.), *Society at War: The Experience of England and France During the Hundred Years War.* New York: Barnes and Noble, 1973.

————, *The Hundred Years War: England and France at War, c. 1300–c. 1450.* Cambridge, Eng.: Cambridge University Press, 1988.

Bachrach, Bernard S., *Merovingian Military Organization, 481–751.* Minneapolis: University of Minnesota Press, 1970.

Barber, Richard, *The Knight and Chivalry.* New York: Macmillan, 1966.

Barnie, John, *War in Medieval English Society: Social Values in the Hundred Years War, 1337–99.* Ithaca, N.Y.: Cornell University Press, 1974.

Beeler, John H., *Warfare in England, 1066–1189.* Ithaca, N.Y.: Cornell University Press, 1966.

————, *Warfare in Feudal Europe, 730–1200.* Ithaca, New York: Cornell University Press, 1972.

Blair, Claude, *European Armour circa 1066 to circa 1700.* London: Batsford, 1958.

Bradbury, Jim, *The Medieval Archer.* New York: St. Martins, 1985.

Burne, A. H., *The Agincourt War: A Military History of the Latter Part of the Hundred Years' War from 1369 to 1453.* New York: Oxford University Press, 1956.

——, *Battlefields of England.* London: Methuen, 1950.

——, *The Crecy War: A Military History of the Hundred Years' War from 1337 to the Peace of Bretigny, 1360.* New York: Oxford University Press, 1955.

Burns, N. T., and C. Reagan, *Concepts of the Hero in the Middle Ages and the Renaissance.* Albany, N.Y.: State University of New York Press, 1975.

Chambers, James, *The Devil's Horsemen: The Mongol Invasion of Europe.* New York: Antheneum, 1979.

Cheyney, E. P., *The Dawn of a New Era, 1250–1453.* New York: Harper, 1936.

Contamine, Philippe, *War in the Middle Ages.* Transl. by Michael Jones. New York: Basil Blackwell, 1986 [1980].

Crone, Patricia, *Slaves on Horses: The Evolution of the Islamic Polity.* New York: Cambridge University Press, 1980.

Dahmus, Joseph, *Seven Decisive Battles of the Middle Ages.* Chicago: Nelson-Hall, 1983.

Ffoulkes, Charles, *Arms and Armament.* London: Harrap, 1945.

Finucane, Ronald C., *Soldiers of the Faith: Crusaders and Moslems at War.* New York: St. Martin's, 1983.

Fowler, Kenneth, *Medieval Mercenaries.* London: Blackwell, 1990.

Gillingham, John, and J. C. Holt (eds.), *War and Government in the Middle Ages.* Totowa, N.J.: Barnes and Noble, 1984.

Glover, R. G., "English Warfare in 1066," *English Historical Review,* LXVII. Jan. 1952.

Glubb, John Bagot, *The Great Arab Conquests.* Englewood Cliffs, N.J.: Prentice-Hall, 1964.

Halperin, Charles J., *Russia and the Golden Horde: The Mongol Impact on Medieval Russian History.* Bloomington, Ind.: Indiana University Press, 1985.

Hewitt, H. J., *The Black Prince's Expedition of 1355–1357.* Manchester: Manchester University Press, 1958.

——, *The Organization of War under Edward III, 1338–62.* New York: Barnes & Noble, 1966.

Heymann, F. G., *John Ziska and the Hussite Revolution.* New York: Russell, 1969.

Hindley, Geoffrey, *Medieval Warfare.* New York: Putnam, 1971.

Hollister, C. W., *Anglo-Saxon Military Institutions.* New York: Oxford University Press, 1962.

——, *The Military Organization of Norman England.* Oxford: Clarendon, 1965.

Jones, Gwynn, *A History of the Vikings.* New York: Oxford University Press, 1968.

Keen, Maurice, *Chivalry.* New Haven, Ct.: Yale University Press, 1984.

——, *The Law of War in the Middle Ages.* London: Routledge & Kegan Paul, 1965.

Koch, H. W., *Medieval Warfare.* London: Bison Books, 1978.

Kwanten, Luc, *Imperial Nomads: A History of Central Asia, 500–1500.* Philadelphia: University of Pennsylvania Press, 1979.

Lewis, A. R., *Naval Power and Trade in the Mediterranean, A.D. 500–1100.* Princeton, N.J.: Princeton University Press, 1951.

Lot, Ferdinand, *L'Art Militaire et les Armées au Moyen Age en Europe et dans le Proche Orient.* Paris: Payot, 1946.

Martin, H. Desmond, *The Rise of Chingis Khan and the Conquest of Northern China.* New York: Octagon Books, 1971.

Mussett, L., *The Germanic Invasions: The Making of Europe, A.D. 400–600.* Transl. by C. James. University Park, Pa.: Pennsylvania State University Press, 1975.

Newark, P., *Medieval Warfare.* London: Bison Books, 1978.

Oman, Sir Charles, "The Art of War in the Fifteenth Century," *Cambridge Medieval History,* VIII. New York: Macmillan, 1936.

——, *The Art of War in the Middle Ages, 378–1515,* rev. ed. Ithaca, N.Y.: Cornell University Press, 1960.

——, *A History of the Art of War in the Middle Ages.* 2 vols. New York: Burt Franklin, 1969.

Painter, Sidney, *A History of the Middle Ages, 284–1500.* New York: Knopf, 1961.

Perroy, Edouard, *The Hundred Years' War.* London: Eyre & Spottiswoode, 1951.

Powicke, Michael, *Military Obligation in Medieval England.* Oxford: Clarendon, 1962.

Riley-Smith, Jonathan. *The Crusades: A Short History.* New Haven, Ct.: Yale University Press, 1987.

Rodgers, W. L., *Naval Warfare under Oars from the Fourth to the Sixteenth Centuries.* Annapolis, Md.: United States Naval Institute, 1939.

Runciman, Steven, *History of the Crusades.* 3 vols. Cambridge: Cambridge University Press, 1951–1955.

Sanders, I. J., *Feudal Military Service in England.* London: Oxford University Press, 1956.

Saunders, J. J., *The History of the Mongol Conquests.* London: Routledge & Kegan Paul, 1971.

Sellman, R. R., *Medieval English Warfare.* London: Roy, 1964.

Smail, R. C., *Crusading Warfare, 1097–1193.* London: Cambridge University Press, 1967.

Thompson, A. Hamilton, "The Art of War to 1400," *Cambridge Medieval History,* VI. New York: Macmillan, 1911–1936.

————, *Military Architecture in England during the Middle Ages.* London: Oxford University Press, 1912.

Unger, Richard, *The Ship in the Medieval Economy, 600–1600.* London: Croom Helm, 1980.

Verbruggen, J. F., *The Art of Warfare in Western Europe During the Middle Ages to 1340.* Transl. by Sumner Willard and S. C. M. Southern. Amsterdam: Elsevier/North Holland, 1977.

White, Lynn, *Medieval Technology and Social Change.* Oxford, Eng.: Clarendon Press, 1962.

Wise, Terrance, *Medieval Warfare.* New York: Hastings House, 1976.

THE RENAISSANCE TO NAPOLEON

GENERAL AND MILITARY

Alden, John R., *The American Revolution.* New York: Harper, 1954.

————, "The Military Side of the Revolution," *Manuscripta,* IX. 1957.

Anderson, M. S., *War and Society in Europe of the Old Regime, 1618–1789.* Fontana War and Society Series. New York: St. Martin's, 1988.

Ashley, M. P., *Crowell's Generals.* New York: St. Martin's, 1954.

————, *The Greatness of Oliver Cromwell.* New York: Crowell Collier & Macmillan, 1966.

Atkinson, C. T., *Marlborough and the Rise of the British Army.* New York: Putnam, 1941.

Baurmeister, Karl, *Revolution in America: Confidential Letters and Journals, 1776–1784.* New Brunswick, N.J.: Rutgers University Press, 1957.

Bayley, C. C., *War and Society in Renaissance Florence: The De Militia of Leonardo Bruni.* Toronto: University of Toronto Press, 1961.

Belloc, Hilaire, *The Tactics and Strategy of the Great Duke of Marlborough.* London: Arrowsmith, 1933.

Berker, T. M., *The Military Intellectual and Battle: Raimundo Montecuccoli and the Thirty Years War.* Albany: State University of New York Press, 1975.

Black, J. (ed.), *The Origins of War in Early Modern Europe.* Edinburgh: Humanities Press, 1987.

Blomfield, Reginald, *Vauban.* London: Methuen, 1938.

Bryant, Arthur, *The Age of Elegance, 1812–1822.* London: Collins, 1950.

————, *Years of Endurance, 1793–1802.* New York: Harper, 1947.

————, *Years of Victory, 1802–1812.* London: Collins, 1944.

Buchan, John, *Oliver Cromwell.* Mystic, Ct.: Verry, 1957.

Burne, A. H., *The Great Civil War: A Military History of the First Civil War, 1642–1646.* London: Eyre & Spottiswoode, 1959.

Burton, I. F., *The Captain-General* (Marlborough). London: Constable, 1969.

Chandler, David, *The Art of War in the Age of Marlborough*. London: Batsford, 1976.

———, *The Campaigns of Napoleon*. New York: Crowell Collier & Macmillan, 1966.

——— (ed.), *Robert Parker and Comte de Merode-Westerloo: The Marlborough Wars*. London: Longmans, 1968.

Churchill, Winston, *Marlborough*. 4 vols. London: Harrap, 1933–1938.

Clark, George, *War and Society in the Seventeenth Century*. Cambridge, Eng.: Cambridge University Press, 1958.

Colby, Elbridge, *Masters of Mobile Warfare* (chapters on Marlborough, Frederick the Great, and Napoleon). Princeton, N.J.: Princeton University Press, 1943.

Colin, J. L. A., *L'Education Militaire de Napoléon*. Paris: Chapelot, 1900.

Craig, G. A., *The Politics of the Prussian Army, 1640–1945*. New York: Oxford University Press, 1964.

Cruickshank, C. G., *Elizabeth's Army*, rev. ed. London: Oxford University Press, 1950.

Davidson, Philip, *Propaganda and the American Revolution*. Chapel Hill, N.C.: University of North Carolina Press, 1967.

Davies, Godfrey, *Wellington and His Army*. Oxford: Oxford University Press, 1954.

Deiss, J. J., *Captains of Fortune: Profiles of Six Italian Condottieri*. New York: T. Y. Crowell, 1966.

Dodge, T. A., *Gustavus Adolphus: A History of the Art of War from the Middle Ages to the War of the Spanish Succession*. Boston: Houghton Mifflin, 1895.

Duffy, Christopher, *Siege Warfare: The Fortress in the Early Modern World, 1494–1660*. London: Routledge & Kegan Paul, 1979.

Duffy, M., *The Military Revolution and the State, 1500–1800*. Exeter, Eng.: University of Exeter Press, 1980.

Dunn, Richard S., *The Age of Religious Wars, 1559–1715*. The Norton History of Modern Europe, 2d ed. New York: Norton, 1979 [1970].

Ergang, R. R., *The Myth of the All-Destructive Fury of the Thirty Years' War*. Pocono Pines, Pa.: Craftsmen, 1956.

———, *The Potsdam Führer*. New York: Cambridge University Press, 1941.

Firth, C. H., *Cromwell's Army*, 2d ed. London: Methuen, 1962.

Fisher, H. A. L., *Napoleon*, 2d ed. New York: Oxford University Press, 1967.

Fortescue, J. W., *Six British Soldiers* (Cromwell and Malborough). London: Williams & Norgate, 1928.

———, *Wellington*. London: Williams & Norgate, 1928.

Frederick II of Prussia, *Instructions for His Generals*, Harrisburg, Pa.: Stackpole, 1951.

Fuller, J. F. C., *British Light Infantry in the Eighteenth Century*. London: Hutchinson, 1925.

———, *Sir John Moore's System of Training*. London: Hutchinson, 1925.

Gershoy, Leo, *From Despotism to Revolution, 1763–1789.* New York: Harper, 1944.

Glover, R. G., *Peninsular Preparation: The Reform of the British Army, 1759–1809.* Cambridge: Cambridge University Press, 1963.

Goodman, David C., *Power and Penury: Government, Technology, and Science in Philip II's Spain.* New York: Cambridge University Press, 1988.

Hale, J. R., *Renaissance Fortification: Art or Engineering?* London: Thames and Hudson, 1977.

————, *War and Society in Renaissance Europe, 1450–1620.* Fonatana History of European War and Society. Baltimore: Johns Hopkins University Press, 1985.

Hale, J. R., and Michael Mallett, *The Military Organization of a Renaissance State: Venice C. 1400 to 1617.* Cambridge, Eng.: Cambridge University Press, 1984.

Hall, A. R., *Ballistics in the Seventeenth Century: A Study in the Relations of Science and War with Reference Principally to England.* New York: Cambridge University Press, 1952.

Hegemann, Werner, *Frederick the Great.* London: Constable, 1929.

Henderson, Nicholas, *Prince Eugen of Savoy: A Biography.* London: Weidenfeld and Nicolson, 1964.

Hibbert, Christopher, *Wolfe at Quebec.* Cleveland: World, 1959.

Higginbotham, R. Don, *The War of American Independence: Military Attitudes, Politics, and Practices.* New York: Macmillan, 1967.

Hill, J. M., *Celtic Warfare, 1595–1763.* Atlantic Highlands, N.J.: Humanities Press, 1986.

Jameson, J. F., *The American Revolution Considered as a Social Movement.* Princeton, N.J.: Princeton University Press, 1926.

Kamen, Henry, *The War of Succession in Spain, 1700–1715.* Bloomington, Ind.: Indiana University Press, 1969.

Kennett, Lee, *The French Armies in the Seven Years' War: A Study of Military Organization and Administration.* Durham, N.C.: Duke University Press, 1967.

Koch, H. W., *The Rise of Modern Warfare, 1618–1815.* Englewood Cliffs, N.J.: Prentice-Hall, 1981.

Leach, D. E., *Flintlock and Tomahawk: New England in King Philip's War.* New York: Norton, 1966.

Lewis, Michael, *Armada Guns: A Comparative Study of English and Spanish Armaments.* New York: Fernhill, 1961.

Liddell Hart, B. H., *The Ghost of Napoleon.* New Haven, Ct.: Yale University Press, 1933.

Luvaas, Jay (ed.), *Frederick the Great on the Art of War.* New York: The Free Press, 1966.

McCardell, Lee, *Ill-Starred General: Braddock of the Coldstream Guards.* Pittsburgh: University of Pittsburgh Press, 1958 (paperback).

Macdonell, A. G., *Napoleon and His Marshals.* London: Macmillan, 1934.

Machiavelli, Nicolo, *The Art of War.* Indianapolis, Ind.: Bobbs-Merrill, 1965.

————, *The Prince*. Harmondsworth, Eng.: Penguin, 1961.

Mackesy, Piers, *The War for America, 1775–1783*. Cambridge, Mass.: Harvard University Press, 1964.

Macleod, W. C., *The American Indian Frontier*. New York: Knopf, 1928.

MacMunn, George, *Gustavus Adolphus*. London: Hodder & Stoughton, 1930.

Mallett, Michael, *Mercenaries and Their Masters: Warfare in Renaissance Italy*. Totowa, N.J.: Rowman and Littlefield, 1974.

Millar, G. J., *Tudor Mercenaries and Auxiliaries, 1485–1547*. Charlottesville, Va.: University of Virginia Press. 1980.

Miller, J. C., *Triumph of Freedom*. Boston: Little Brown, 1948 (paperback).

Montross, Lynn, *Rag, Tag, and Bobtail: The Story of the Continental Army, 1775–1783*. New York: Harper, 1952.

Nicholson, G. W. L., *Marlborough and the War of the Spanish Succession*. Ottawa: Queen's Printer, 1955.

Nickerson, Hoffman, *The Armed Horde, 1793–1939: A Study of the Rise, Survival, and Decline of the Mass Army*. New York: Putnam's, 1940.

Oman, Carola, *Sir John Moore*. Mystic, Ct.: Verry, 1953.

Oman, Charles, *History of the Art of War in the Sixteenth Century*. London: Methuen, 1937.

————, *A History of the Peninsular War*. 5 vols. Oxford: Clarendon, 1902.

————, *Studies in the Napoleonic Wars*. London: Methuen, 1929.

————, *Wellington's Army, 1809–1814*. New York: Longmans Green, 1912.

Omond, J. S., *Parliament and the Army*. Cambridge: Cambridge University Press, 1933.

O'Neil, B. H., *Castles and Cannon: A Study of Early Artillery Fortifications in England*. Oxford: Clarendon, 1960.

Pargellis, S. M., *Lord Loudoun in North America*. Hamden, Ct.: Shoe String Press, 1968.

Parker, Geoffrey, *The Army of Flanders and the Spanish Road: 1567–1659*. Cambridge, Eng.: Cambridge University Press, 1972.

————, *The Military Revolution: Military Innovation and the Rise of the West, 1500–1800*. Cambridge, Eng.: Cambridge University Press, 1988.

————, *The Thirty Years' War*. New York: Military Heritage Press, 1988 [1987].

Parker, H. M. D., *The Thirty Years War*. London: Routledge and Kegan Paul, 1967.

Parker, Harold, *Three Napoleonic Battles*. Durham, N.C.: Duke University Press, 1973 [1944].

Parkman, Francis, *Montcalm and Wolfe*. 2 vols. Boston: Little, Brown, 1905–1907.

Peckham, H. H., *The War for Independence: A Military History*. Chicago: Chicago University Press, 1958.

Pepper, Simon, and Nicholas Adams, *Firearms and Fortifications: Military Architecture and Siege Warfare in Sixteenth-century Siena*. Chicago: University of Chicago Press, 1986.

Phipps, R. W., *The Armies of the First French Republic and the Rise of the Marshals of Napoleon.* 5 vols. London: Oxford University Press, 1926–39.

Prebble, John, *Culloden.* Harmondsworth, Eng.: Penguin, 1967.

Quimby, R. S., *The Background of Napoleonic Warfare: The Theory of Military Tactics in Eighteenth Century France.* New York: AMS Press, 1957.

Roberts, Michael, *The Military Revolution, 1560–1660.* Belfast: Belfast University Press, 1956.

Roberts, Penfield, *The Quest for Security, 1715–1740.* New York: Harper, 1947.

Rose, J. Holland, *The Personality of Napoleon.* London: G. Bell, 1912.

Rothenberg, Gunther E., *The Art of War in the Age of Napoleon.* Bloomington, Ind.: Indiana University Press, 1980.

Russell, C. P., *Guns on the Early Frontiers: A History of Firearms from Colonial Times through the Years of the Western Fur Trade.* Berkeley, Calif.: University of California Press, 1957.

Savory, Reginald, *His Britannic Majesty's Army in Germany during the Seven Years' War.* New York: Oxford University Press, 1966.

Saxe, Maurice de, *Reveries on the Art of War.* Harrisburg, Pa.: Military Service, 1944.

Scheer, G. F., and H. F. Rankin, *Rebels and Redcoats.* Cleveland: World, 1957.

Scott, Samuel F., *The Response of the Royal Army to the French Revolution, 1787–1793.* Oxford, Eng.: Clarendon Press, 1978.

Scouller, R. E., *The Armies of Queen Anne.* New York: Oxford University Press, 1966.

Snyderman, G. S., *Behind the Tree of Peace: A Sociological Analysis of Iroquois Warfare.* Philadelphia: University of Pennsylvania Press, 1948.

Stacey, C. P., *Quebec, 1759: The Siege and the Battle.* New York: St. Martin's, 1959.

Stoye, John, *The Siege of Vienna.* London: Collins, 1964.

Taylor, F. L., *The Art of War in Italy, 1494–1529.* London: Cambridge University Press, 1921.

Taylor, Frank, *The Wars of Marlborough, 1702–1709.* 2 vols. Oxford: Blackwell, 1921.

Thompson, I. A. A., *War and Government in Habsburg Spain, 1560–1620.* London: Athlone Press, 1976.

Thompson, J. W., *The Wars of Religion in France, 1559–1576.* New York: Ungar, 1964.

Trevelyan, G. M., *England under Queen Anne.* 3 vols. New York: Longmans Green, 1932–1934.

Vaillant, G. C., *The Aztecs of Mexico,* 2d ed. Harmondsworth, Eng.: Penguin, 1962.

Ward, Christopher, *The War of the Revolution,* 2 vols. Ed. by John R. Alden. New York: Macmillan, 1952.

Watson, Francis, *Wallenstein: Soldier Under Saturn.* London: Chatto and Windus, 1938.

Wedgwood, C. V., *The Common Man in the Great Civil War.* Leicester: Leicester University Press, 1957.

Wedgwood, C. V., *The King's War, 1641–1647.* New York: Macmillan, 1959.

———, *The Thirty Years' War.* Harmondsworth, Eng.: Penguin, 1961.

Weller, Jac, *Wellington in the Peninsula, 1808–1814.* London: N. Vane, 1962.

White, J. M., *Marshal of France: The Life and Times of Maurice, Comte de Saxe.* New York: Rand McNally, 1962.

Wise, S. F., "The American Revolution and Indian History," *Character and Circumstance: Essays in Honour of D. G. Creighton.* Toronto: Macmillan, 1970.

Woodhouse, A. S. P. (ed.), *Puritanism and Liberty: The Army Debates, 1647–49.* Chicago: University of Chicago Press, 1951.

Yorck von Wartenburg, Maximilian, *Napoleon as a General.* 2 vols. London: Paul Trench, Trübner, 1902.

NAVAL

Albion, R. G., *Forests and Sea Power: The Timber Problem of the Royal Navy, 1652–1862.* Hamden, Ct.: Archon, 1965.

Bamford, P. W., *Forests and French Sea Power, 1660–1789.* Toronto: University of Toronto Press, 1956.

Boxer, Charles R., *The Dutch Seaborne Empire, 1600–1800.* New York: Knopf, 1965.

———, *The Portuguese Seaborne Empire, 1415–1825.* New York: Knopf, 1970.

Bryant, Arthur, *Samuel Pepys.* 3 vols. Cambridge: Cambridge University Press, 1933–1936.

Cipolla, Carlo, *Guns and Sails in the Early Phase of European Expansion, 1400–1700.* London: Collins, 1965.

Clark, G. N., *The Dutch Alliance and the War against French Trade, 1688–1697.* New York: Longmans Green, 1923.

Corbett, Julian, *The Campaign of Trafalgar.* London: Longmans Green, 1910.

———, *Drake and the Tudor Navy.* 2 vols. New York: Burt Franklin, 1965.

———, *England in the Seven Years' War.* 2 vols. London: Longmans Green, 1907.

———, (ed.), *Fighting Instructions, 1530–1816.* London: Navy Records Society, 1905.

———, *The Successors of Drake.* New York: Burt Franklin, 1969.

Ehrman, John, *The Navy in the War of William III, 1689–1697.* Cambridge: Cambridge University Press, 1953.

Graham, G. S. *Empire of the North Atlantic: The Maritime Struggle for North America,* 2d ed. Toronto: University of Toronto Press, 1958.

Great Britain, Admiralty, *Evidence Relating to the Tactics . . . at Trafalgar.* London: HMSO, 1913.

Grenfell, Russell, *Nelson, the Sailor.* London: Faber & Faber, 1949.

Guilmartin, John F., Jr., *Gunpowder and Galleys: Changing Technology and Mediterranean Warfare at Sea in the Sixteenth Century.* New York: Cambridge University Press, 1974.

James, William, *The British Navy in Adversity.* New York: Longmans Greens, 1926.

Laughton, J. K. (ed.), *State Papers Relating to the Defeat of the Spanish Armada.* London: Navy Records Society, 1894.

Lewis, Michael, *A Social History of the Navy, 1793–1815.* London: Allen & Unwin, 1960.

Mackesy, Piers, *The War in the Mediterranean, 1803–1810.* Cambridge, Mass.: Harvard University Press, 1957.

——, *The Life of Nelson.* 2 vols. New York: Haskell, 1968.

——, *The Major Operations of the Navies in the War of American Independence.* New York: Greenwood, 1968.

Mahan, A. T., *Sea Power in its Relations to the War of 1812.* 2 vols. New York: Greenwood, 1969.

Malone, J. J., *Pine Trees and Politics: Naval Stores and Forest Policy in Colonial New England.* Seattle: University of Washington Press, 1964.

Martin, Colin, and Geoffrey Parker, *The Spanish Armada.* New York: W. W. Norton, 1988.

Mattingly, Garrett, *The Defeat of the Spanish Armada.* New York: Houghton Mifflin, 1962.

Morison, S. E., *Admiral of the Ocean Sea.* Boston: Little, Brown, 1942.

Oakeshott, W. F., *Founded upon the Seas.* Cambridge: Cambridge University Press, 1942.

Oman, Carola, *Nelson.* Mystic, Ct." Verry, 1967.

Owen, J. H., *War at Sea under Queen Anne.* Cambridge, Eng.: Cambridge University Press, 1938.

Parry, J. H., *The Age of Reconnaissance.* New York: Praeger, 1969.

Perrin, W. G., *Nelson's Signals: The Evolutions of the Signal Flags.* London: HMSO, 1908.

Phillips, C. R., *Six Galleons for the King of Spain: Imperial Defence in the Early Seventeenth Century.* Baltimore: Johns Hopkins University Press, 1986.

Pool, Bernard, *Navy Board Contracts, 1660–1832.* London: Longmans, 1966.

Powley, E. B., *The English Navy in the Revolution of 1688.* Cambridge: Cambridge University Press, 1928.

Richmond, Herbert, *The Navy as an Instrument of Policy, 1558–1727.* Cambridge, Eng.: Cambridge University Press, 1953.

Rose, J. Holland, "Napoleon and Sea Power," *The Indecisiveness of Modern War.* Port Washington, N.Y.: Kennikat, 1968.

Rowse, A. L., *Sir Richard Grenville.* London: Cape, 1937.

Taylor, A. H., "The Battle of Trafalgar," *Mariners' Mirror,* XXXVI. 1950.

Tunstall, Brian, *Admiral Byng and the Loss of Minorca.* London: P. Allan, 1928.

————, *Nelson.* New York: Dufour, 1950.

Williamson, J. A., *The Age of Drake.* New York: Barnes & Noble, 1960.

————, *Hawkins of Plymouth,* 2d ed. New York: Barnes & Noble, 1969.

Woodrooffe, Thomas, *The Enterprise of England: An Account of Her Emergence as an Oceanic Power.* London: Faber & Faber, 1958.

THE NINETEENTH CENTURY

GENERAL AND MILITARY

Angell, Norman, *The Great Illusion: A Study of the Relation of Military Power in Nations to their Economic and Social Advantage.* New York: Putnam's, 1911.

Black, R. C., *The Railroads of the Confederacy.* Chapel Hill, N.C.: University of North Carolina Press, 1952.

Bourne, Kenneth, *Britain and the Balance of Power in North America, 1815–1908.* Berkeley, Calif.: University of California Press, 1967.

Bruce, R. V., *Lincoln and the Tools of War.* Urbana: University of Illinois Press, 1989 [1956].

Catton, Bruce, *The Centennial History of the Civil War.* 3 vols. Garden City, N.Y.: Doubleday, 1961–1965.

Challener, Richard D., *The French Theory of the Nation in Arms, 1866–1939.* New York: Russell, 1955.

Clausewitz, Carl von, *On War.* Ed. by Michael Howard and Peter Paret. Princeton, N.J.: Princeton University Press, 1986.

Coggins, Jack, *Arms and Equipment of the Civil War.* Garden City, N.Y.: Doubleday, 1962.

Coles, Harry L., *The War of 1812.* Chicago: University of Chicago Press, 1965.

Curtiss, John S., *The Russian Army under Nicholas I, 1825–1855.* Durham, N.C.: Duke University Press, 1965.

Dowdey, Clifford, *Death of a Nation: The Story of Lee and his Men at Gettysburg.* New York: Knopf, 1958.

Downey, F. D., *The Guns at Gettysburg.* New York: Crowell Collier & Macmillan, 1962.

Du Picq, A., *Battle Studies.* Harrisburg, Pa.: Military Service, 1947.

Falls, C. B., *A Hundred Years of War, 1850–1950.* New York: Crowell Collier & Macmillan, 1962.

Fite, E. D., *Social and Industrial Conditions in the North during the Civil War.* New York: Ungar, 1963.

Fortescue, J. W., *History of the British Army,* XI–XIII. New York: Macmillan, 1902–1930.

Freeman, D. S., *R. E. Lee.* 4 vols. New York: Scribner's, 1951.

Greene, Francis Vinton, *The Russian Army and its Campaigns in Turkey in 1877–78.* New York: Appleton, 1879.

Henderson, G. F. R., *Stonewall Jackson and the American Civil War.* New York: Longmans Green, 1963.

Hilton, Richard, *The Indian Mutiny: A Centenary History.* London: Hollis & Carter, 1957.

Howard, Michael, *The Franco-Prussian War: The German Invasion of France, 1870–1871.* New York: Crowell Collier & Macmillan, 1969.

———, "The French and Prussian Staff Systems before 1870," *Journal of the American Military History Foundation,* II. 1938.

———, "The French Discovery of Clausewitz and Napoleon," *Journal of the American Military Institute,* IV. 1940.

Irvine, D. D., "Origins of Capital Staffs," *Journal of Modern History,* X. 1938.

Jessup, P. C., *et al., Neutrality: Its History, Economics and Law.* 4 vols. New York: Columbia University Press, 1935–36.

Jomini, Henri, *Summary of the Art of War.* Harrisburg, Pa.: Stackpole, 1952.

Kitchen, Martin, *The German Officer Corps, 1890–1914.* Oxford: Clarendon, 1968.

Livermore, T. L., *Numbers and Losses in the Civil War in America, 1861–65.* New York: Kraus, 1968.

Luvaas, Jay, *The Civil War: A Soldier's View: A Collection of Civil War Writings.* Chicago: University of Chicago Press, 1958.

———, *The Education of an Army: British Military Thought, 1815–1940.* Chicago: University of Chicago Press, 1964.

———, *The Military Legacy of the Civil War: The European Inheritance.* Chicago: University of Chicago Press, 1959.

Luard, C. E., "Field Railways and their General Application in War," *R.U.S.I. Journal,* XVII. 1873.

McElwee, William, *The Art of War: Waterloo to Mons.* Bloomington, Ind.: Indiana University Press, 1974.

McPherson, James, *Battle Cry of Freedom: The Civil War Era.* New York: Oxford University Press, 1988.

McWhiney, Grady, and D. Jamieson Perry, *Attack and Die: Civil War Military Tactics and the Southern Heritage.* University, Ala.: University of Alabama Press, 1982.

Maurice, J. F., *The System of Field Manoeuvres Best Adapted for Enabling our Troops to Meet a Continental Army.* Edinburgh: Blackwood, 1872.

Moltke, H. K. B. von, *The Franco-German War of 1870–71.* London: Harper, 1907.

Nelson, O. L., *National Security and the General Staff.* Washington: Infantry Journal Press, 1946.

Nickerson, Hoffman, *The Armed Horde, 1793–1939: A Study of the Rise, Survival, and Decline of the Mass Army.* New York: Putnam's, 1940.

Oppenheim, L. F. L., *International Law, II. Disputes: War and Neutrality,* 7th ed. New York: Longmans Green, 1963.

Paret, Peter, *Clausewitz and the State: The Man, His Theories, and His Times.* Princeton, N.J.: Princeton University Press, 1985 [1976].

Pratt, E. A., *The Rise of Rail Power in War and Conquest, 1833–1914*. London: King, 1916.

Preston, R. A., *Canada and "Imperial Defense": A Study of the Origins of the British Commonwealth's Defense Organization, 1867–1919*. Durham, N.C.: Duke University Press, 1967.

Randall, J. G., and David Donald, *The Civil War and Reconstruction*, 2d ed. Boston: Heath, 1961.

Ritter, E. A., *Shaka Zulu: The Rise of the Zulu Empire*. New York: Longmans Green, 1964.

Schellendorff, Paul von, *The Duties of the General Staff*. London: HMSO, 1907.

Sen, Surendra Nath, *Eighteen Fifty-Seven*. Delhi: Government of India, 1957.

Stacey, C. P., *Canada and the British Army: A Study in the Practice of Responsible Government*, rev. ed. Toronto: University of Toronto Press, 1963.

——, "The Myth of the Unguarded Frontier, 1815–1871," *American Historical Review*, LVI. 1950.

Stackpole, E. J., *Chancellorsville: Lee's Greatest Battle*. Harrisburg, Pa.: Stackpole, 1958.

——, *Drama on the Rappahannock: The Fredericksburg Campaign*. Harrisburg, Pa.: Military Service, 1957.

Steele, M. F., *American Campaigns*. 2 vols. Washington: U.S. Infantry Association, 1939–43.

Trask, David, *The War with Spain in 1898*. New York: Macmillan, 1981.

U.S. Military Academy, Department of Military Art and Engineering, *Jomini, Clausewitz, and Schlieffen*. West Point: Government Printing Office, 1951.

Upton, Emory, *The Military Policy of the United States*. New York: Greenwood, 1968.

Vandiver, Frank, *Mighty Stonewall*. New York: Mc-Graw-Hill, 1957.

Webster, Charles, *The Congress of Vienna*. New York: Barnes & Noble, 1963.

Whitton, Frederick E., *Moltke*. London: Constable, 1921.

William, K. P., *Lincoln Finds a General*. 5 vols. New York: Macmillan, 1949–1959.

Williams, T. H., *Lincoln and His Generals*. New York: Random House, 1952 (paperback).

Woodham-Smith, Cecil, *The Reason Why*. New York: Dutton, 1952.

NAVAL

Albion, R. G., and J. B. Pope, *Sea Lanes in Wartime: The American Experience: 1775–1942*. Hamden, Ct.: Shoe String Press, 1968.

Anderson, Bern, *By Sea and by River: The Naval History of the Civil War*. New York: Knopf, 1963.

Bartlett, C. J., *Great Britain and Sea Power, 1815–1853*. Oxford: Clarendon, 1963.

Baxter, J. P., "The British Government and Neutral Rights," *American Historical Review,* XXXIV. Oct. 1928.

———, *The Introduction of the Ironclad Warship.* Hamden, Ct.: Shoe String Press, 1968.

Callendar, G. A. R., and F. H. Hinsley, *The Naval Side of British History, 1485–1945.* London: Chatto, 1960.

Chapelle, Howard I. *The Search for Speed under Sail, 1700–1855.* New York: Norton, 1967.

Cotter, C. H., *A History of Nautical Astronomy.* New York: American Elsevier, 1968.

Daly, Robert W., *How the 'Merrimac' Won: The Strategic Story of the C.S.S. Virginia.* New York: Crowell, 1957.

Gardiner, C. Harvey, *Naval Power in the Conquest of Mexico.* Austin, Tex.: University of Texas Press, 1956.

Hearnshaw, F. J. C., *Sea Power and Empire.* London: Harrap, 1940.

James, W. M., *The Influence of Sea Power on the History of the British People.* Cambridge: Cambridge University Press, 1948.

Lewis, Michael, *The Navy in Transition: 1814–1864.* Mystic, Ct.: Verry, 1965.

Lloyd, Christopher, *The Navy and the Slave Trade: The Suppression of the African Slave Trade in the Nineteenth Century.* New York: Barnes & Noble, 1968.

Marder, A. J., *The Anatomy of British Sea Power: A History of British Naval Power in the Pre-dreadnought Era, 1880–1905.* London: F. Cass, 1964.

Marder, A. J., "From Jimmu Tenno to Perry," *American Historical Review.* Oct. 1945.

Penn, Geoffrey, *"Up Funnel, Down Screw!": The Story of the Naval Engineer.* London: Hollis & Carter, 1955.

Preston, Antony, and John Major, *Send a Gunboat: A Study of the Gunboat and its Role in British Policy, 1854–1904.* London: Longmans Green, 1967.

Richmond, Herbert, *Statesmen and Sea Power.* Oxford: Clarendon, 1946.

Robertson, F. L., *The Evolution of Naval Armament.* London: Constable, 1921.

Savage, Carlton, *The Policy of the United States toward Maritime Commerce in War, 1776–1918.* 2 vols. New York: Kraus, 1969.

Seager, Robert, II, *Alfred Thayer Mahan: The Man and His Letters.* Annapolis, Md.: Naval Institute Press, 1977.

Schurman, D. M., *The Education of a Navy: The Development of British Naval Strategic Thought, 1867–1914.* Chicago: University of Chicago Press, 1965.

Smith, D. B., and A. C. Dewar (eds.), *The Russian War, 1854–55.* 3 vols. London: Navy Records Society. 1943–1945.

Sprout, H., and M. T. Sprout, *The Rise of American Naval Power.* Princeton, N.J.: Princeton University Press, 1943.

Sprout, M. T., "Mahan," E. M. Earle (ed.), *Makers of Modern Strategy.* Princeton, N.J.: Princeton University Press, 1948.

Wescott, A. F. (ed.), *American Sea Power Since 1775.* Philadelphia: Lippincott, 1947.

Westcott, A. F. (ed.), *Mahan on Naval Warfare: Selections from the Writings of Rear Admiral Alfred T. Mahan.* Boston: Little, Brown, 1918.

Wilson, H. W., *Ironclads in Action.* 2 vols. London: S. Low & Marston, 1896.

THE TWENTIETH CENTURY TO 1939

Albertini, Luigi, *The Origins of the War of 1914.* 3 vols. New York: Oxford University Press, 1952–1957.

Aron, Raymond, *The Century of Total War.* Boston: Beacon, 1955.

Ashworth, Tony, *Trench Warfare, 1914–1918.* New York: Holmes & Meier, 1980.

Aston, George, *The Study of War for Statesmen and Citizens.* New York: Longmans Green, 1927.

Baldwin, Hanson W., *World War I: An Outline History.* New York: Harper & Row, 1962.

Banse, Ewald, *Germany Prepares for War.* New York: Harcourt Brace, 1934.

Barnett, Correlli, *The Swordbearers: Supreme Command in the First World War.* New York: Morrow, 1964.

Bliss, Tasker, "The Evolution of the Unified Command," *Foreign Affairs,* I. Dec. 1922.

Bloch, I. S., *The Future of War in its Technical, Economic and Political Relations.* Boston: Ginn, 1902.

Boyle, Andrew, *Trenchard.* London: Collins, 1962.

Bruntz, G. F., *Allied Propaganda and the Collapse of the German Empire in 1918.* Stanford, Calif.: Stanford University, Hoover War Library Publications, 1938.

Buell, R. L., *The Washington Conference.* New York: Putnam's, 1922.

Chambers, F. P., *The War Behind the War, 1914–1918.* London: Faber & Faber, 1939.

Chatterton, E. K., *The Big Blockade.* London: Hurst & Blackett, 1932.

Churchill, W. L. S., *The Unknown War: The Eastern Front.* London: Butterworth, 1931.

Churchill, W. L. S., *The World Crisis, 1911–1918.* 6 vols. New York: Scribner's, 1923–1931.

Coffman, Edward M., *The War to End all Wars: The American Military Experience in World War I.* New York: Oxford University Press, 1969.

Cruttwell, C.R.M.T. *A History of the Great War, 1914–1918.* Oxford: Oxford University Press, 1940.

Dawson, R. M., "The Cabinet Minister and Administration: Asquith, Lloyd George, Curzon," *Political Science Quarterly,* LV. Sept. 1940.

——, "The Cabinet Minister and Administration: Winston Churchill at the Admiralty," *Canadian Journal of Economics and Political Science,* VI. Aug. 1940.

De Gaulle, Charles, *The Army of the Future.* Philadelphia: Lippincott, 1941.

Dunlop, J. K., *The Development of the British Army, 1899–1914.* London: Methuen, 1938.

Edmonds, J. E., *A Short History of World War One.* London: Oxford University Press, 1951.

Ehrman, John, *Cabinet Government and War, 1890–1940.* Hamden, Ct.: Shoe String Press, 1969.

Erickson, John, *The Soviet High Command: A Military-Political History, 1918–1941.* New York: St. Martin's, 1962.

Falkenhayn, Erich von, *General Headquarters, 1914–1916, and Its Critical Decisions.* New York: Dodd Mead, 1919.

Falls, C. B., *The First World War.* London: Longmans, 1964.

Fay, S. B., *The Origins of the World War,* 2d ed. New York: The Free Press, 1966.

Fedotoff-White, Dimitri, *The Growth of the Red Army.* Princeton, N.J.: Princeton University Press, 1944.

Foerster, Wolfgang, *La Stratégie Allemande . . . 1914–1918.* Paris: Payot, 1929.

Fredette, R. H., *The Sky on Fire: The First Battle of Britain, 1917–18 and the Birth of the Royal Air Force.* New York: Holt, Rinehart & Winston, 1966.

Frost, H. H., *The Battle of Jutland.* London: Stevens & Brown, 1936.

Fuller, J. F. C., *Armoured Warfare.* Harrisburg, Pa.: Military Service, 1943.

——, *Lectures on Field Service Regulations II.* London: Sifton Praed, 1931.

——, *Memoirs of an Unconventional Soldier.* London: Nicholson & Watson, 1936.

——, *The Reformation of War.* London: Hutchinson, 1923.

Fussell, Paul, *The Great War and Modern Memory.* New York: Oxford University Press, 1975.

Garthoff, R. L., *Soviet Military Doctrine.* Glencoe, Ill.: The Free Press, 1953.

Gibson, Langhorne, and J. E. T. Harper, *The Riddle of Jutland.* New York: Coward-McCann, 1934.

Golovine, N. N., *The Russian Campaign of 1914: The Beginning of the War and Operations in East Prussia.* London: Rees, 1933.

Goodspeed, D. J., *Ludendorff: Genius of World War I.* Toronto: Macmillan, 1966.

Gordon, H. J., *The Reichswehr and the German Republic, 1919–1926.* Princeton, N.J.: Princeton University Press, 1957.

Gottman, Jean, "The Background of Geopolitics," *Military Affairs,* VI. 1942.

Hankey, M. P., *The Supreme Command, 1914–1918.* 2 vols. New York: Fernhill, 1961.

Harper, J. E. T., *The Truth about Jutland.* London: Murray, 1927.

Higham, R., *Armed Forces in Peacetime: Britain, 1918–1940.* London: Foulis, 1963.

———, *The British Rigid Airship, 1908–1931.* London: Foulis, 1961.

———, *The Military Intellectuals in Britain, 1918–1939.* New Brunswick, N.J.: Rutgers University Press, 1966.

Hoffman, Max., *War Diaries and Other Papers.* 2 vols. London: M. Secker, 1929.

Holley, I. B., *Ideas and Weapons: Exploitation of the Aerial Weapon by the United States during World War I: A Study in the Relationship of Technological Advance, Military Doctrine and the Development of Weapons.* New Haven, Ct.: Yale University Press, 1953.

Holt, Edgar, *The Boer War.* London: Putnam, 1958.

Hunter, T. M., *Marshal Foch: A Study in Leadership.* Ottawa: Queen's Printer, 1961.

Johnson, R. M., *First Reflections on the Campaign of 1918.* New York: Holt, 1920.

Larson, Robert H., *The British Army and the Theory of Armored Warfare, 1918–1940.* Newark, Del.: University of Delaware Press or Associated University Press, Toronto, 1984.

Levine, I. D., *Flying Crusader: The Story of General William Mitchell.* London: Davies, 1943.

Liddell Hart, B. H., *The Defence of Britain.* London: Faber & Faber, 1939.

———, *Europe in Arms.* London: Faber & Faber, 1937.

———, *The Future of Infantry.* London: Faber & Faber, 1933.

———, *The Real War, 1914–1918,* rev. ed. Boston: Little Brown, 1964.

———, *The Remaking of Modern Armies.* London: Murray, 1927.

———, *The Revolution in Warfare.* New Haven, Ct.: Yale University Press, 1927.

———, *The Tanks.* London: Cassell, 1959.

Lloyd George, David, *War Memoirs.* 6 vols. London: Nicholson & Watson, 1933–1936.

Ludendorff, Erich, *My War Memories, 1914–1918.* 2 vols. New York: Harper, 1919.

Mackinder, Halford, *The Scope and Methods of Geography and the Geographical Pivot of History.* London: Murray, 1951.

Madariaga, Salvador de, *Disarmament.* New York: Coward-McCann, 1929.

Marder, A. J., *Fear God and Dread Nought: The Correspondence of Admiral of the Fleet Lord Fisher of Kelverstone.* 3 vols. London: Cape, 1952–1959.

———, *From Dreadnought to Scapa Flow,* 5 vols. London: Oxford University Press, 1961–1970.

Maurice, F. B., *Governments and War.* London: Heinemann, 1926.

Maurice, F. B., *Lessons of Allied Cooperations, Naval, Military and Air, 1914–1918.* London: Oxford University Press, 1942.

Miksche, F. O., *Blitzkrieg.* London: Faber & Faber, 1942.

Millis, Walter, *The Road to War: America, 1914–1917.* New York: Fertig, 1935.

Moorehead, Alan, *Gallipoli.* New York: Harper, 1956.

Morgan, J. H., *Assize of Arms: The Disarmament of Germany and Her Rearmament, 1919–1939.* New York: Oxford University Press, 1946.

Morton, Louis, "The Origins of American Military Policy," *Military Affairs,* XXII. 1958.

Moyse-Bartlett, Hubert, *The King's African Rifles: A Study in the Military History of East and Central Africa, 1890–1945.* Aldershot, Eng.: Gale & Polden, 1956.

Obermann, Emil, *Soldaten, Bürger, Militaristen: Militär und Demokratie in Deutschland.* Stuttgart, W. Germ.: Cotta, 1958.

O'Callaghan, Sean, *The Easter Lily: The Story of the I.R.A.* New York: Roy, 1956.

Parkes, Oscar, *British Battleships: Warrior 1860 to Vanguard 1950: A History of Design, Construction and Armament.* London: Seeley, 1958.

Pollard, A. F., *A Short History of the Great War.* London: Methuen, 1950.

Posen, Barry R., *The Sources of Military Doctrine: France, Britain, and Germany Between the World Wars.* Cornell Studies in Security Affairs. Ithaca, N.Y.: Cornell University Press, 1984.

Puleston, W. D., *High Command in the World War.* London: Scribner's, 1934.

Raleigh, Walter, and H. A. Jones, *The War in the Air.* 6 vols. Oxford: Clarendon, 1922–1937.

Read, J. M., *Atrocity Propaganda, 1914–1919.* New Haven, Ct.: Yale University Press, 1941.

Renouvin, Pierre, *The Forms of War Government in France.* New Haven, Ct.: Yale University Press, 1927.

Ritter, Gerhard, *The Schlieffen Plan: Critique of a Myth.* New York: Dufour, 1968.

Robertson, William, *Soldiers and Statesmen.* London: Cassell, 1926.

Robinson, D. H., *The Zeppelin in Combat,* rev. ed. London: Foulis, 1966.

Root, Elihu, *The Military and Colonial Policy of the United States.* Cambridge, Mass.: Harvard University Press, 1916.

Rosinski, Herbert, "The Role of Sea Power in Global Warfare of the Future," *Brassey's Naval Annual.* 1947.

Royal Institute of International Affairs, *International Sanctions.* London: Oxford University Press, 1938.

Scammel, J. M., "Spenser Wilkinson and the Defense of Britain," *Journal of the American Military Institute,* IV, 1940.

Sikorski, Wladyslaw, *Modern Warfare.* New York: Roy, 1943.

Siney, M. C., *The Allied Blockade of Germany, 1914–1916.* Ann Arbor, Mich.: University of Michigan Press, 1957.

Slessor, J. C., *Air Power and Armies.* London: Oxford University Press, 1936.

Smithers, A. J., *A New Excalibur: The Development of the Tank, 1909–1939.* London: Leo Cooper in association with Secker & Warburg, 1986.

Spaight, J. M., *Air Power and War Rights.* New York: Longmans Green, 1924.

———, *Air Power Can Disarm.* London: Pitman, 1948.

———, *The Beginnings of Organized Air Power.* London: Longmans Green, 1927.

Spears, E. L., *Prelude to Victory.* London: Cape, 1939.

Sprout, H. M., *Toward a New Order of Sea Power: American Naval Policy and the World Scene, 1918–1922.* Princeton, N.J.: Princeton University Press, 1946.

Strausz-Hupé, Robert, *Geopolitics.* New York: Putnam's, 1942.

Sueter, M. F., *Airmen or Noahs.* London: Pitman, 1928.

———, *The Evolution of the Tank.* London: Hutchinson, 1937.

Sykes, F. H., *Aviation in Peace and War.* London: Arnold, 1922.

Taylor, Edmond, *The Strategy of Terror.* Boston: Houghton Mifflin, 1940.

Thomas, Hugh, *The Spanish Civil War.* New York: Harper, 1961.

Tschuppik, Karl, *Ludendorff: The Tragedy of a Military Mind.* Boston: Houghton Mifflin, 1932.

Tuchman, Barbara, *The Guns of August.* New York: Macmillan, 1962.

Vagts, Alfred, "Geography in War and Geopolitics," *Military Affairs,* VII. 1943.

———, *The Military Attaché.* Princeton, N.J.: Princeton University Press, 1967.

Wavell, Archibald, *Allenby.* 2 vols. London: Harrap, 1940–1943.

Wheeler-Bennett, J.W., *The Nemesis of Power: The Germany Army in Politics, 1918–1945.* New York: Viking Press (Compass), 1964.

Wilkinson, Spenser, *The Brain of an Army: The German General Staff.* London: Constable, 1895.

Winter, Denis, *Death's Men: Soldiers of the Great War.* Harmonsworth, Middlesex: Penguin, 1985.

Woodward, E. L., *Great Britain and the German Navy.* London: E. L. Cass, 1964.

WORLD WAR II

Allport, G. W., and Leo Postman, *The Psychology of Rumor.* New York: Russell, 1965.

Andrews, Marshall, *Disaster through Air Power.* New York: Rinehart, 1950.

Bacon, Reginald, *Modern Naval Strategy.* London: Muller, 1950.

Badoglio, Pietro, *Italy in the Second World War.* New York: Oxford University Press, 1948.

Baxter, J. P., *Scientists against Time.* Cambridge, Mass.: MIT Press, 1965.

Benoist-Mechin, J., *Sixty Days that Shook the West: The Fall of France, 1940.* New York: Putnam, 1963.

Bloc, Marc, *Strange Defeat: A Statement of Evidence Written in 1940.* New York: Norton, 1968.

Bradley, Omar, *A Soldier's Story.* New York: Holt, 1951.

Brodie, Bernard, *Strategic Air Power in World War II*. Santa Monica, Calif.: Rand Corporation, 1957.

Bryant, Arthur, *Triumph in the West, 1943–1946: Based on the Diaries and Autobiographical Notes of Field Marshal the Viscount Alanbrooke*. Garden City, N.Y.: Doubleday, 1959.

Bryant, Arthur, *The Turn of the Tide: A History of the War Years Based on the Diaries of Field Marshal Lord Alanbrooke, Chief of the Imperial General Staff*. Garden City, N.Y.: Doubleday, 1957.

Buckmaster, M. J., *They Fought Alone: The Story of British Agents in France*. London: Odhams, 1958.

Burne, A. H., *Strategy as Exemplified in World War II*. Cambridge: Cambridge University Press, 1946.

Canada, Department of National Defence, *Official History of the Canadian Army in the Second World War*. 4 vols. Ottawa: Queen's Printer, 1955–1970.

Carroll, Wallace, *Persuade or Perish*. Boston: Houghton Mifflin, 1948.

Chuikov, Vasilii, *The Battle for Stalingrad*. New York: Ballantine, 1968.

Churchill, W. L. S., *The Second World War*. 6 vols. Boston: Houghton Mifflin, 1948-1953.

Ciano, Galeazzo, *The Ciano Diaries, 1939–1943*. New York: Fertig, 1946.

Clark, Alan, *Barbarossa: The Russian-German Conflict, 1941–1945*. New York: Morrow, 1965.

Clark, Mark, *From the Danube to the Yalu*. New York: Harper, 1950.

Cookridge, E. H., *Inside SOE: The Story of Special Operations in Western Europe, 1940–45*. London: Arthur Barker, 1966.

Craven, W. F., and J. L. Cate (eds.), *The Army Air Forces in World War II*. 6 vols. Chicago: University of Chicago Press 1948–1955.

Creswell, John, *Sea Warfare, 1939–1945*, rev. ed. Berkeley, Calif.: University of California Press, 1967.

Dallin, Alexander, *German Rule in Russia, 1941–1945: A Study of Occupation Policies*. New York: St. Martin's, 1957.

De Guingand, Francis, *Operation Victory*. London: Hodder & Stoughton, 1947.

De Seversky, A. P., *Air Power: Key to Survival*. New York: Simon & Schuster, 1950.

———, *Victory through Air Power*. New York: Simon & Schuster, 1942.

De Weerd, H. A., *Great Soldiers of the Two World Wars*. New York: Norton, 1941.

Dickens, G. C., *Bombing and Strategy*. London: S. Low, Marston, 1947.

Draper, Theodore, *The Six Weeks' War*. London: Methuen, 1946.

Eisenhower, D. D., *Crusade in Europe*. New York: Doubleday, 1948.

Erickson, John. *The Road to Stalingrad: Stalin's War Against Germany*. New York: Harper & Row, 1975.

Falls, C. B., *The Second World War: A Short History,* 3d ed. London: Methuen, 1950.

Feis, Herbert, *Churchill, Roosevelt, Stalin: The War They Waged and the Peace They Sought.* Princeton, N.J.: Princeton University Press, 1957.

Fleming, Peter, *Invasion, 1940: An Account of the German Preparations and the British Counter-measures.* London: Hart-Davies, 1957.

Fuller, J. F. C., *The Second World War.* London: Eyre & Spottiswoode, 1954.

Galland, Adolf, *The First and the Last: The Rise and Fall of the German Fighter Forces, 1938–1945.* New York: Ballantine, 1969.

Garthoff, R. L., *Soviet Military Policy.* New York: Praeger, 1966.

Gilbert, Felix (ed.), *Hitler Directs His War.* New York: Oxford University Press, 1950.

Goebbels, Joseph, *The Goebbels Diaries, 1942–43,* New York: Doubleday, 1948.

Goerlitz, W., *Paulus and Stalingrad.* London: Methuen, 1963.

Great Britain, Privy Council, *History of the Second World War.* 55 vols. to date. London: HMSO, 1952–.

Guderian, Heinz, *Panzer Leader.* London: Michael Joseph, 1952; abridged ed., New York: Ballantine, 1967.

Harris, Arthur, *Bomber Offensive.* London: Collins, 1947.

Higgins, Trumbull, *Winston Churchill and the Second Front, 1940–1943.* New York: Oxford University Press, 1957.

Hinsley, F. H., *Hitler's Strategy: The Naval Evidence.* Cambridge: Cambridge University Press, 1951.

Irving, T. A., "Psychological Analysis of Wartime Rumour Patterns in Canada," *Bulletin of the Canadian Psychological Association,* III. 1943.

Jacobsen, H. A., and J. Rohwer (eds.), *Decisive Battles of World War II: The German View.* New York: Putnam's, 1965.

James, William, *The British Navies in the Second World War.* London: Longmans Green, 1947.

Keegan, John, *The Second World War.* New York: Viking, 1990.

Kilmarx, R. A., *A History of Soviet Air Power.* New York: Praeger, 1962.

Kris, Ernst, and Hans Speier, *German Radio Propaganda.* New York: Oxford University Press, 1944.

Liddell Hart, B. H., *Defence of the West.* London: Cassell, 1950.

———, *The Other Side of the Hill,* 3d ed. London: Cassell, 1956.

———, (ed), *The Rommel Papers.* London: Collins, 1953.

———, *Strategy,* 2d rev. ed. New York: Praeger, 1967.

Lundin, C. L., *Finland in the Second World War.* Bloomington, Ind.: Indiana University Press, 1957.

MacDonald, Charles B., *The Mighty Endeavor: American Armed Forces in the European Theater in World War II.* New York: Oxford University Press, 1969.

Manstein, Erich von, *Lost Victories.* Chicago: Regnery, 1958.

Marshall, G. C., *The Winning of the War in Europe and the Pacific.* Washington: Simon & Schuster, 1945.

Martienssen, A. K., *Hitler and His Admirals.* London: Secker & Warburg, 1948.

Maund, L. E. H., *Assault from the Sea.* London: Methuen, 1949.

Mellenthin, F. W. von., *Panzer Battles, 1939–1945.* London: Cassell, 1955; Norman, Okla.: University of Oklahoma Press, 1964.

Montgomery, Bernard, *El Alamein to the River Sangro.* Germany: British Army of the Rhine, 1946.

———, *Memoirs.* Cleveland, Ohio: World, 1958.

———, *Normandy to the Baltic.* London: Hutchinson, 1946.

Morison, S. E., *History of United States Naval Operations in World War II.* 15 vols. Boston: Little Brown, 1947–1962.

Morton, Louis, "The Decision to Use the Atomic Bomb," *Foreign Affairs,* XXXV. 1957.

Morton, Louis, "Pacific Command: A Study in Interservice Relations," *Harmon Memorial Lectures in Military History,* no. 3. Colorado Springs: USAF Academy, 1961.

Nickerson, Hoffman, *Arms and Policy, 1939–1944.* New York: Putnam's, 1945.

Puleston, W. D., *The Influence of Sea Power in World War II.* New Haven, Ct.: Yale University Press, 1947.

Rohwer, Jürgen, *Die U-boot-Erfolge der Achsenmächte, 1939–1945.* Munich: Lehmann, 1968.

Rohwer, Jürgen, and G. Hümmelchen, *Chronik des Seekrieges, 1939–1945.* Oldenburg, Germ.: Stalling, 1968.

Ruge, Friedrich, *Sea Warfare, 1939–1945: A German Viewpoint.* London: Cassell, 1957.

Schull, J. J., *The Far Distant Ships: An Official Account of Canadian Naval Operations in the Second World War.* Ottawa: King's Printer, 1950.

Sherwood, R. E., *Roosevelt and Hopkins.* New York: Harper, 1950.

Shulman, Milton, *Defeat in the West.* London: Secker and Warburg, 1947; rev. ed., New York: Ballantine, 1968.

Slessor, J. C., *The Central Blue: Recollections and Reflections.* New York: Praeger, 1956.

Slim, William, *Defeat into Victory.* New York: McKay, 1956.

Spector, Ronald H., *Eagle Against the Sun: The American War with Japan.* New York: Vintage Books, 1985 [1984].

Speidel, Hans, *Invasion 1944.* Chicago: Regnery, 1950.

Stacey, C. P., *The Canadian Army, 1939–1945.* Ottawa: King's Printer, 1948.

Sykes, Christopher, *Orde Wingate.* London: Collins, 1961.

Taylor, Edmond, *The Strategy of Terror.* Boston: Houghton Mifflin, 1940.

Tedder, A. W., *Air Power in War.* London: Hodder & Stoughton, 1948.

Terrell, Edward, *Admiralty Brief: The Story of the Inventions That Contributed to Victory in the Battle of the Atlantic.* London: Harrap, 1958.

Tuker, Francis, *The Pattern of War.* London: Cassell, 1948.

U.S. Army, Chief of Military History, *The United States Army in World War II.* 75 vols. to date. Washington: Government Printing Office, 1947–1983.

U.S. Marine Corps Historical Branch, *History of U.S. Marine Corps Operations in World War II.* 5 vols. Washington: USMC Headquarters, 1958–1968.

U.S. Military Academy, *The War in Eastern Europe (June 1941 to May 1945).* West Point: U.S. Military Academy, 1949.

United States Strategic Bombing Survey, *Summary Report (European War).* Washington: Government Printing Office, 1945.

Webster, Charles, and Noble Frankland, *The Strategic Air Offensive against Germany, 1939–1945.* 4 vols. London: HMSO, 1961.

Werth, Alexander, *Russia at War, 1941–1945.* New York: Dutton, 1964.

Westphal, Siegfried, *The German Army in the West.* London: Cassell, 1951.

Wheatley, Ronald, *Operation Sea Lion: German Plans for the Invasion of England, 1939–1942.* Oxford: Clarendon, 1958.

Wheeler-Bennett, J. W., *The Nemesis of Power: The German Army in Politics, 1918–1945.* New York: St. Martin's, 1956.

Wilmot, Chester, *The Struggle for Europe.* London: Collins, 1952.

Wright, Gordon. *The Ordeal of Total War, 1939–1945,* The Rise of Modern Europe. New York: Harper and Row, 1968.

Young, Desmond, *Rommel.* London: Collins, 1951.

THE WORLD SINCE 1945

Ambler, John S., *The French Army in Politics, 1945–62.* Columbus, Ohio: Ohio State University Press, 1966. (Reissued by Doubleday in paperback as *Soldiers Against the State.*)

Aron, Raymond, *Peace and War: A Theory of International Relations.* London: Weidenfeld & Nicholson, 1967.

Barclay, C. N., *The First Commonwealth Division . . . in Korea 1950–53.* Aldershot, Eng.: Gale & Polden, 1951.

Bator, Victor, *Viet Nam, a Diplomatic Tragedy: The Origins of the United States Involvement.* Dobbs Ferry, N.Y.: Oceana, 1965.

Beaton, Leonard, *The Struggle for Peace.* New York: Praeger, 1967.

Berger, Carl, *The Korean Knot: A Military-Political History,* rev. ed. Philadelphia: University of Pennsylvania Press, 1964.

Billings-Yun, Melanie, *Decision Against War: Eisenhower and Dien Bien Phu, 1954.* Contemporary American History Series. New York: Columbia University Press, 1988.

Blackett, P. M. S., *Fear, War, and the Bomb: Military and Political Consequences of Atomic Energy.* New York: McGraw-Hill, 1948.

Bloomfield, Lincoln P., et al., *International Military Forces: The Question of Peacekeeping in an Armed and Dearming World.* Boston: Little, Brown, 1964.

Bowett, D. W., *United Nations Forces: A Legal Study.* New York: Praeger, 1965.

Brodie, Bernard, *Strategy in the Missile Age.* Princeton, N.J.: Princeton University Press, 1959.

Brogan, Patrick, *The Fighting Never Stopped: A Comprehensive Guide to World Conflict since 1945.* New York: Vintage Books, 1990.

Brown, Weldon A., *The Last Chopper: The Denouement of the American Role in Vietnam, 1963–1975.* Port Washington, N.Y.: Kennikat Press, 1976.

Bull, Hedley, *The Control of the Arms Race: Disarmament and Arms Control in the Missile Age.* New York: Praeger, 1961.

Butterworth, Robert Lyle, *Managing Interstate Conflict, 1945–74: Data with Synopses.* Pittsburgh, Pa.: University Center for International Studies, University of Pittsburgh, 1976.

Byers, R. B. (ed.), *Deterrence in the 1980s: Crisis and Dilemma.* New York: St. Martin's Press, 1985.

Caidin, Martin, *Spaceport U.S.A.: The Story of Cape Canaveral and the Air Force Missile Center.* New York: Dutton, 1959.

Devillers, Philippe, and Jean LaCouture, *End of a War: Indochina, 1954.* New York: Praeger, 1969.

Dunn, Peter M., and Bruce W. Watson (eds.), *American Intervention in Grenada: The Implications of Operation "Urgent Fury".* Boulder, Colo.: Westview Press, 1985.

Dinerstein, H. S., *War and the Soviet Union: Nuclear Weapons and the Revolution in Soviet Military and Political Thinking.* New York: Praeger, 1959.

Fall, Bernard, *The Two Viet Nams: A Political and Military Analysis,* 2d ed. New York: Praeger, 1967.

Fallows, James, *National Defense.* New York: Random House, 1981.

Fehrenbach, T. R., *This Kind of War: A Study in Unpreparedness.* New York: Macmillan, 1963.

Foot, M. R. D., *Men in Uniform: Military Manpower in Modern Industrial Societies.* New York: Praeger, 1961.

Foot, Rosemary, *The Wrong War: American Policy and the Dimensions of the Korean Conflict, 1950–1953.* Ithaca, N.Y.: Cornell University Press, 1985.

Freedman, Lawrence, *The Evolution of Nuclear Strategy,* 2d ed. London: Macmillan, 1989 [1983].

———, *U.S. Intelligence and the Soviet Strategic Threat,* 2d ed. Princeton, N.J.: Princeton University Press, 1986 [1977].

Gaddis, John Lewis, *Strategies of Containment: A Critical Appraisal of Postwar American National Security Policy.* New York: Oxford University Press, 1982.

Gallois, Pierre, *The Balance of Terror: Strategy for the Nuclear Age.* Boston: Houghton Mifflin, 1961.

Garthoff, R. L., *The Soviet Image of Future War.* Washington: Public Affairs Press, 1959.

Gavin, J. M., *War and Peace in the Space Age.* New York: Harper, 1958.

Goodspeed, D. J., *A History of the Defence Research Board of Canada.* Ottawa: Queen's Printer, 1958.

Gordenker, Leon, *The United Nations Secretary General and the Maintenance of Peace.* New York: Columbia University Press, 1967.

Halle, Louis J., *The Cold War as History.* New York: Harper and Row, 1967.

Halperin, M. H., *Limited War: An Essay on the Development of the Theory and an Annotated Bibliography.* Cambridge, Mass.: Harvard University Center for International Affairs, Occasional Papers in International Affairs, No. 3, May, 1962.

Hannah, Norman B., *The Key to Failure: Laos and the Vietnam War.* Lanham, Md.: Madison Books, 1990 [1987].

Herken, Greg, *The Winning Weapon: The Atomic Bomb in the Cold War, 1945–1950.* Princeton, N.J.: Princeton University Press, 1988 [1981].

Herring, George C., *America's Longest War: The United States and Vietnam, 1950–1975,* 2d ed. New York: Knopf, 1986.

Higgins, Rosalyn, *United Nations Peacekeeping Operations: Documents and Commentary.* London: Royal Institute of International Affairs, 1969.

Hitch, C. J., and R. N. McKean, *Economics of Defense in the Nuclear Age.* Cambridge, Mass.: Harvard University Press, 1960; New York: Atheneum, 1965 (paperback).

Howard, Michael, and Robert Hunter, *Israel and the Arab World: The Crisis of 1967.* London: Institute for Strategic Studies, 1967.

Horowitz, Irvine L., *The War Game: Studies in the New Civilian Militarism.* New York: Ballantine, 1963.

Irmscher, W. F. (ed.), *Man and Warfare: Thematic Readings for Composition.* Boston: Little Brown, 1964.

Jessup, P. C., and H. J. Taubenfeld, *Control for Outer Space and the Arctic Analogy.* New York: Columbia University Press, 1959.

Kahn, Herman, *On Escalation: Metaphors and Scenarios.* New York: Praeger, 1965.

——, *On Thermonuclear War.* Princeton, N.J.: Princeton University Press, 1960; New York: The Free Press, 1969.

——, *Thinking about the Unthinkable.* New York: Horizon Press, 1962; Avon, 1969.

Kingston-McCloughry, E. J., *The Direction of War: A Critique of the Political Direction and High Command in War.* New York: Praeger, 1955.

Kissinger, Henry A., *The Necessity for Choice: Prospects of American Foreign Policy.* New York: Harper, 1961; Norton, 1969.

Kissinger, Henry A., *Nuclear Weapons and Foreign Policy.* New York, Harper, 1957; Norton, 1969.

Klare, Michael T., and Peter Kornbluh (eds.), *Low Intensity Warfare: Counterinsurgency, Proinsurgency, and Antiterrorism in the Eighties.* New York: Pantheon Books, 1988.

Knorr, Klaus (ed.), *NATO and American Security.* Princeton, N.J.: Princeton University Press, 1959.

Knorr, Klaus, and Thornton Read (eds.), *Limited Strategic War*. New York: Praeger, 1962.

Krepinevich, Andrew F., Jr., *The Army and Vietnam*. Baltimore: Johns Hopkins University Press, 1986.

Kreppon, Michael, *Strategic Stalemate: Nuclear Weapons and Arms Control in American Politics*. New York: St. Martin's Press; Council on Foreign Relations, 1984.

LaFeber, Walter, *America, Russia, and the Cold War, 1945–1971,* 5th ed. New York: Knopf, 1985.

Levant, Victor, *Quiet Complicity: Canadian Involvement in the Vietnam War*. Toronto: Between the Lines, 1986.

Lewy, Guenter, *America in Vietnam*. Oxford, Eng.: Oxford University Press, 1978.

Liddell Hart, B. H., *Deterrent or Defense*. New York: Praeger, 1960.

Mandelbaum, Michael, *The Nuclear Question: The United States and Nuclear Weapons, 1946–1976*. Cambridge, Eng.: Cambridge University Press, 1979.

Mao Tse-tung, *On Protracted War,* Peking: Foreign Language Press, 1954.

———, *Strategic Problems of China's Revolutionary War*. Peking: Foreign Language Press, 1954.

Martin, Andrew, *Collective Security: A Progress Report*. Paris: UNESCO, 1952.

F. M. Mickolus, et al., *International Terrorism in the 1980s*. 2 vols. Ames, Ia.: Iowa State University Press, 1989.

Middlebrook, Martin, *Operation Corporate: The Falklands War, 1982*. New York: Viking, 1985.

Middleton, Drew, *The Defense of Western Europe*. New York: Appleton, 1952.

Miksche, F. O., *The Failure of Atomic Strategy—And a New Proposal for the Defense of the West*. New York: Praeger, 1959.

Miksche, F. O., and E. Combaux, *War between Continents*. London: Faber & Faber, 1948.

Miller, Linda B., *World Order and Local Disorder: The United States and Internal Conflicts*. Princeton, N.J.: Princeton University Press, 1967.

Morison, Samuel E., *Strategy and Compromise*. Boston: Little, Brown, 1958.

Noel-Baker, Philip, *The Arms Race*. New York: Oceana, 1960.

O'Ballance, Edgar, *The Arab-Israeli War, 1948*. London: Faber & Faber, 1956.

Osgood, Robert E., *The Nuclear Dilemma in American Strategic Thought*. Boulder, Colo.: Westview Press. 1988.

Paschall, Rod, *LIC 2010: Special Operations & Unconventional Warfare in the Next Century*. Future Warfare Series, vol. 5, ed. by Perry M. Smith. Washington: Brassey's (US), 1990.

Pearson, L. B., *Diplomacy in the Nuclear Age*. Cambridge, Mass.: Harvard University Press, 1959.

Possony, S. T., *Strategic Air Power: A Pattern of Dynamic Security*. Washington: Infantry Journal Press, 1949.

Prados, John, *The Soviet Estimate: U.S. Intelligence and Soviet Strategic Forces*. Princeton, N.J.: Princeton University Press, 1986 [1982].

Preston, Richard A., "The Great Debate on Strategy: A Survey of the Literature," *R.C.A.F. Staff College Journal*. 1961.

Pruitt, D. G., and R. C. Snyder, *Theory and Research on the Causes of War*. Englewood Cliffs, N.J.: Prentice-Hall, 1969.

Pustay, J. S., *Counter-insurgency Warfare*. New York: Free Press of Glencoe, 1965.

Rapoport, Anatol, *Strategy and Conscience*. New York: Harper & Row, 1964; Schocken, 1969.

Rees, David, *Korea: The Limited War*. New York: St. Martin's, 1964.

Rosen, Stephen (ed.), *Testing the Theory of the Military Industrial Complex*. Lexington, Mass.: D. C. Heath, 1973.

Rovere, Richard, and Arthur Schlesinger, *The General and the President*. New York: Farrar, Straus, & Young, 1951.

Royal Institute of International Affairs, *Defence in the Cold War*. London: R.I.I.A., 1950.

Sarkesian, Sam C., *The New Battlefield: The United States and Unconventional Conflicts*. Contributions in Military Studies, No. 54. New York: Greenwood Press, 1978.

Saunders, M. C. (ed.), *The Soviet Navy*. New York: Praeger, 1958.

Schelling, Thomas C., *Arms and Influence*. New Haven, Ct.: Yale University Press, 1966.

————, *Strategy and Arms Control*. New York: Twentieth Century Fund, 1961 (paperback).

————, *The Strategy of Conflict*. Cambridge, Mass.: Harvard University Press, 1960; New York: Oxford University Press, 1963.

Schlesinger, James R., *The Political Economy of National Security*. New York: Praeger, 1960.

Shackley, Theodore, *The Third Option: An American View of Counterinsurgency Operations*. New York: Readers' Digest Co., 1981.

Shafer, D. Michael, *Deadly Paradigms: The Failure of U.S. Counterinsurgency Policy*. Princeton, N.J.: Princeton University Press, 1988.

Shultz, Richard H., et al. (eds.), *Guerrilla Warfare and Counterinsurgency: U. S. and Soviet Policy in the Third World*. Lexington, Mass.: D.C. Heath, 1989.

Singer, J. D., *Deterrence, Arms Control, and Disarmament*. Columbus, Ohio: Ohio State University Press, 1962.

Sivard, Ruth Leger, *World Military and Social Expenditures*. Leesburg, Va.: WMSE Publications, 1986.

Slessor, J. C., *The Great Deterrent: A Collection of Lectures, Articles, and Broadcasts on the Development of Strategic Policy in the Nuclear Age*. New York: Praeger, 1957.

Slessor, J. C., *Strategy for the West*. London: Cassell, 1954.

Smith, Charles D., *Palestine and the Arab–Israeli Conflict*. New York: St. Martin's, 1988.

Smoke, Richard, *National Security and the Nuclear Dilemma: An Introduction to the American Experience.* New York: Random House, 1984.

Smyth, J. G., *The Western Defences.* London: Wingate, 1951.

Sokolovsky, Vasilii D. (ed.), *Military Strategy: Soviet Doctrine and Concepts.* New York: Praeger, 1963.

Spanier, J. W., *The Truman-MacArthur Controversy and the Korean War.* Cambridge, Mass.: Harvard University Press, 1959.

Stern, F. M., *The Citizen Army: Key to Defense in the Atomic Age.* New York: St. Martin's, 1957.

Stubbing, Richard A., *The Defense Game: An Insider Explores the Astonishing Realities of America's Defense Establishment.* Assisted by Richard A. Mendel. New York: Harper & Row, 1986.

Summers, Harry G., *On Strategy: A Critical Analysis of the Vietnam War.* New York: Dell, 1984 [1982].

Taylor, Maxwell D., *The Uncertain Trumpet.* New York: Harper, 1960.

Thomas, R. C. W., *The War in Korea, 1950–1953: A Military Study.* Aldershot, Eng.: Gale & Polden, 1954.

Tucker, Robert W., *The Just War: Exposition of the American Concept.* Baltimore: Johns Hopkins University Press, 1960.

Turner, Gordon B., and R. D. Challener, *National Security in the Nuclear Age: Basic Facts and Theories.* New York: Praeger, 1960.

United Nations Secretariat, *The Blue Helmets: A Review of United Nations Peacekeeping.* New York: United Nations Department of Public Information, 1985.

United States, Department of the Army, *Korea—1950.* Washington: Office of the Chief of Military History, 1952.

Ward, Barbara, *Policy for the West.* London: Allen & Unwin, 1951.

Welles, Sumner, *Where Are We Heading?* New York: Harper, 1946.

Whelan, Richard, *Drawing the Line: The Korean War, 1950–1953.* Boston: Little, Brown, 1990.

Wise, S. F., "The Balance of Nuclear Terror," *Queen's Quarterly,* LXVII. 1960.

Wood, Herbert F., *Strange Battleground: The Operations in Korea and their Effects on the Defence Policy of Canada.* Ottawa: Queen's Printer, 1966.

Young, Oran R., *The Intermediaries: Third Parties in International Crises.* Princeton, N.J.: Princeton University Press, 1969.

Reference added in proof:

Dresziger, N. F. (ed.), *Ethnic Armies: Polyethnic Armed Forces from the Time of the Hapsburgs to the Age of the Superpowers,* Waterloo Ont., Can.: Wilfid Laurier University Press, 1990.

INDEX